59 Advances in Biochemical Engineering Biotechnology

Managing Editor: T. Scheper

Springer-Verlag Berlin Heidelberg GmbH

Bioprocess and Algae Reactor Technology, Apoptosis

With Contributions by
M. Al-Rubeai, J. F. Cornet, C. G. Dussap,
C. B. Elias, J. Gomes, J.-B. Gros, D. C. Hill,
J. B. Joshi, A.S. Menawat, L. J. Nisbet, O. Pulz,
K. Scheibenbogen, S. R. Wrigley

With 56 Figures and 1 Color Figure

Springer

This series represents critical reviews on the present and future trends in Biochemical Engineering/Biotechnology, including microbiology, genetics, biochemistry, chemistry, computer science and chemical engineering. It is addressed to all scientists at universities and in industry who wish to keep up-to-date in this extremly fast developing area of science.

In general, special volumes are dedicated to selected topics and are edited by well known guest editors. The managing editor and publisher will however always be pleased to receive suggestions and supplementary information. Manuscripts are accepted in English.

In references Advances in Biochemical Engineering/Biotechnology is abbreviated as Adv. Biochem. Engin./Biotechnol. as a journal.

ISBN 978-3-662-14820-4 ISBN 978-3-540-69541-7 (eBook)
DOI 10.1007/978-3-540-69541-7

Library of Congress Catalog Card Number 72-152360

ISSN 0724-6145

Originally published by Springer-Verlag Berlin Heidelberg New York in 1998
Softcover reprint of the hardcover 1st edition 1998

Typesetting: Macmillan India Ltd., Bangalore-25
SPIN: 10573576 02/3020 - 5 4 3 2 1 0 - Printed on acid-free paper

Attention all "Enzyme Handbook" Users:

A file with the complete volume indexes Vols. 1 through 11 in delimited ASCII format is available for downloading at no charge from the Springer EARN mailbox. Delimited ASCII format can be imported into most databanks.

The file has been compressed using the popular shareware program "PKZIP" (Trademark of PKware INc., PKZIP is available from most BBS and shareware distributors).

This file distributed without any expressed or implied warranty.

To receive this file send an e-mail message to:
SVSERV@DHDSPRI6.BITNET.
The message must be: "GET/ENZHB/ENZ_HB.ZIP".

SPSERV is an automatic data distribution system. It responds to your message. The following commands are available:

HELP	returns a detailed instruction set for the use of SVSERV,
DIR (*name*)	returns a list of files available in the directory "name",
INDEX (*name*)	same as "DIR"
CD <*name*>	changes to directory "name",
SEND <*filename*>	invokes a message with the file "filename"
GET <*filename*>	same as "SEND".

Table of Contents

Fed-Batch Bioproduction of Spectinomycin

J. Gomes* and A.S. Menawat**
* Department of Biochemical Engineering & Biotechnology,
Indian Institute of Technology, Delhi, Hauz Khas, New Delhi 110016, India
** Advanced Solutions & Know-how, 1426, Hidden Creek North, Saline,
MI-48176, USA

Actinomycetes produce about 67% of the known antibiotics covering a wide range of chemical structures. However, their filamentous growth present several problems during industrial processes. Among these problems oxygen transfer limitation is critical. In this chapter we present the role of oxygen in spectinomycin production by a *Streptomyces* species. Spectinomycin, a broad spectrum antibiotic effective against penicillin resistant gonorrhea, is an aminoglycoside constituted from two glucose moieties. Its bioproduction is strongly influenced by glucose and oxygen. We have shown that for a fixed dissolved oxygen concentration, there are two specific glucose concentrations which give maximum final titers of spectinomycin. The bi-modal maximum indicates the influence of two intermediate metabolites in spectinomycin biosynthesis. We propose a mechanism for spectinomycin biosynthesis and subsequently develop a model based on this mechanism. The proposed mechanism for spectinomycin biosynthesis is validated by successfully reconstructing the air flow rate profiles. A nonlinear systems theory technique termed External Differential Representation, is implemented to reconstruct the spectinomycin bioconversion process which then predicts the spectinomycin concentration from the air flow rate profile. This signifies that spectinomycin titers in industrial fed-batch processes can be controlled if a priori information about the air flow rate profile yielding maximum spectinomycin is available.

Advances in Biochemical Engineering/
Biotechnology, Vol. 59
Managing Editor: Th. Scheper

List of Symbols and Abbreviations

a	constant for activation term
A^*	saturated dissolved oxygen concentration ($g\,l^{-1}$)
A	dissolved oxygen concentration ($g\,l^{-1}$)
b	constant for activation term
C	cell mass concentration ($g\,l^{-1}$)
g	glucose concentration
K_e	exponential model constant ($g\,l^{-1}$)
K_{e1}	exponential model constant for product 1 ($g\,l^{-1}$)
K_{e2}	exponential model constant for product 2 ($g\,l^{-1}$)
K_i	inhibition constant for Haldane-Monod or exponential structures ($g\,l^{-1}$)
K_{i1}	inhibition constant for product 1 in exponential model ($g\,l^{-1}$)
K_{i2}	inhibition constant for product 2 in exponential model ($g\,l^{-1}$)
k_{la}	mass transfer coefficient (h^{-1})
K_m	Monod constant ($g\,l^{-1}$)
P	product concentration ($g\,l^{-1}$)
S	substrate concentration ($g\,l^{-1}$)
S_f	feed concentration ($g\,l^{-1}$)
T_{f0}	time of glucose feed initiation (hr)
v	volumetric air flow rate ($1\,min^{-1}$)
V	volume (l)
Y	yield coefficient [(cell mass concentration ($g\,l^{-1}$))/(substrate concentration ($g\,l^{-1}$))]
Y_{CA}	yield coefficient of cell mass based on oxygen (g cell mass)/(g oxygen)
Y_{CS}	yield coefficient of cell mass from substrate (g cell mass)/(g substrate)
Y_{P1A}	yield coefficient of product 1 based on oxygen (g product 1)/(g oxygen)
Y_{P2A}	yield coefficient of product 2 based on oxygen (g product 2)/(g oxygen)
Y_{P1S}	yield coefficient of product 1 based on substrate (g product 1)/(g substrate)
Y_{P2S}	yield coefficient of product 2 based on substrate (g product 2)/(g substrate)
FIA	flow injection analysis
HPLC	high pressure liquid chromatography
PHB	poly-β-hydroxybutyric
GC	gas chromatography
EDR	external differential representation
OUR	oxygen uptake rate

Greek Letters

α	activation term
μ	specific growth rate (Monod Kinetics) (h^{-1})
$\hat{\mu}$	medium viscosity
μ_m	maximum specific growth rate
$\sum$	summation
ρ	density ($g\,ml^{-1}$) medium viscosity

1 Introduction

Spectinomycin is a broad spectrum antibiotic produced primarily by Abbott Laboratories and Upjohn Company in the USA. Since the initial isolation of the antibiotic by researchers from both companies technology has come a long way to large scale industrial production. We found no literature relating to fundamental research, such as the application of genetic engineering, in improving the yields of spectinomycin. Most of the existing literature relates to the determination of the structure of spectinomycin and its measurement by various analytical techniques. Production technology is well established and various improvements have been made from engineering contributions.

This work investigates the relation between glucose and oxygen in spectinomycin bioproduction. Previous research at Abbott indicate that spectinomycin production is favored at two different concentrations of glucose. This phenomenon is examined in detail and the significance of the results are discussed.

Following a general overview of antibiotics spectinomycin is discussed in some detail, including methods of production and analysis. Results of the experimental optimization of glucose feed concentration and analysis of the results are presented. The mechanism of spectinomycin biosynthesis is described by a model employing an exponential structure. We show conclusively that biosynthesis depends on the allocation of glucose and oxygen resource by the microorganism and that yield can be controlled by manipulating these two variables. Finally, nonlinear systems theory is applied to predict the concentration from air flow rate values. A control strategy for optimal spectinomycin production is also presented.

1.1 Physiology of Antibiotics

Among the diverse secondary metabolites available a group of biologically active compounds called antibiotics occupy a special status. Since the 1940's, discoveries of new and potent molecules have contributed significantly to advances in medicine and agriculture. However, the search continues for new antibiotics and other secondary metabolites to combat naturally resistant fungi and bacteria, as well as those microbes which have acquired resistance.

It is remarkable that microorganisms producing antibiotics remain viable in their own environment. In spite of this apparent insensitivity, addition of antibiotic to the culture during active growth undermines the viability of the microorganism. Organisms usually start to produce antibiotics after passing through the active growth phase. The initiation of antibiotic production varies from organism to organism and on the environmental conditions. Normally, enzymes directly involved in the antibiotic production are repressed or inhibited during the growth phase.

The resistance mechanisms developed by the antibiotic-producing microorganism against its own antibiotic are similar to those of clinically resistant bacteria. Some of the more common mechanisms found in resisting antibiotics are [1]:

a) Permeability modifications: modifications of the cell wall structure during the antibiotic production phase reduce permeability and protect the microorganism against the high extracellular concentration of its antibiotic.
b) Compartmentation: special compartments developed during the production phase, allow the microorganism to hold the antibiotic before excretion into the environment. This protects the cell from internal antibiotic build-up.
c) Enzymatic deactivation: synthesis of special enzymes which convert the antibiotic into inactive or weaker derivatives helps in resisting the biological activity of the molecules.
d) Modification of target sites: modification of the target site of the antibiotic within its own cellular machinery, e.g., in the ribosomes, negates its effect on the microorganism.
e) Feedback inhibition and repression: shutting off the biosynthetic pathway for the antibiotic prevents the concentration from becoming lethal. Feedback inhibition and repression almost always accompanies other modes of resistance.

The beginning of secondary metabolism is strictly regulated due to the bioactive nature of antibiotics. The precise nature of the molecular mechanisms of the various regulatory factors are still not known. However, activation of transcription of many different operons having the same physiological functions regulates the start of secondary metabolism for most producing microorganisms. These functions may involve synthesis of certain proteins or control cellular activities not directly related to growth. If such functions affect antibiotic synthesis directly or indirectly, their control mechanisms become essential regulatory phenomena. Among the various regulatory factors, carbon, nitrogen and phosphate play key roles in antibiotic synthesis.

The nutrients present in the environment exert a strong influence on the production of antibiotics. General observations indicate that the onset of antibiotic production occurs when the nutritional conditions limit growth. Often the specific growth rate reflects regulatory phenomena because it depends on the type and quantity of nutrients present in the environment. For example, in the case of the steady-state production of β-1,3-glucanase, an exoenzyme secondary metabolite, by an actinomycete in a glycerol-limited chemostat culture occurs only at growth rates below $0.2\,h^{-1}$ [2]. Tylosin production by steady-state chemostat cultures of *Streptomyces fradiae* also depends on the growth rate [3]. Likewise, there is a direct correlation between growth rate and gramicidin-S production by *Bacillus brevis* in chemostat cultures [4].

Nutritional limitations of carbon, nitrogen and phosphorus exert the strongest influence on regulatory patterns [5–9]. There is sufficient evidence indicating the absence of activity of enzymes, specifically involved in the biosynthetic pathway of the antibiotic, before production begins. These critical enzymes become active at low nutritional levels only under growth limiting conditions. This clearly suggests that catabolite repression is one of the mechanisms of regulation. However it is still not known whether repression occurs at the transcription or at the translation level.

1.1.1 Carbon Regulation

Glucose, an excellent carbon source for growth, interferes with the biosynthesis of many antibiotics. In antibiotic production facilities where glucose and a more slowly metabolizable carbon source (citrate, glycerol) is used, glucose is consumed during growth while the second carbon source is utilized for antibiotic production. Depending on the microorganism, the carbon regulation may be triggered by a carbon source other than glucose. For example, *Streptomyces niveus* prefers citrate as the carbon source over glucose in the synthesis of novobiocin [10].

Often the specific growth rate determines the onset of antibiotic synthesis. In such situations, controlling the growth rate in the optimum ranges for either growth or production eliminates the interference from glucose. For example, maintaining the growth rate within the range of 0.009 and 0.014 h^{-1}, prevents the decay of penicillin [11,12]. The residual glucose concentration also affects the commencement of secondary metabolism in the fungal culture. Consequently, the feeding rate of glucose determines the productivity of penicillin by *Penicillium chrysogenum*. Martin and Demain [1] have presented a detailed study of the interfering effects of glucose and other carbon sources on various antibiotic synthetic processes.

1.1.2 Nitrogen Regulation

Nitrogen atoms constitute an important part of many antibiotic molecules and cellular materials and therefore participates in important regulatory mechanisms. Usually, microorganisms integrate nitrogen containing primary metabolites into the final antibiotic molecule. Nitrogen regulates the production of proteins and other nitrogenous biomolecules by microorganisms. It also regulates the activity of enzymes which utilize nitrogen resources. For example, ammonia represses nitrite and nitrate reductase, glutamate dehydrogenase and extracellular protease. Microorganisms also show distinct preference for the type of nitrogen nutrient they need for growth or synthesis of secondary metabolites.

The literature provides sufficient evidence that nitrogen significantly influences the production of antibiotics. *Streptomyces griseus* produced the highest titer of Streptomycin when the media contained soya bean meal, a rich source of nitrogen [13]. Novobiocin production by *Streptomyces niveus* requires proline as the nitrogen source. When the defined media contained both ammonium salts and proline, the microorganism preferentially utilized the ammonium salts for growth and proline for novobiocin production. Clearly, a nitrogen regulatory phenomena controlled the antibiotic production. Similarly, production of gramicidin S by *Bacillus brevis* specifically required phenylanaline. In another report, highest yields of cephalosporin by *Streptomyces clavuligerus* [6] required supplementation of medium with certain amino acids and urea.

These examples illustrate two possible circumstances of nitrogen incorporation – (i) including a direct precursor into the antibiotic molecule and (ii) utilizing the nitrogen atom or a nitrogen group by an antibiotic-specific reaction on a specific intermediate. In the first case, supplying the production medium with the direct precursor or a source from which the molecule is easily available significantly enhances the yield. In the second case, optimizing the medium with several suitable sources of nitrogen is necessary.

Several routes of nitrogen regulatory phenomena include:

a) Control at the catabolic level.
b) Regulation at the biosynthetic level.
c) Transport of metabolites.
d) Modulation of intracellular concentration of free metabolites.
e) Protein turnover and degradation.
f) Regulation of antibiotic-synthetase formation.

All regulatory mechanisms are manifestations of expression at the genetic level. Expression of the antibiotic genes depend on many factors, both external and internal. Nitrogen is just one of the many external factors influencing these mechanisms. Antibiotic biosynthesis control occurs at the transcription, the translational and as well as the post translational levels. Regulatory signals from nitrogen metabolism exert influence on each of these levels. However, the type and degree of influence for most industrial antibiotics are only partially known.

1.1.3 Phosphate Regulation

Phosphate is the crucial growth-limiting nutrient in several antibiotic biosynthetic processes. In candicidin bioproduction with *Streptomyces griseus*, phosphate levels are depleted about 2 h before the commencement of the production phase [14]. Phosphate concentration remains low during the entire production phase of candicidin bioproduction. Addition of 10 mM phosphate at the start of a candicidin bioproduction prevents depletion of extracellular phosphate. Consequently, growth continues throughout the bioproduction with no antibiotic production. Phosphate levels are depleted during the growth of *Streptomyces aureofaciens* before the production of tetracycline begins. Phosphate in concentrations ranging from 0.3–300 mM usually supports extensive cell growth, but concentrations of 10 mM and above suppresses biosynthesis of many antibiotics [1].

Intracellular ATP levels affect the production of antibiotic. Martin and Demain [15] observed that extracellular phosphate concentration directly affected the intraceullar concentration of ATP. They observed that *Streptomyces griseus* synthesized antibiotic only under phosphate limited conditions. Phosphate addition increased the concentration of phosphoribosyl pyrophosphate which stimulates the uptake of adenine.

Biosynthetic mechanisms for growth are directly related to the available energy and hence the intracellular ATP concentration. The correlation between ATP concentration or adenylate charge in cells and regulatory enzymes is still controversial. Energy charge defined by Atkinson and Walton [16] as

$$\frac{ATP + 0.5ADP}{ATP + ADP + AMP}$$

provides a measure for the amount of energy stored in the adenylate system. During the exponential growth phase of *Escherichia coli* the energy charge varies between 0.8–0.9 [17]. The microorganism remains viable for energy charge values between 0.8–0.85. For energy charge values below 0.5 the cells die. These values are true for most microorganisms [18]. Energy charge numbers and their correlation with antibiotic production are still not universally accepted [19]. Martin et al. [20] established that at an energy charge of about 0.8, *Streptomyces griseus* produces antibiotic. Their studies show that an addition of 10 mM phosphate increases the energy charge number to 0.85 with a simultaneous switch to active growth. Although the energy charge number increases only by 0.5, the intracellular ATP concentration triples. Clearly, both extracellular and intracellular phosphate strongly regulate antibiotic biosynthesis.

In the case of spectinomycin it is not phosphorus or nitrogen but carbon which regulates its biosynthesis. It has been established conclusively [21,22] that the ultimate precursor for the two carbon rings of spectinomycin is glucose. Therefore, production of spectinomycin begins with limiting levels of glucose in the medium.

2 Spectinomycin Biosynthesis

2.1 *Organism and Metabolism*

Spectinomycin is a broad spectrum antibiotic used primarily for the treatment of penicillin resistant gonorrhea. The antibiotic is an aminoglycoside comprising two glucose moieties joined together by two oxygen bridges in a tri-ring structure (Fig. 1). The formation of the aminoglycoside group of antibiotics triggers off directly from glucose-6-phosphate, i.e., intracellular glucose. Two possible pathways exist for the production of spectinomycin from intracellular glucose: (i) by aldolase conversion to myo-inositol and (ii) by oxidoreductase

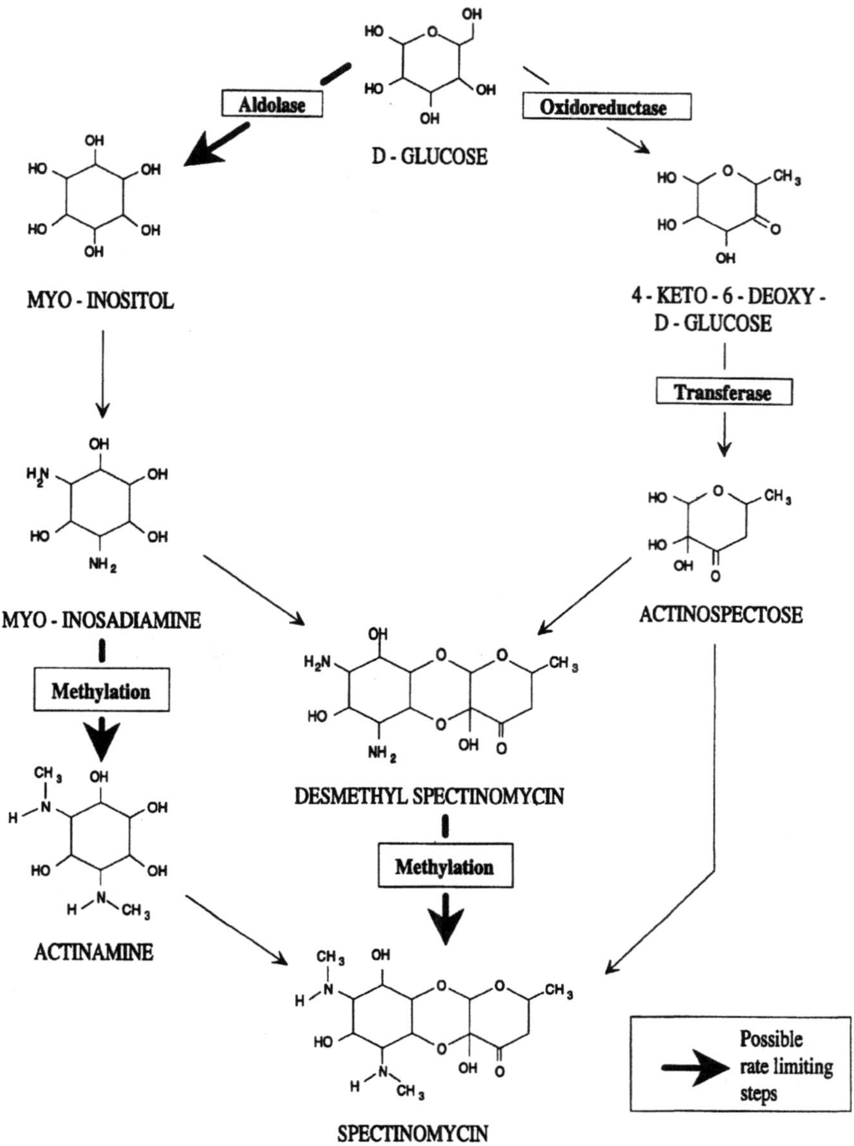

Fig. 1. Biosynthetic pathway for spectinomycin

conversion to 4-keto-6-deoxy-D-glucose. The first pathway continues from myo-inositol to form 1,3-myo-inosadiamine, then actinamine and finally spectinomycin. The intermediate 4-keto-6-deoxy-D-glucose of the second pathway forms actinospectose and then spectinomycin. Two intermediates of these alternate pathways, 1,3-myo-inosadiamine and actinospectose, also combine to give the unmethylated spectinomycin which when methylated produces spectinomycin.

Spectinomycin is produced by several species[1] of *Streptomyces*, such as *Streptomyces flavopersicus* and *Streptomyces spectabilis*. They belong to a group of branching microorganisms called actinomycetes. They are closely related to the true bacteria but are considered as higher, filamentous bacteria. Actinomycetes occupy a distinct position between the true bacteria and the true fungi. They thrive on a variety of substrates and produce a wide spectrum of antibiotics.

The Abbott Laboratories [23] and the Upjohn Company [24] pioneered the research on spectinomycin. It was known as actinospectacin when first discovered in the early sixties. It blocks protein synthesis by selectively inhibiting initiating ribosomes on bacterial and viral messengers although it fails to block ribosomes already engaged in chain elongation [25]. A series of preliminary research projects conducted by Upjohn [26, 27] culminated in establishing the structure of spectinomycin [28]. Later studies [21] further explained the biosynthesis. In particular, Otsuka et al. [22] studies the stereochemical aspects of the biosynthesis and explained the mode of conversion of glucose into the actinamine moiety and the formation of the cyclitol portion.

Cotton seed flour, soybean meal and brewers yeast, which are common constituents of industrial medium formulations, provide a rich source of amino acids for the organism. Ammonium hydroxide, the base used to control pH, also acts as a source of nitrogen during the growth phase. Between the growth phase (primary metabolism) and the production phase (secondary metabolism), there is a distinct switch from acidic to basic pH. Hence during the production phase only sulfuric acid needs to be added in requisite amounts to control the pH, whereas, during the growth phase only ammonium hydroxide needs to be added to control pH. Since the microorganism is capable of using amino acids directly from the medium, active anaploretic pathways must be present. However, it is the carbon source which plays a decisive role in spectinomycin biosynthesis. Glucose is the limiting substrate for growth and spectinomycin production by *Streptomyces*. Since glucose-6-phosphate is the precursor both for spectinomycin biosynthesis and for growth and proliferation, allocation of this resource is critical for its production.

2.2 Role of Glucose and Oxygen

Microorganisms producing spectinomycin utilize glucose for growth and maintenance and for the synthesis of the antibiotic. Since spectinomycin is an aminoglycoside and is synthesized from the intermediate glucose-6-phosphate,

[1] The species name will not be disclosed on confidential grounds.

there is a high probability of carbon catabolite regulation. Hence, it is not surprising that the level of residual glucose in the medium determines the start of spectinomycin production.

The microorganism can channel glucose in two possible ways: (i) utilize glucose predominantly for growth until the stationary phase, then predominantly for synthesis of the antibiotic and (ii) utilize glucose simultaneously for growth and production. The first method of glucose utilization corroborates carbon catabolite regulation whereas the second method does not. Evidence in the literature suggests that antibiotic-producing microorganisms usually follow the first method of glucose utilization.

The biosynthetic pathway presented in Fig. 1 shows that either the conversion of D-glucose to myo-inositol or the methylation reactions control the rate of antibiotic synthesis. The conversion of D-glucose to myo-inositol involves the reduction of NAD^+ to NADH and its reoxidization to NAD^+ in a subsequent reaction. Active methionine donates the methyl group during the methylation step. In a cyclical mechanism, the methyl group originates with the conversion of L-homocysteine to L-methionine utilizing methyl tetrahydrofolate in the presence of methyl cobalamine as catalyst. L-Methionine donates a methyl group when activated with ATP (Fig. 2).

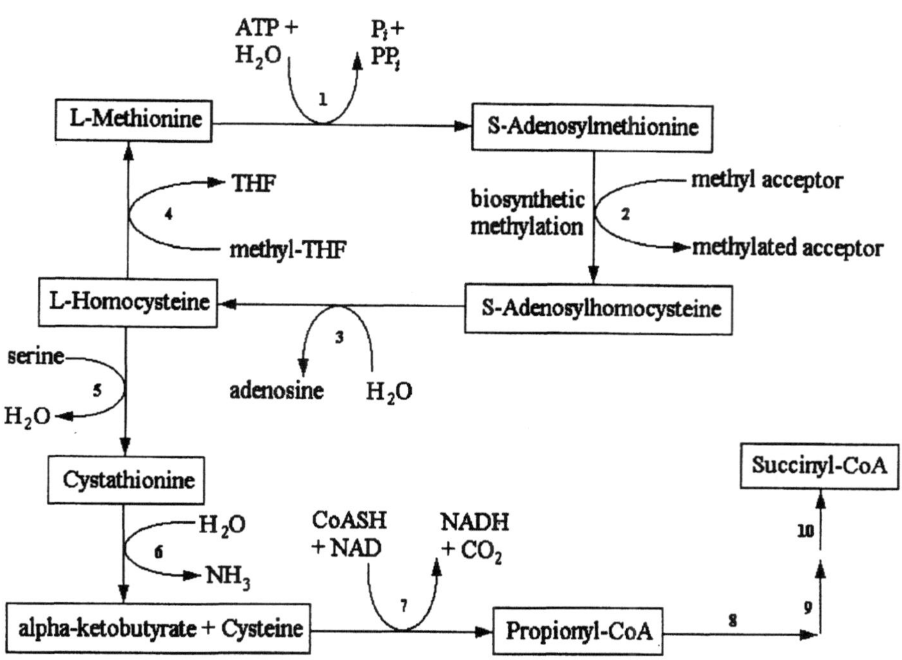

Fig. 2. L-Methionine as the biosynthetic methylating agent and its degradation to succinyl-CoA. (1) Methionine adenosyl transferase; (2) Methylase; (3) Adenosyl homocysteinase; (4) Homocysteine methyltransferase; (5) Cystathionine β-synthase; (6) Cystathionine γ-lyase; (7) α-keto acid dehydrogenase; (8) Propionyl-CoA mutase carboxylase; (9) Methylmalonyl-CoA racemase; (10) Methylmalonyl-CoA mutase

Since ATP synthesis depends directly on the respiratory action, a bioenergetic consideration indicates that both glucose and oxygen requirements require proper balancing for high yields in spectinomycin bioproduction. Further, on examining the mechanism of the two possible rate limiting steps, and considering that the microorganism possesses an enhanced anaploretic pathway for assimilating amino acids for growth, we find that methylation controls the rate of antibiotic synthesis. This implies that spectinomycin biosynthesis begins after the growth phase. However, since the organism requires nutrients for maintenance and survival even during production, an optimum glucose concentration exists for maximal antibiotic production. These aspects should be considered in formulating a scheme for spectinomycin bioproduction.

3 Bioproduction of Spectinomycin

3.1 Preparation of Inoculum

The inoculum for spectinomycin bioproduction is started from a spore suspension. The spore suspension obtained either from 20% glycerol suspensions or sporulating petri dishes, is transferred aseptically into the seed culture vessel. For laboratory scale bioproduction, the inoculation medium contains dextrose, cotton seed flour, brewers yeast, potassium phosphate and antifoam agent. In larger scale productions the dextrose may be replaced with a more economical source of carbon such as cerelose. Cotton seed flour and brewers yeast are excellent sources of amino acids and essential minerals for healthy growth of the microorganism. The initial pH is normally about 5.7 and is not adjusted before inoculation. The seed culture vessel is normally incubated at 32°C for about 33–36 h to obtain a heavy inoculum. This duration of incubation is sufficient to ensure that the seed culture is in the mid-exponential phase of growth. The amount of dextrose and potassium phosphate is critical in determining the quality of the seed culture in terms of spectinomycin productivity.

The seed culture vessel may be simply an Erlenmeyer flask for laboratory scale or a small bioreactor for pilot scale bioproduction. In the case of large scale industrial bioproduction the seed culture vessel is a well equipped reactor with options for in-situ sterilization and compressed air transfer facility to the main production bioreactor. In both cases the inoculum is grown in the batch mode.

Alternatively, two or three petri dishes may be plated from a healthy slant and incubated at 32°C. The dishes sporulate in about 3–4 days. This essential extra time requires planning and may not be acceptable for certain production facilities. However, for laboratory scale operations this technique is often followed to ensure that the microorganism is still good for antibiotic production. The spores may then be dry-scraped and suspended with sterile water and used for inoculating the seed culture. Direct inoculation from the slant is not recommended because only sporulating mycelia are good for spectinomycin production.

3.2 *Bioproduction Media*

Satisfactory yields of spectinomycin are obtained from a variety of media utilizing soybean meal as the principal nitrogen source. A typical media of the following composition was satisfactory: soybean meal 20 $g\,l^{-1}$, cerelose 25 $g\,l^{-1}$, soybean oil 5 $g\,l^{-1}$, NaCl 5 $g\,l^{-1}$, $CaCO_3$ 1 $g\,l^{-1}$ and silicone antifoam 1 $g\,l^{-1}$ with pH unadjusted [23]. Another media formulation contains 0.5% phytone, 0.5% yeast extract, 1% glucose, 0.1% Bacto-Casetone, and 0.5% NaCl [22]. However, media formulations used in industry employ cheaper ingredients whose composition are optimized to give higher final titers of the antibiotic and are a closely guarded secret. Some variations in the production media formulations include the utilization of cotton seed flour as the principal source of nitrogen and amino acids. The other essential nutrients are obtained from brewers yeast as well as from cotton seed flour. The initial pH in this case is about 5.7 and is not adjusted before inoculation.

The carbon requirement is usually met with cerelose or a similar commercial sugar source. In general, the seed culture formulation is not very different from the final production medium recipe to ensure a minimum loss in time for adaptation. Of course, the seed culture will almost always contain larger amounts of the more easily metabolizable ingredients as compared to the bioproduction medium. For laboratory scale spectinomycin production, the seed culture medium contains twice the amount of carbon source as in production medium. This enables the microorganism to increase its cell mass rapidly in the seed culture, while in plant production the amount is calculated such that carbon source exhausts within 24 h when the idiophase or production phase sets in. To maintain the productivity and to supply the maintenance energy during the production phase continuous glucose feeding is necessary. Therefore, spectinomycin production is carried out in the fed-batch mode.

3.3 *On-line and Off-line Analysis*

On-line measurements are necessary in the implementation of effective process control. Recent advances in the field of fiber optics and its application in biosensors technology resulted in the development of miniature sensors for glucose [29]. Although these glucose sensors are sensitive, selective and stable, they are not durable enough for industrial bioprocesses. Only recently has the development of an autoclavable glucose sensor been reported in the literature [30]. A fluorescence based fiber optic biosensor for the flow injection analysis (FIA) of penicillin reports excellent performance on pure samples [31]. Recently,

Park et al. [32] have developed a FIA system for glucose, fructose, and sucrose. They used a biosensor with permeabilized *Zymomonas mobilis* and invertase. Cells of *Zymomonas mobilis* were permeabilized with toluene and co-immobilized with invertase within a porcine-skin gelatin membrane. The response time of the FIA system is approximately 10 s. However, the actual implementation of most biosensors in industrial processes have not been demonstrated. Although a wide range of biosensors exist for measuring various biomolecules [33–36] their direct successful applications are few. Restriction in autoclaving and interference from components of complex medium formulations has discouraged a wider application of biosensors.

The wide scope of on-line and off-line measurements in process analysis, monitoring and control has been reviewed pragmatically by Royce [37]. Although most of the process variables such as the pH, temperature, dissolved oxygen, pressure, redox potential, gas and liquid flow rates, and power input are routinely measured on-line, they provide only a segment of the total information required to characterize the state of the bioreactor. With the exception of dissolved oxygen, these measurements give little or no direct insight into the physiological state of the process. For example, the cell mass (cell number or viability), substrate and the product concentration, which give a better understanding of the physiological state of the process are usually measured off-line.

It is possible to relate the off-gas measurements to the physiological state of the process. Consequently, off-gas is measured routinely for monitoring growth, especially in aerated bioreactors. Off-gas can be measured using gas chromatography although it takes upto 10–30 min to complete an analysis by this method. This restricts the use of gas chromatography in on-line optimization and control. More commonly, oxygen and carbon dioxide are measured by paramagnetic and infrared analyzers, respectively. The major problem with an infrared carbon dioxide analyzer is its non specificity. It suffers from interference from other compounds having similar radiation absorption characteristics, especially water. Hence, elaborate arrangements are necessary to remove moisture from the sample. Paramagnetic oxygen analyzers are extremely sensitive to pressure, flow rate and temperature, so that precision accessories are required to maintain temperature and control the gas flow rate. However, since these analyzers require only 5 s or less to complete an analysis and are accurate up to 2% or better, they are more suited than gas chromatography (GC) for on-line implementations.

Mass spectrometry has been widely used in bioprocess monitoring and control, and the cost is affordable then it is the method of choice. Mass spectrometers are capable of making rapid and accurate estimates of multiple species in bioreactor off-gas with linear response ranging over several orders of magnitude. Recently, due to considerable improvement in analysis techniques, there has been a resurgence in its application in bioprocess monitoring and control [38]. Mass spectrometry in combination with separation techniques such as high pressure liquid chromatography (HPLC) and capillary electrophoresis has opened a new front in biomolecule analysis. Mass spectrometers are capable of giving high precision measurements of bioreactor off-gas. This enables the application of elemental balancing or related techniques to obtain information about biological rates directly. A comparison of off-line measurements with on-line estimation of state variables (poly-β-hydroxybutyric (PHB) acid, residual biomass, nitrogen source and glucose) using an elemental balancing technique [38] shows excellent reconstruction.

The only disadvantage with mass spectrometry is its high cost. The design and cost of the mass spectrometer varies considerably with the species to be analyzed and the resolution desired. Consequently, research has continued in the development of new instruments for measuring off-gas. Christensen et al. [39] have developed an acoustic off-gas analyzer with performance comparable to that of a mass spectrometer. Carbon dioxide and oxygen can be analyzed continuously with this instrument with a relative accuracy of 0.01–0.33% in 0.2 s. Sipior et al. [40] have designed a phase fluorometric optical carbon dioxide gas sensor that can be made at a cost less than US$1000. The sensor is based upon the change in lifetime of a donor (sulforhodamine 101) fluorophore induced by fluorescence resonance energy transfer to a pH sensitive non fluorescent acceptor, *m*-cresol purple. On-line carbon dioxide measurements every 6 s with this sensor compared well with off-line measurements by GC requiring 20 min analysis time. The authors indicate that this sensor has potential *in vivo* applications.

In industrial scale spectinomycin production, temperature, power input, dissolved oxygen and pH are measured on-line. Off-gas analysis is normally performed using paramagnetic oxygen analyzers and infrared carbon dioxide analyzers. Glucose and spectinomycin are however measured off-line. Cell mass concentration is not reported since it cannot be measured accurately in complex medium, especially because *Streptomyces* exhibits filamentous growth. Sometimes viscosity measurements are correlated to the physiological state of the bioreactor and used to determine the start of spectinomycin production.

For the data presented here, temperature, dissolved oxygen and pH were measured on-line. The residual glucose concentration was measured off-line by an enzymatic method. Samples were taken from the bioreactor and first centrifuged in 50 ml sterile tubes at 3000 rcf for 15 min. Then 10 ml of the resulting supernatant was spun down at 14 000 g in a microcentrifuge to remove the finer suspended particles. Finally, the resulting clear supernatant was used for measuring glucose in a YSI model 27 industrial analyzer. Spectinomycin was assayed using a turbidimetric method. Problems associated with its measurements are discussed in the following section.

3.4 Spectinomycin Estimation

The absence of UV absorbing chromophores and the inherently weak chemical structure of spectinomycin makes it a difficult molecule to assay. Consequently, many different types of assay methods have been tried [23, 41–45]. The spectinomycin molecule derived from two glucose moieties also contains two secondary amines, three hydroxyl groups and a carbonyl group (Fig. 3). Consequently, this molecule is structurally weak with the α-ketohemiketal (4a position) being the most unstable. Spectinomycin easily degrades into actinospectinoic acid and actinamine. The absence of UV absorbing chromophores in spectinomycin and its primary degradation products makes its assay difficult.

The lack of UV absorbing chromophores necessitates the derivatization of spectinomycin for detection. Myers and Rindler [44] developed an HPLC method which utilizes reverse phase ion-pairing chromatography followed by post column oxidative degradation and derivatization for fluorometric detection. Among the HPLC techniques, the one developed by Tsuji and Jenkins [45] achieved the highest degree of success. The method involves pre-column derivatization of secondary amines of spectinomycin with 2-naphthalenesulfonyl chloride using lincomycin as catalyst. A normal phase silica column gives adequate separation for detecting the derivatized spectinomycin at 254 nm. This method applies to both chloride and sulphate salts. A gas-liquid chromatography [41] method involves derivatization with hexamethyl disilazane (HMDS). The polyfunctional nature mandates careful standardization of the silylation conditions to form the single tetrakis-TMS-spectinomycin derivative. The instability of this TMS derivative results in the decrease of its response during chromatographic runs. Unfortunately, these methods work relatively well only with pure samples or in defined medium cultivation. They fail or show many problems when the sample is taken from a complex medium cultivation.

Considerable literature exists on microbiological methods [46, 47]. Several bioassay techniques presented in the literature vary from colony counting to zonal diameter measurement. Among these, the turbidimetric method is the most accurate and yields results with a high degree of confidence. Furthermore, the turbidimetric method is dependable and also amenable to automation. Several antibiotic industries implement automated turbidimetric machines to

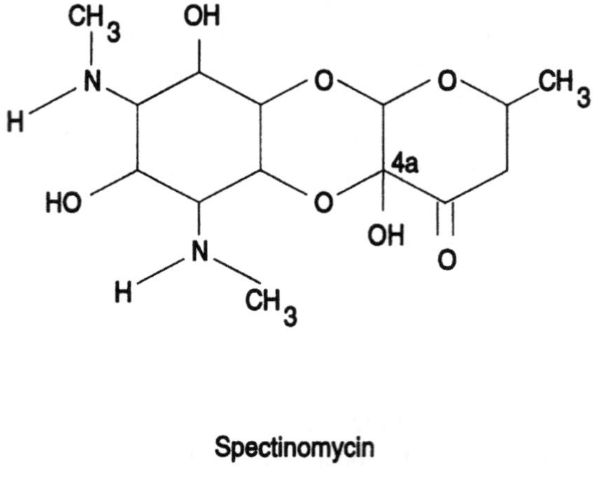

Fig. 3. The unstable feature of the spectinomycin molecule – α-ketohemiketal 4a position

sample, process the sample and measure the antibiotic concentration. Microbiological techniques employ *Escherichia coli* for assaying the chloride and sulphate salts respectively. The turbidimetric methods developed for spectinomycin require a substantial amount of time and are subject to interference from a variety of factors. Moreover, the assay reflects the potency of the sample and does not provide information about the degradation products and intermediates.

4 Fed-Batch Cultivation

Spectinomycin bioproduction proceeds in three steps. First, a spore suspension is thawed for inoculating the seed culture. Alternatively, the slants are incubated and plated on petri dishes to generate spores to be used in seed culture preparation. Petri dishes incubated at 32°C require 4 days to produce spores. Next, the seed culture is grown at 32°C for 33–36 h to obtain a rich and heavy inoculum in exponential growth phase. Finally, the production bioreactor is inoculated with 10% of the final medium volume of seed culture for growth and production. The age of the seed culture is critical in determining the quality of growth and production, and should not cross 36 h. In the case of inoculum preparation, some degree of variability is expected. Although small variations in the inoculum size will not significantly affect the bioproduction, a reasonable degree of control is necessary.

For the fed-batch cultivation, the bioproduction medium is prepared according to a standardized composition and procedure as described in Sect. 3.1 and 3.2. The bioreactor operates with an agitation of 200 rpm, at 28°C and pH adjusted to 6.4 before inoculation, dissolved oxygen is preset to some defined initial value. The pH, temperature, dissolved oxygen, air flow rate and agitation speed were monitored on-line with a computer using a software developed for data acquisition and control. The dissolved oxygen was controlled by manipulating only the air flow rate, and the pH was controlled by adding acid or base. Whereas temperature and agitation were controlled with B-Braun control modules, glucose was measured as described in Sect. 3.3, and spectinomycin was assayed by a turbidimetric method using *Escherichia coli.*

The fed-batch production was conducted for glucose feed concentrations of 100, 125, 150, 175, 200 and 400 $g\,l^{-1}$. The dissolved oxygen in the medium was controlled at 50% of the saturation value for all bioproduction runs. In this manner the glucose feed concentration for the highest yield and the correlation between glucose and oxygen demand were determined.

4.1 Glucose Feed Optimization

This optimization accounts only for the carbon and oxygen resource requirements for production in a 15 l bioreactor. Since these two variables are critical for biosynthesis we did not expect significant variations from these results for more extended optimizations. The results are presented comprehensively in Table 1. The actual spectinomycin concentration values are not revealed and have been presented in new scaled units. The time-weighted averages of the residual glucose concentration and the glucose uptake rate are also presented. Although the time weighted averages are useful in interpreting the data, the nature of the actual profiles must also be considered for a more complete analysis. We present a typical data set in Figs. 4 and 5. Fig. 4 presents the on-line

Table 1. Characteristics of fed-batch bioproduction of spectinomycin

Glucose feed concentration ($g l^{-1}$)	100	125	150	175	200	400
Maximum Spectinomycin concentration in 5 days (arbitrary units)	207	253	690	193	450	173
Spectinomycin concentration at intermediate saturation (arbitrary units)	95	100	300	75	196	none
Duration of Intermediate Span (h)	30	36	12	30	30	none
†Average residual glucose concentration ($g l^{-1}$)	0.71	0.53	1.67	1.16	1.58	31.30
†Average glucose uptake rate ($g h^{-1}$)	3.3	3.75	4.45	5.0	5.17	9.3
Maximum air flow rate ($l min^{-1}$)	8	11	5.5	11.5	10.5	5

† Time weighted average value during production phase (from 24 h to end of run)

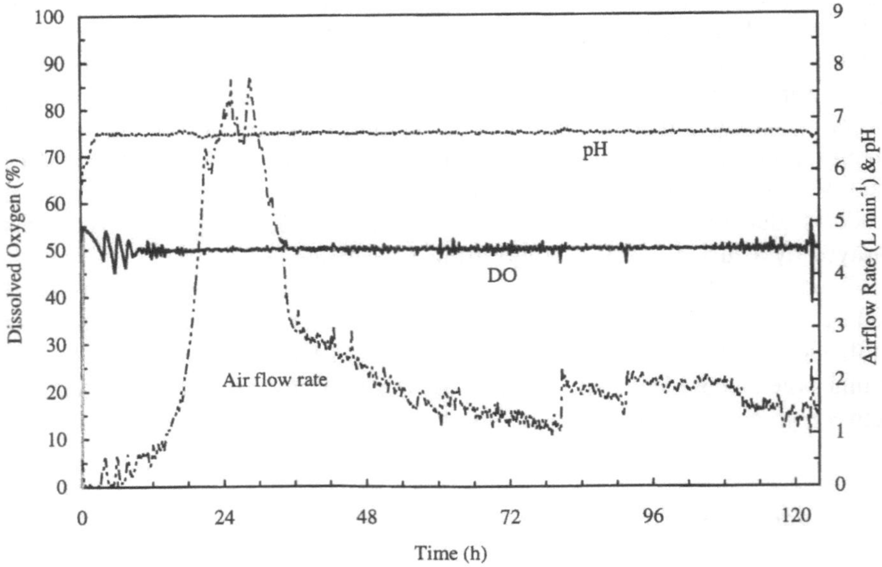

Fig. 4. Dissolved oxygen (DO) and pH control of spectinomycin bioproduction with glucose feed concentration of $100 g l^{-1}$

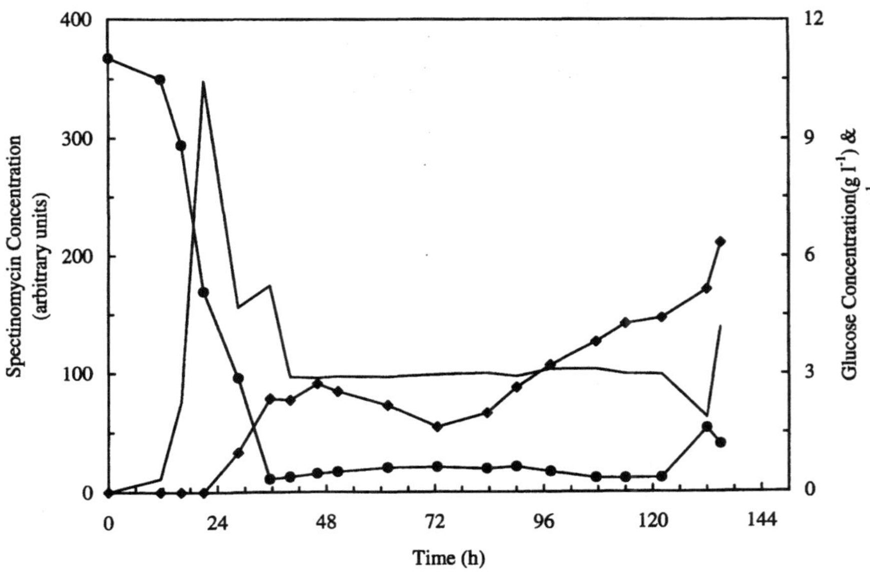

Fig. 5. Spectinomycin concentration (◆), glucose concentration (●) and glucose uptake rate (—) data for spectinomycin bioproduction with glucose feed concentration of $100\,g\,l^{-1}$

measurements of dissolved oxygen, pH and air flow rate, and the computed value of the projected dissolved oxygen concentration for fed-batch production with a glucose feed concentration of $100\,g\,l^{-1}$. Fig. 5 presents the corresponding off-line measurements of spectinomycin and glucose concentrations. The complete optimization data is presented later in Sect. 6.

From the information presented in Table 1 the following trends are observed with increasing glucose feed concentration.

a) The final spectinomycin concentrations show dual maxima at 690 units for glucose feed concentration of $150\,g\,l^{-1}$ and 450 units for glucose feed concentration of $200\,g\,l^{-1}$. Also, for a given glucose feed concentration, the spectinomycin concentration shows two levels of saturation.
b) The duration of the plateau period of spectinomycin concentration between the two saturation levels is minimum for the glucose feed concentration of $150\,g\,l^{-1}$. There was no distinct intermediate level for the glucose feed concentration of $400\,g\,l^{-1}$.
c) The average residual glucose concentration of $1.67\,g\,l^{-1}$ was observed for the feed concentration of $150\,g\,l^{-1}$. Although the highest average concentration occurs for the feed concentration of $400\,g\,l^{-1}$, it is ignored because entirely different mechanisms govern the observed trends.
d) The glucose uptake rate shows an increasing trend from 3.3–9.3 $g\,h^{-1}$ respectively for the glucose feed concentration from 100–400 $g\,l^{-1}$.
e) The maximum air flow rate values follow a trend opposite to that of the spectinomycin concentrations, excluding the $400\,g\,l^{-1}$ data based on (c).

A qualitative analysis of these observations are presented in the following sections. A better understanding of spectinomycin biosynthesis would be possible if cell mass measurements were available, unfortunately, accurate cell mass measurements in complex media with *Streptomyces* was not possible.

4.2 Characteristics of Spectinomycin Bioproduction

A typical profile for spectinomycin synthesis is represented diagrammatically in Fig. 6. Synthesis begins when the glucose concentration in the medium falls below a critical value (G_1). The lag phase is approximately 24 h under the given conditions. In the results of the fed-batch optimization conducted with glucose feed concentrations ranging from 100–400 $g\,l^{-1}$, two distinct saturation levels were observed in the synthesis profile as shown in Fig. 6. The first saturation (S_1) occurs between 2–3 days of operation, whereas the second saturation (S_2) occurs after the fourth day of operation.

The two different saturation levels indicate the presence of separate pathways for spectinomycin. This is clearly depicted in Fig. 1 where intracellular glucose can be utilized along separate pathways for production. The pathway selected depends predominantly on the residual glucose concentration and oxygen supply. It was observed that as the glucose feed concentration and hence

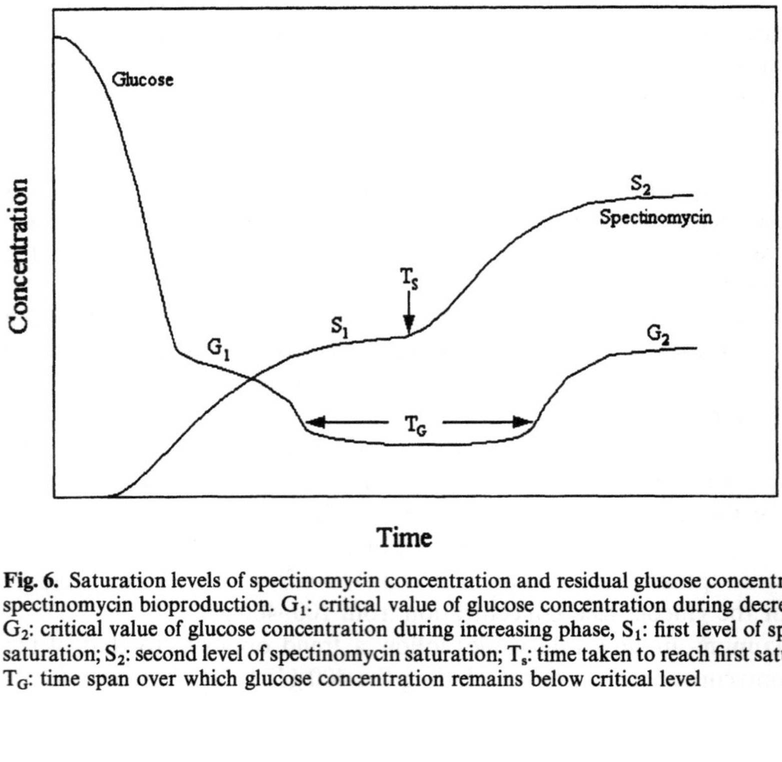

Fig. 6. Saturation levels of spectinomycin concentration and residual glucose concentration during spectinomycin bioproduction. G_1: critical value of glucose concentration during decreasing phase; G_2: critical value of glucose concentration during increasing phase, S_1: first level of spectinomycin saturation; S_2: second level of spectinomycin saturation; T_s: time taken to reach first saturation level; T_G: time span over which glucose concentration remains below critical level

the residual glucose concentration approaches the optimum, the span (T_G) of the first level of saturation reduces. As the glucose feed concentration approaches the optimum, the allocation of the carbon and oxygen resource is such that the microorganism preferentially chooses the high yield path. Consequently, it utilizes the low yield path to a lesser extent and hence the duration of the intermediate saturation level reduces. We have not determined which branch of glucose-6-phosphate leads to higher yields of spectinomycin. However, we have observed that this span is 36 h and the longest for the glucose feed concentration of 125 $g l^{-1}$. In the case of glucose feed concentrations of 100 $g l^{-1}$ and 175 $g l^{-1}$, it is approximately 30 h. The smallest span of 12 h occurs when the glucose feed concentration is 150 $g l^{-1}$, which is the optimum. However, two different saturation levels were not observed in the last case where the glucose feed concentration was 400 $g l^{-1}$. Hence, for high glucose feed concentrations such as 400 $g l^{-1}$ a different mechanism of resource allocation and uptake exists. Unfortunately, there is insufficient information to make further comments on this phenomenon.

4.3 *Influence of Residual Glucose Concentration*

The residual glucose concentration is a good indicator of spectinomycin productivity in a production schedule. In Fig. 6 a typical profile is presented. There are two different levels of the possible residual glucose concentration. The higher level of about 3.5 $g l^{-1}$ corresponds to a higher yield of spectinomycin, whereas, the lower residual glucose concentration level of about 0.5 $g l^{-1}$ corresponds to a lower yield. As the glucose feed concentration approaches the optimum value, the residual glucose concentration remains at the higher level for a longer duration. This indicates that near the optimum, the glucose feed is sufficient to meet the needs of maintenance and spectinomycin biosynthesis without becoming inhibitory or exhausting [48]. However, at glucose feed concentrations both higher and lower than the optimum, the residual glucose concentration remains at the lower level for most of the production phase. Also, in these cases, the residual glucose concentration increases towards the later part of the bioproduction. At the same time a finite glucose uptake rate exists. Possibly, a switch between biosynthetic pathways occurs to counter the sub-optimal conditions and enables the microorganism to survive.

The residual glucose concentrations for all cases are presented in Table 1. A proper analysis would require the consideration of both the residual glucose concentrations and the glucose uptake rates. Note that the glucose uptake rate has an increasing trend from the glucose feed concentration of 100–400 $g l^{-1}$. For glucose feed concentrations of 100 $g l^{-1}$ and 125 $g l^{-1}$ the residual glucose concentrations are, 0.71 $g l^{-1}$ and 0.53 $g l^{-1}$, respectively. In these cases the corresponding glucose uptake rates are 3.3 $g h^{-1}$ and 3.75 $g h^{-1}$. Thus, although the glucose feed concentration and the glucose uptake rate increased, the residual glucose concentration failed to increase. Therefore, even at a glucose feed concentration of 125 $g\,l^{-1}$, the carbon resource supply is below the quantity need to sustain high spectinomycin productivity.

In the diametrically opposite case, for glucose feed concentrations of 175 $g\,l^{-1}$ and 200 $g\,l^{-1}$, the residual glucose concentrations are, 1.16 $g\,l^{-1}$ and 1.58 $g\,l^{-1}$, respectively, whereas, the corresponding glucose uptake rates are 5.0 $g\,h^{-1}$ and 5.17 $g\,h^{-1}$. An examination of the residual glucose concentration profiles show that an increasing trend towards the later part of the bioproduction contributes to the higher residual glucose concentrations. Although the glucose uptake rate increased significantly, the residual glucose concentration remained nearly the same. Therefore, it appears that a glucose feed concentration in excess of that required for the production of spectinomycin increases the flux of the carbon resource in alternative pathways. In this process the microorganism would utilize more energy in alternate pathways thereby resulting in a lower yield of spectinomycin.

An analysis of the profiles indicates that the increase in the residual glucose concentration towards the end of production contributes significantly to the value of 31.3 $g\,l^{-1}$ for the 400 $g\,l^{-1}$ glucose feed concentration. Table 1 also shows that the dual saturation profile in spectinomycin synthesis observed for other glucose feed concentrations are not apparent in this case. However, the glucose uptake rate is as high as 9.3 $g\,h^{-1}$. Since the yield of spectinomycin in this case is only 173 units, the carbon could be channelized towards the synthesizing of cell material or towards the accumulation of intermediates. We speculate that a distinctly different metabolic behavior occur for high glucose feed concentrations.

Therefore excluding the results for the 400 $g\,l^{-1}$ glucose feed concentration, the highest time-weighted average residual glucose concentration of 1.67 $g\,l^{-1}$ occurs for the optimum glucose feed concentration of 150 $g\,l^{-1}$. However, the absolute value of 3–3.5 $g\,l^{-1}$ is more important. Menawat [48] found that the residual glucose concentration of 2.1 $g\,l^{-1}$ gives the highest productivity of spectinomycin. From the residual glucose profile it appears that if the glucose concentration in the bioreactor is maintained at the higher level throughout the production phase, the highest yield of spectinomycin will be obtained. To achieve this it may be necessary to implement a variable glucose feeding strategy during the production phase.

4.4 Variations in Air Flow Rate

The air flow rate is continuously manipulated to meet the changing oxygen needs of the microorganism. Consequently, the air flow rate profile reflects the variations in oxygen demand during the course of bioproduction. The oxygen requirement is the highest during the exponential growth phase. During the transition from growth to production of spectinomycin, the air flow rate gradually decreases. It reaches a new value which remains nearly constant during the production phase. Compressed air was supplied to the bioreactor at a constant pressure head of 20 psig through a linear trim Badger control valve. The valve actuator required air to open and its maximum flow rate at 20 psig was $12\,l\,min^{-1}$.

The maximum air flow rate attained during exponential phase was different for the various glucose feed concentrations used. A maximum air flow rate of $11.5\,l\,min^{-1}$ was required when the glucose feed concentration was $175\,g\,l^{-1}$. This is similar to the maximum air flow rate of $11\,l\,min^{-1}$ for the glucose feed concentration of $125\,g\,l^{-1}$. These two glucose feed concentrations are equidistant from the optimum value of $150\,g\,l^{-1}$. In fact, even the nature of the air flow rate profiles are similar. Notice that the final and intermediate values of spectinomycin concentrations are also in the same range. Clearly, the route for the biosynthesis of spectinomycin followed by the microorganism is the same in these two cases. Consequently, the energy requirements and hence the oxygen demand are comparable. In the case of glucose feed concentrations of $100\,g\,l^{-1}$ and $200\,g\,l^{-1}$ the maximum air flow rates are $8\,l\,min^{-1}$ and $10.5\,l\,min^{-1}$, respectively. Here also the energy requirements and hence the oxygen demand are comparable.

The lowest value for the maximum air flow rate of $5\,l\,min^{-1}$ was obtained for the glucose feed concentration of $400\,g\,l^{-1}$ followed by $5.5\,l\,min^{-1}$ for the glucose feed concentration of $150\,g\,l^{-1}$. The glucose feed concentration of $150\,g\,l^{-1}$ corresponds to the highest final spectinomycin concentration of 690 units. Also, the time-weighted average glucose uptake rate of $4.45\,g\,h^{-1}$ is intermediate, while its time-weighted average residual glucose concentration of $1.67\,g\,l^{-1}$ is the highest. These values indicate that the oxygen and carbon utilization during both the growth phase and the production phase is more balanced. Hence, the high spectinomycin yield is also energetically favored.

However, the lowest spectinomycin yield was obtained for the $400\,g\,l^{-1}$ feed concentration. In this case, the other variables such as intermediate spectinomycin saturation and the residual glucose concentration also do not follow the trends observed in the other feed concentration data. A more detailed study is necessary before this phenomenon can be explained.

5 Two Metabolite Hypothesis

There are two routes for the synthesis of spectinomycin starting from glucose-6-phosphate as shown in Fig. 1. Along the first route 1,3-myo-inosadiamine and actinospectose combine to give desmethyl spectinomycin which on methylation gives spectinomycin. Whereas the second route involves the methylation of 1,3-myo-inosadiamine first to actinamine. Actinamine then combines with actinospectose to produce spectinomycin. Depending on the amount of glucose and oxygen present, production via one of these routes predominates. At low glucose feed concentrations the carbon resource is not sufficient to sustain high spectinomycin productivity. Whereas, at glucose feed concentrations higher than the optimum the residual glucose concentration represses spectinomycin synthesis. As a result, the final spectinomycin titers obtained shows a dual maxima as indicated in Fig. 7. These maxima occur for glucose feed concentrations of 150 $g\,l^{-1}$ and 200 $g\,l^{-1}$. Figure 8 shows similar results obtained in an independent investigation [48].

Figure 1 illustrates that the methylation of 1,3 myo-inosadiamine is the energetically critical step. Methylation occurs by the donation of a methyl group by L-methionine in the presence of ATP. This ATP is generated by the amphibolic utilization of glucose. Hence, the microorganism will need to channel the energy so that it is subjected to minimum strain. Therefore, two situations are possible:

(a) For a given amount of glucose resource, there may be insufficient energy to spare for methylation of 1,3-myo-inosadiamine. Thus, 1,3-myo-inosadiamine accumulates and results in the desmethyl spectinomycin formation. When the energetics are more favorable, spectinomycin is produced by the methylation of desmethyl spectinomycin.

(b) The glucose resource available may be such that the microorganism is able to spare sufficient energy for the methylation of 1,3 myo-inosadiamine to actinamine. In this case both actinamine and actinospectose would yield spectinomycin.

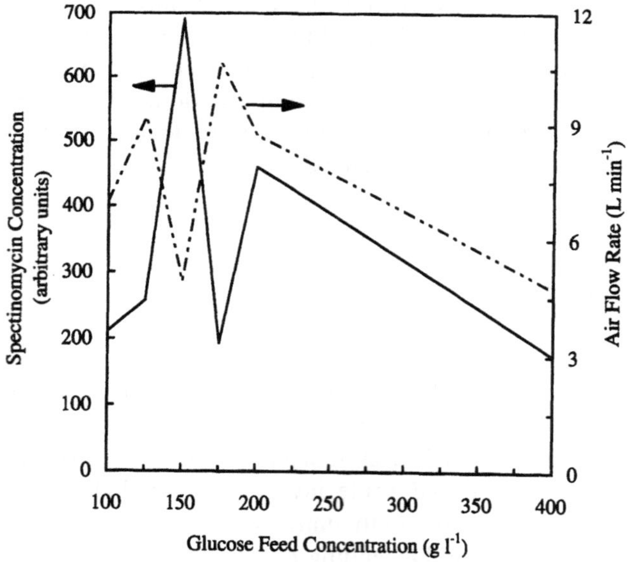

Fig. 7. Variation in spectinomycin yields and maximum air flow rate with glucose feed concentrations

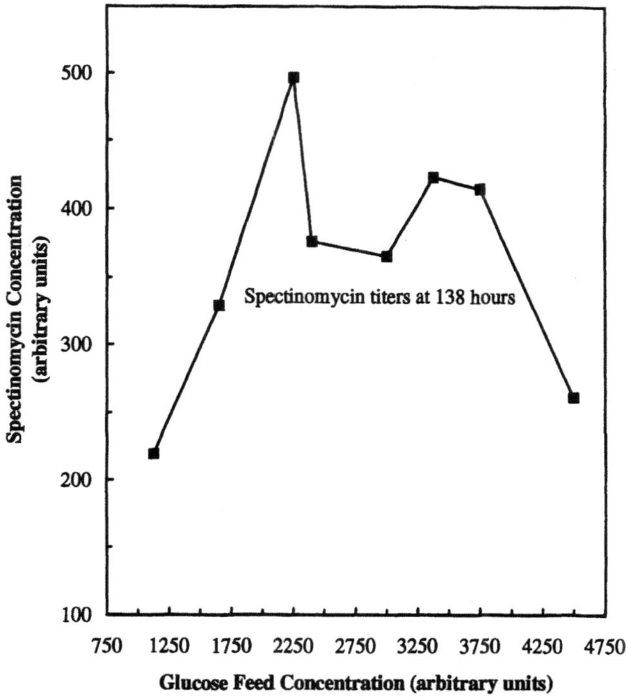

Fig. 8. Evidence of bi-modal maxima in spectinomycin bioproduction [48]

Therefore, depending on the glucose feed concentration, the two intermediate metabolites, 1,3 myo-inosadiamine and actinamine, or both, determine the rate and the total amount of spectinomycin synthesized. Only a more detailed investigation will reveal which pathway is favored for a given glucose feed concentration. Nevertheless, certain qualitative inferences can be made about the mechanism of spectinomycin biosynthesis based on the available information.

6 Exponential Structure for Mechanistic Model

Spectinomycin synthesis begins after a lag phase of about 24 h. As synthesis continues its concentration increases quickly and saturates at the first level. Later its concentration increases once again and saturates at a second higher level. We interpret this profile in the following simplistic manner. When the microorganism is moving out its growth phase, some bio-signal triggers or *activates* the beginning of spectinomycin synthesis. The metabolic activity of the microorganism then gradually switches over to the synthesis of spectinomycin synthesis through a series of enzymatic *reactions*. Spectinomycin synthesis continues until it saturates at first level. The deactivation or *inhibition* of the first route of biosynthesis is followed by the activation of spectinomycin synthesis through an alternate route. This happens when the environmental conditions can no longer support the synthesis of spectinomycin by the first pathway. As before, the synthesis of spectinomycin continues until it saturates at some new level. Hence, spectinomycin biosynthesis follows an *activation-reaction-inhibition* sequence of events. Clearly two separate routes are involved in which two different key intermediates, 1,3 myo-inosadiamine and actinospectose, participate. Therefore, the *activation-reaction-inhibition* sequence holds for each of the two intermediates participating in the biosynthesis of spectinomycin.

A mathematical description of the process should also incorporate parameter sensitivity, both geometric and numerical, for implementation in process control. Geometric sensitivity enables a close tracking of the actual experimental data profile, whereas numerical sensitivity reduces computation time and allows accurate parameter evaluation. This is vital for spectinomycin biosynthesis because the lag phase is considerable. The subsequent increase in concentration is steep and there are two saturation levels. To describe these rapid changes in the profile an exponential variation of the Monod kinetics is developed here.

Assuming that the overall rate is inhibited according to first order kinetics, and applying an exponential variation of the Monod equation (see Appendix), the following exponential inhibition kinetics is obtained.

$$\mu(S) = \mu_m \exp\left(-\frac{K_e}{S}\right) \exp\left(-\frac{S}{K_i}\right) \tag{1}$$

where μ_m is the maximum specific growth rate, K_e is the exponential equivalent of the Monod constant K_m, K_i is the inhibition constant and S is the substrate concentration. The first exponential component $\exp(-K_e/S)$ is the reaction term and the second exponential component $\exp(-S/K_i)$ is the inhibition term. This is the exponential structure used to describe the biosynthesis of spectinomycin. The model equations (Eq. A.7) are presented in the appendix.

6.1 Model Simulation

The parameter values of the model Eq. A.7 obtained along with the sum of the squares of the error per data point are also presented in Table 2. These errors are the combined sum of errors for spectinomycin and residual glucose concentrations per data point. T_{f0}, which denotes the starting times for glucose feeding is known from the recorded data. The parameters presented in Table 2 were computed by nonlinear regression and excluded the last two data points in each of the data sets. The last two data points were excluded because the regulatory mechanism governing this phenomenon is not accounted in the model. Such

Table 2. Exponential model parameters

	$100\,g\,l^{-1}$	$125\,g\,l^{-1}$	$150\,g\,l^{-1}$	$175\,g\,l^{-1}$	$200\,g\,l^{-1}$	$400\,g\,l^{-1}$
Titer (units)	207	253	690	193	450	173
μ_m (h^{-1})	0.350	0.350	0.350	0.400	0.350	0.170
K_e ($g\,l^{-1}$)	3.000	3.000	3.000	3.000	3.000	5.300
K_i ($g\,l^{-1}$)	4.200	4.200	4.000	3.850	4.200	11.5125
μ_{P1} (h^{-1})	0.500	0.500	0.500	0.500	0.500	0.250
K_{e1} ($g\,l^{-1}$)	0.150	0.150	0.150	0.150	0.150	0.150
K_{i1} ($g\,l^{-1}$)	0.220	0.220	0.275	0.275	0.275	0.220
μ_{P2} (h^{-1})	0.250	0.250	0.250	0.250	0.150	0.100
K_{e2} ($g\,l^{-1}$)	2.000	2.000	2.000	2.000	2.000	8.000
K_{i2} ($g\,l^{-1}$)	1.500	1.500	3.200	2.200	2.500	10.0985
Y_{CS} ($gC\,g^{-1}S$)	0.400	0.400	0.400	0.270	0.400	0.490
Y_{P1S} ($gP_1\,g^{-1}S$)	0.300	0.300	0.450	0.280	0.400	0.135
Y_{P2S} ($gP_2\,g^{-1}S$)	0.543	0.543	0.543	0.523	0.543	0.090
K_d (h^{-1})	3.23E-02	3.28E-02	2.13E-02	5.33E-02	3.54E-02	3.325E-02
T_{f0} (h)	18	18	18	22	20	22
C_0 ($g\,l^{-1}$)	2.5	3.5	2.5	3.4	3.4	3.0
$\sum e^2/N$	0.394	0.446	1.774	0.438	1.285	5.300

a phenomena definitely requires a more structured model for describing the mechanisms and cannot be achieved by the proposed model. However, the simulated profiles of the model over the other data points in each case are accurate and thus demonstrate their ability to describe spectinomycin production.

The model parameters determined from the data when the glucose feed concentration is $100\,g\,l^{-1}$ also adequately describes the data from bioproduction with other glucose feed concentrations. However, calculations show that slightly different values of some of the parameters give a better fit of the data (Table 2). The only exception occurs for a glucose feed concentration of $400\,g\,l^{-1}$.

Remarkably, the parameters for one fed-batch bioproduction describes the other data for the other production runs, where the glucose feed concentrations are different in each case. Therefore the model is a good representation of the mechanism of bioproduction. Further, the premise of the model, that two metabolites contribute to the final formation of spectinomycin, is supported by the satisfactory match between the simulated profiles and data supports this hypothesis. Figures 9–14 show how the model fits the bioproduction data.

It would be useful if an invertible relationship exists between spectinomycin and the air flow rate (oxygen demand) in the proposed model. If such a relationship exists then it would be possible to control spectinomycin biosynthesis by manipulating the air flow rate. However, one stepping stone remains, namely, the cell mass concentration which appears in the model. As mentioned before, it is difficult to accurately determine the cell mass concentration on-line in complex medium for filamentous microorganisms. To overcome this problem a nonlinear systems tool called the external differential representation (EDR) [49–52]

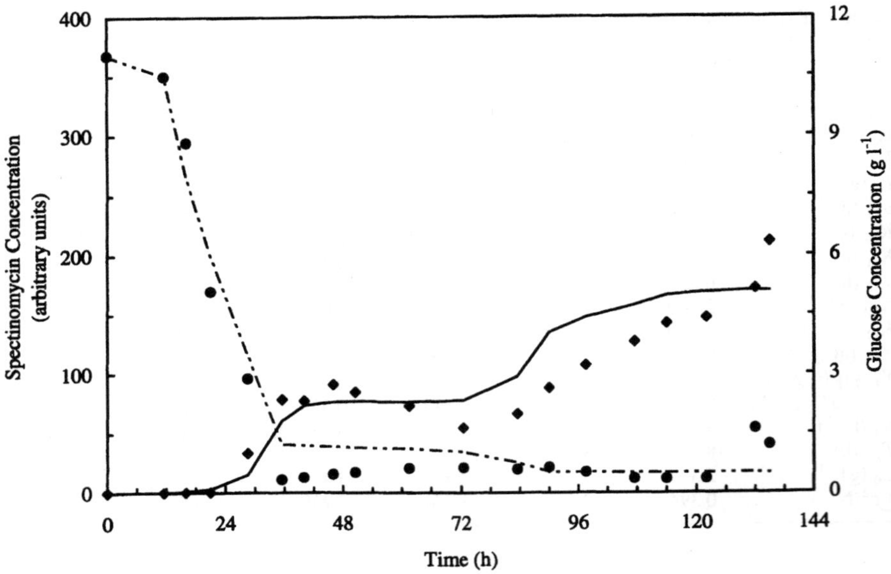

Fig. 9. Fed-batch bioproduction of spectinomycin with glucose feed concentration of 100 $g l^{-1}$ ● glucose concentration; ◆ spectinomycin concentration; — model fit for spectinomycin concentration; ---- model fit for glucose concentration

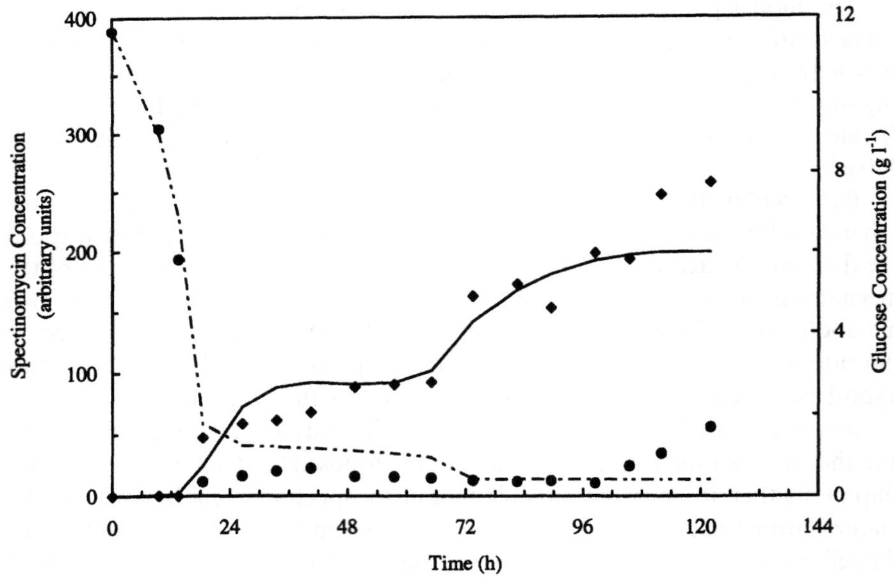

Fig. 10. Fed-batch bioproduction of spectinomycin with glucose feed concentration of 125 $g l^{-1}$. ● glucose concentration; ◆ spectinomycin concentration; — model fit for spectinomycin concentration; ---- model fit for glucose concentration

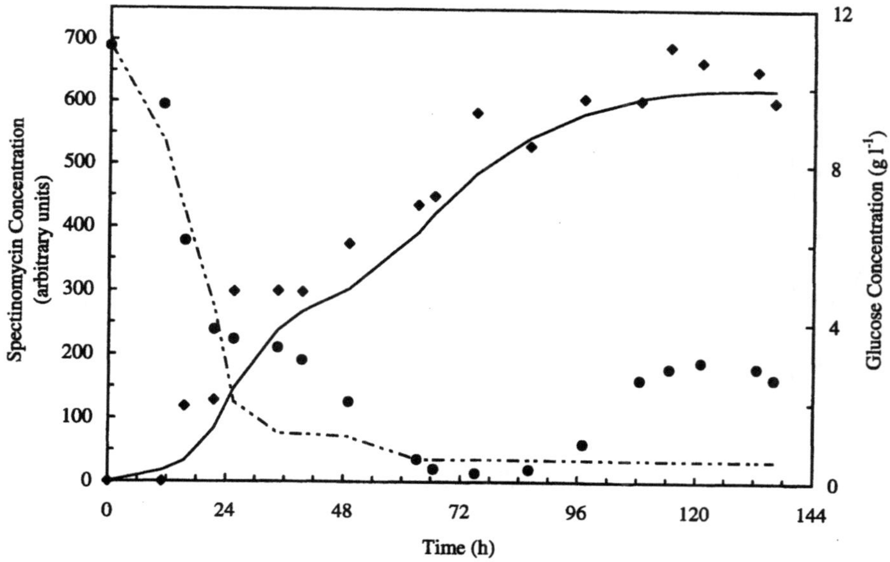

Fig. 11. Fed-batch bioproduction of spectinomycin with glucose feed concentration of 150 g l^{-1}. ● glucose concentration; ◆ spectinomycin concentration; — model fit for spectinomycin concentration; ---- model fit for glucose concentration

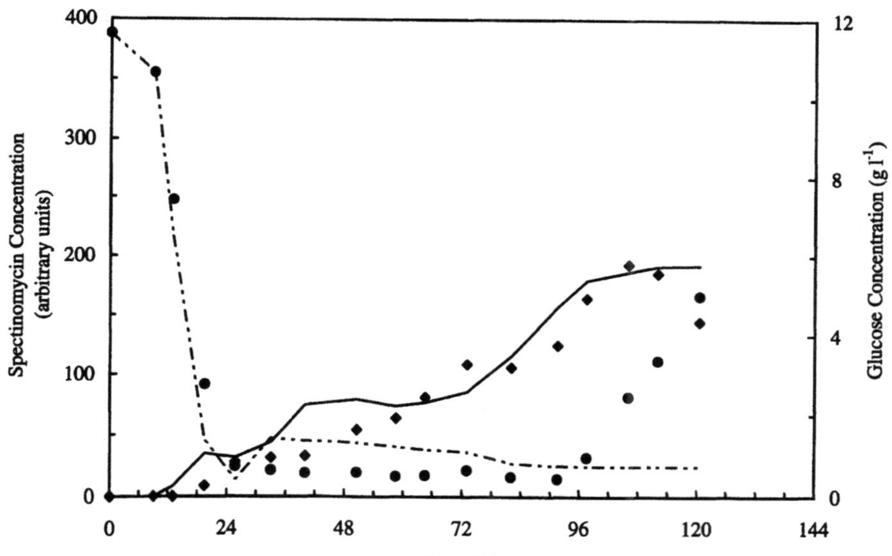

Fig. 12. Fed-batch bioproduction of spectinomycin with glucose feed concentration of 175 g l^{-1}. ● glucose concentration; ◆ spectinomycin concentration; — model fit for spectinomycin concentration; ---- model fit for glucose concentration

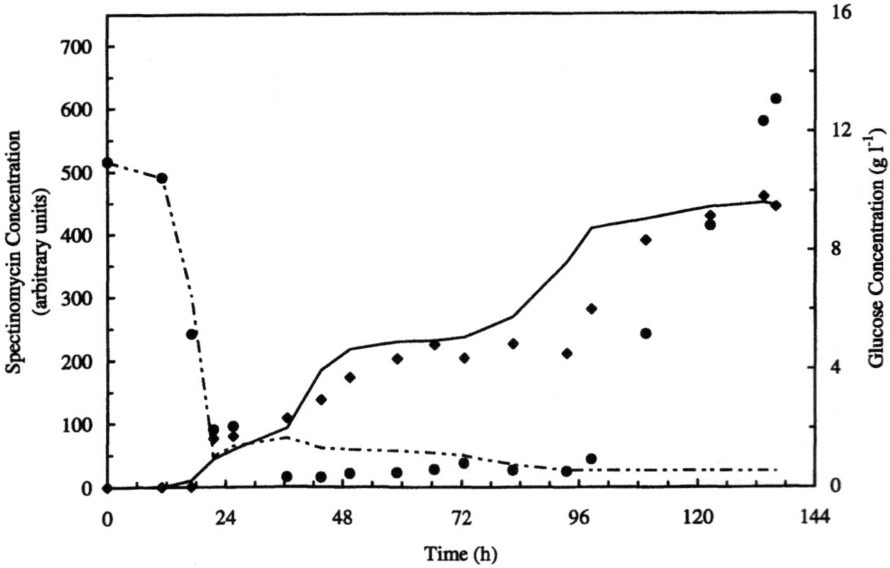

Fig. 13. Fed-batch bioproduction of spectinomycin with glucose feed concentration of 200 g l^{-1}. ● glucose concentration; ◆ spectinomycin concentration; — model fit for spectinomycin concentration; ---- model fit for glucose concentration

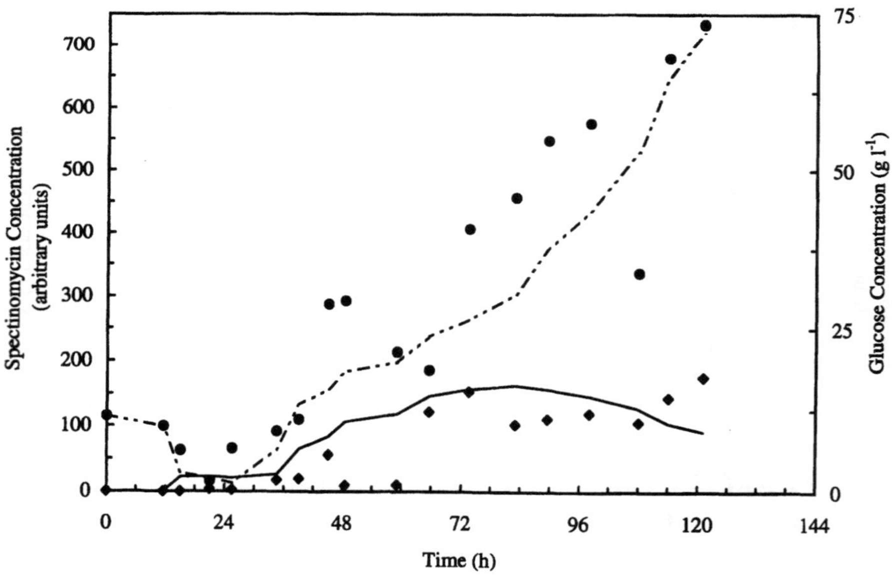

Fig. 14. Fed-batch bioproduction of spectinomycin with glucose feed concentration of 400 g l^{-1}. ● glucose concentration; ◆ spectinomycin concentration; — model fit for spectinomycin concentration; ---- model fit for glucose concentration

was employed. The EDR essentially provided the missing link between spectinomycin and the air flow rate to enable the prediction of one from the other.

7 Prediction

7.1 External Differential Representation

The EDR is the mathematical relationship or mapping between the inputs and the outputs of a system. It correlates the outputs to some or all the system inputs. Figure 15 gives a geometric interpretation of this mapping. Input-output structures, which are analogous to transfer functions in process control, have many applications. For example, the EDR can be used in parameter estimation, trajectory tracking and process control.

The system described by Eq. (A.7) is first checked for observability with the product and substrate concentrations [53–56]. Since controlling the dissolved oxygen concentration at 50% saturation eliminates its dynamics, the oxygen

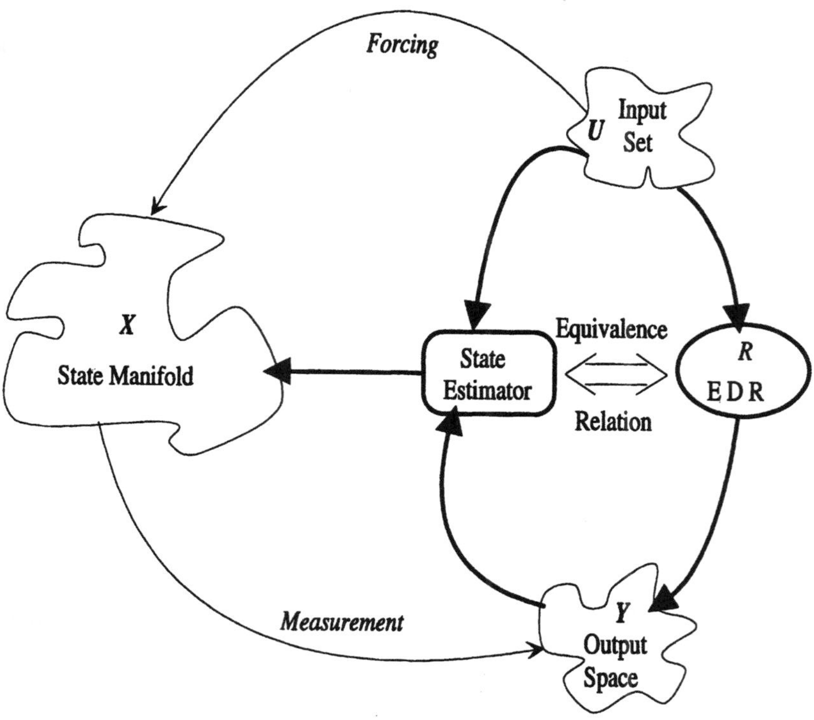

Fig. 15. Geometry of the External Differential Representation (EDR)

balance equation can be used to compute the product concentration provided the observability condition holds. If the observability criterion is satisfied, an equation for computing the cell mass concentration from the glucose and spectinomycin concentrations can be derived. Following a step-by-step procedure [50–52], it is easy to determine that S and P measurements make the process observable. The EDR for the cell mass concentration is given by

$$C = -\dot{S} + \frac{S_F}{V} - \frac{S}{V}\frac{dV}{dt} + \alpha \frac{\mu_m}{Y_{CS}} \exp\left(-\frac{K_e}{S} - \frac{S}{K_i}\right) + \alpha_1 \frac{\mu_{P_1}}{U_{P1S}} \exp\left(-\frac{K_{e1}}{S} - \frac{S}{K_{i1}}\right) + \alpha_2 \frac{\mu_{P_2}}{Y_{P2S}} \exp\left(-\frac{K_{e2}}{S} - \frac{S}{K_{i2}}\right) \tag{2}$$

7.2 Air Flow Rate

The activation-reaction-inhibition model describes the mechanism of spectinomycin biosynthesis as shown in the previous section. Furthermore, the simulated profiles of the model match the data accurately (Table 2). However, the oxygen dynamics do not appear in the model (Eq. A.7) because the dissolved oxygen is controlled at 50%. The air flow rate which varies to meet the oxygen demand contains physiological information otherwise present in dissolved oxygen measurements. The following equation approximates the oxygen utilized during the fermentation when dissolved oxygen is constant.

$$\begin{aligned} OUR = &\frac{1}{Y_{CA}} \alpha \hat{\mu}_m \exp\left(-\frac{K_e}{S} - \frac{S}{K_i}\right) C \\ &+ \frac{1}{Y_{P1A}}\left(\alpha_1 \hat{\mu}_{P_1} \exp\left(-\frac{K_{e1}}{S} - \frac{S}{K_{i1}}\right) C - K_d P_1\right) \\ &+ \frac{1}{Y_{P2A}}\left(\alpha_2 \hat{\mu}_{P_2} \exp\left(-\frac{K_{e2}}{S} - \frac{S}{K_{i2}}\right) C - K_d P_2\right) \\ = &k_{la}(A^* - A) \end{aligned} \tag{3}$$

Where Y_{CA}, Y_{P1A} and Y_{P2A} are the yield coefficients for cell, product 1 and product 2 respectively. The above equation can be implemented to estimate the oxygen uptake rate (OUR) measurements of dissolved oxygen in the bioreactor provided the value of the mass transfer coefficient k_{la} is available for every data point. Since the k_{la} is a function of the hydrodynamic conditions of the bioreactor it plays a significant role in its operation and control. In general k_{la} is a function of the volumetric air flow rate v, the medium viscosity μ, the medium density ρ and the power to volume ratio P/V used by the bioreactor. Under similar hydrodynamic conditions and agitation, a direct correlation exists between the air flow rate v and the mass transfer coefficient k_{la}. Physiologically, oxygen demand varies according to the metabolic state of the microorganism. Consequently, the air flow rate is linked to the rate of antibiotic production.

The oxygen balance contains the same activation-reaction-inhibition terms as the model but with the yield coefficients defined for oxygen utilization. Since the air flow rate reflects the oxygen demand of the process, it should be possible to predict the airflow rate from the oxygen balance equation. For this prediction of the air flow rate to be meaningful, the parameters evaluated from the data by nonlinear regression should be used. If this prediction is possible, it would demonstrate that the energetic relationship which exists between spectinomycin and oxygen utilization is also described by the model. Hence, if the air flow rate profile for maximum spectinomycin productivity is known a priori, then the maximum spectinomycin productivity can be obtained for every fed-batch run by controlling the air flow rate profile.

It is indeed possible to reconstruct the air flow rate profile using the parameters evaluated (Table 2) by nonlinear regression from the spectinomycin and glucose data. The air flow rate data points used for demonstrating the prediction are each the mean values of 20 on-line sample points. The yield coefficients Y_{CA}, Y_{P1A}, and Y_{P2A} were calculated from the data for the substrate uptake for growth and product formation. Here k_{la} is treated as a parameter. Once k_{la} is evaluated, the air flow rate can be computed using a suitable expression relating air flow rate to the mass transfer coefficient k_{la}. Figures 16–20 show the predicted air flow rate profiles along with the actual air flow rate data.

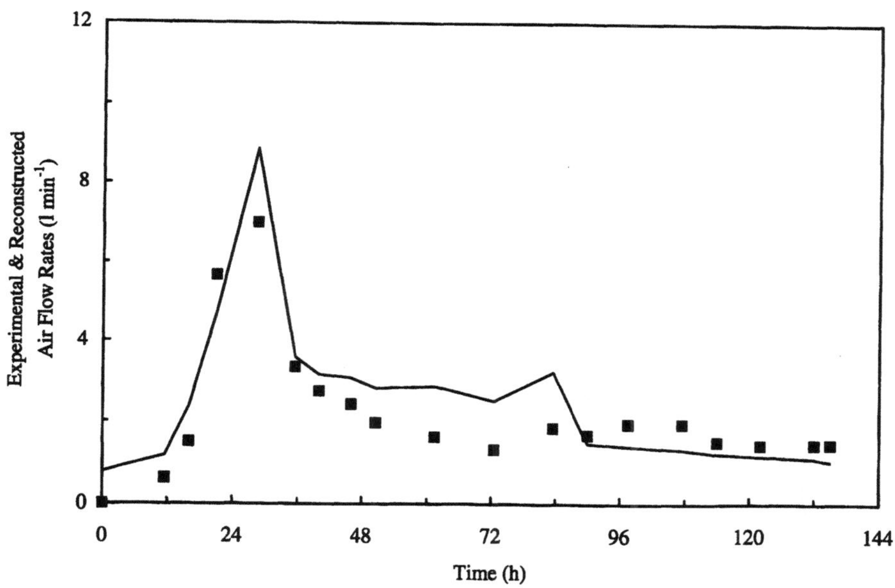

Fig. 16. Reconstruction of the air flow rate profile for spectinomycin bioproduction with glucose feed concentration of 100 g l^{-1}. ■ air flow rate data; — predicted air flow rate

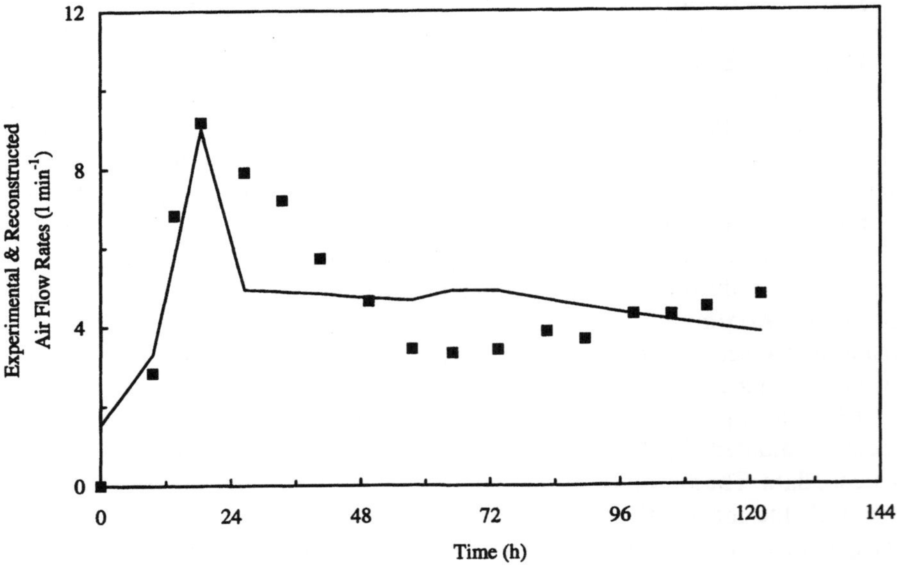

Fig. 17. Reconstruction of the air flow rate profile for spectinomycin bioproduction with glucose feed concentration of 125 $g l^{-1}$. ■ air flow rate data; — predicted air flow rate

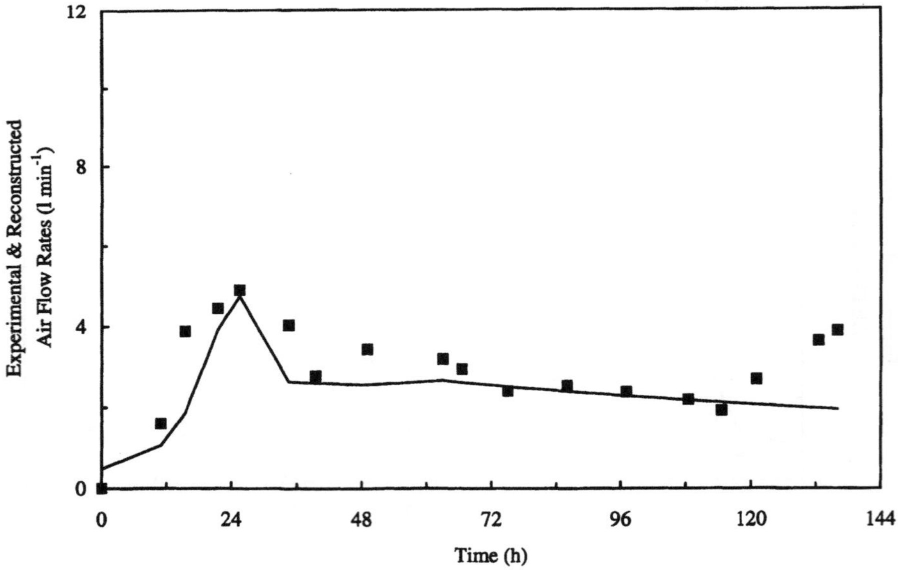

Fig. 18. Reconstruction of the air flow rate profile for spectinomycin bioproduction with glucose feed concentration of 150 $g l^{-1}$. ■ air flow rate data; — predicted air flow rate

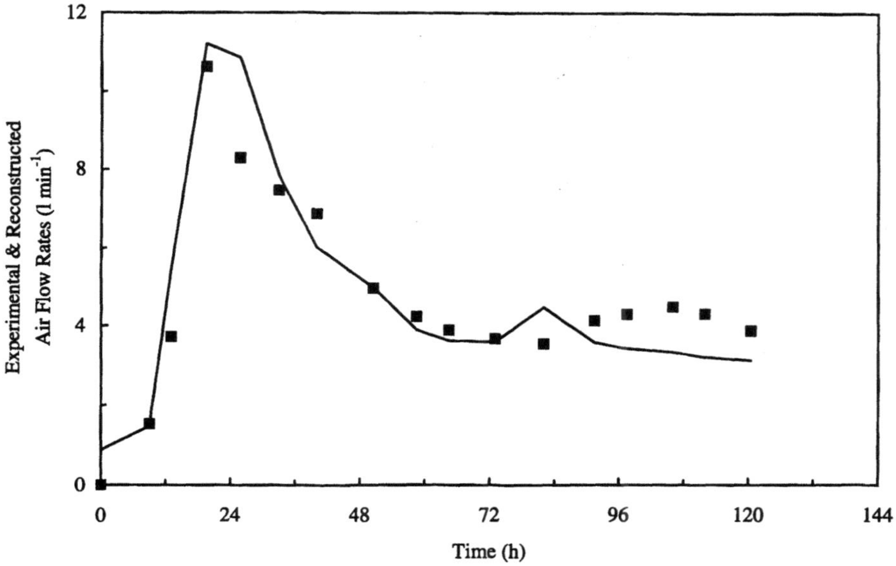

Fig. 19. Reconstruction of the air flow rate profile for spectinomycin bioproduction with glucose feed concentration of 175 $g\,l^{-1}$. ■ air flow rate data; — predicted air flow rate

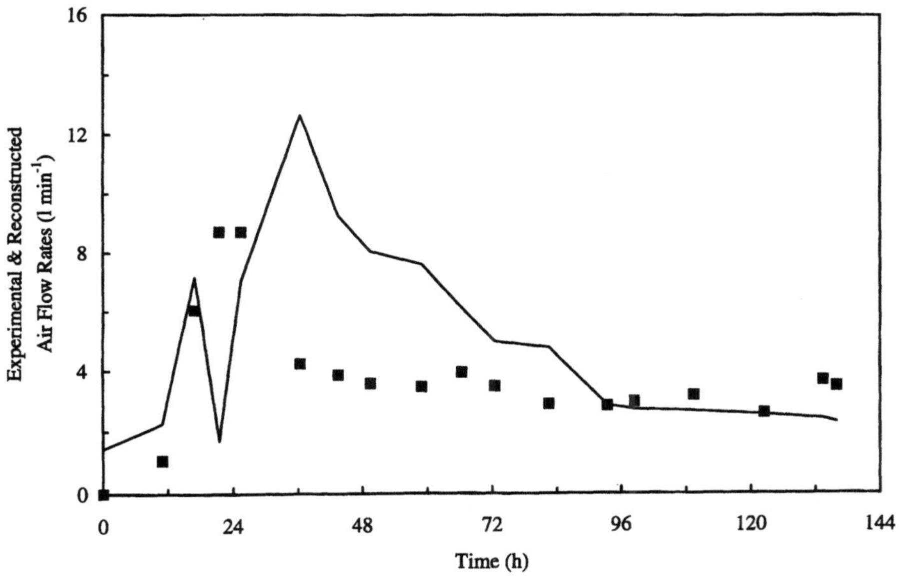

Fig. 20. Reconstruction of the air flow rate profile for spectinomycin bioproduction with glucose feed concentration of 200 $g\,l^{-1}$. ■ air flow rate data; — predicted air flow rate

7.3 Spectinomycin

Prediction of spectinomycin from the air flow rate data is the industrially more important reconstruction problem. Since the hydrodynamic conditions in all the bioproduction runs were maintained by controlling the operational conditions, the relationship between air flow rate and spectinomycin is not dependent on the physical conditions. Hence, for simplicity and to demonstrate that this prediction is possible, it is assumed that spectinomycin varies linearly with the air flow rate. The other variables are related according to the model. Thus, Eq. (3) after some manipulation, and noting that $(A^* - A)$ is constant when dissolved oxygen is controlled at 50% becomes

$$\begin{aligned} P &= P_1 + P_2 \\ &= \kappa\,\alpha\,\mu_m \exp\left(-\frac{K_e}{S} - \frac{S}{K_i}\right)C + \kappa_1\,\alpha_1\,\mu_{P_1} \exp\left(-\frac{K_{e1}}{S} - \frac{S}{K_{i1}}\right)C \\ &\quad + \kappa_2\,\alpha_2\,\mu_{P_2} \exp\left(-\frac{K_{e2}}{S} - \frac{S}{K_{i2}}\right)C - \kappa_3 v \end{aligned} \tag{4}$$

The parameters obtained to give best fit of the bioproduction runs (Table 2) are the same parameters used in the prediction of the antibiotic concentration. Only the coefficients κ, κ_1, κ_2 and κ_3 need to be determined. These coefficients are determined by minimizing the error in the predicted values of spectinomycin concentration.

Figures 21–23 present the predicted profiles for spectinomycin and correspond to glucose feed concentrations of 125 gl^{-1}, 150 gl^{-1} and 175 gl^{-1}, respectively. The trends obtained are excellent. The prediction reflects the different characteristics of the data, such as the dual saturation feature. Further refinement of the prediction is possible but has not been carried out in this study.

It is evident that the relationship between the air flow rate (oxygen demand) and the spectinomycin concentration is invertible. Hence, by implementing a suitable control strategy it is possible to control the spectinomycin productivity by controlling the bioproduction along a desired air flow rate trajectory. However, this control strategy should be defined within the overall strategy to maximize the production of spectinomycin. It must be clear how and why the fed-batch bioproduction of spectinomycin is to be conducted. These ideas are discussed in the following section.

8 Control Strategy for Bioproduction

The fed-batch bioproduction of spectinomycin is governed by the central question as to how the microorganism allocates glucose towards spectinomycin

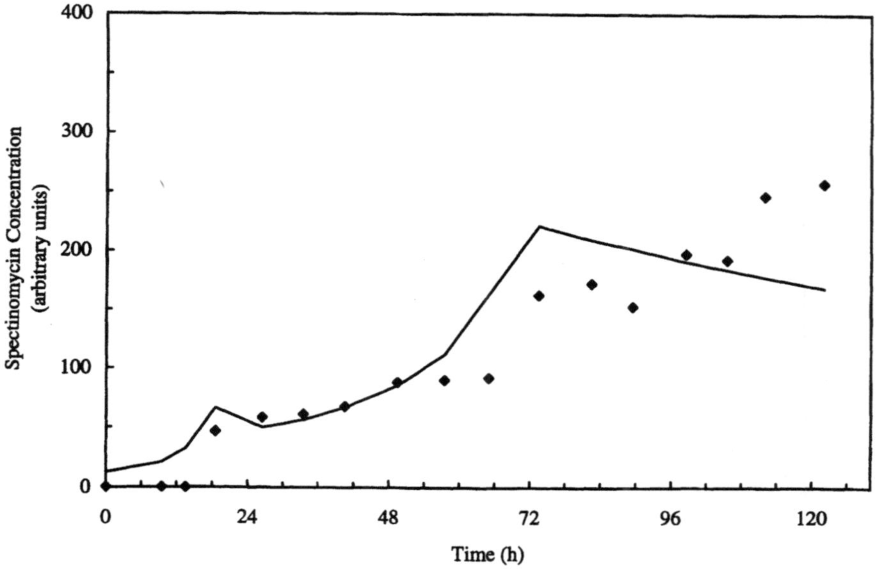

Fig. 21. Prediction of spectinomycin concentration from air flow rate data using the External Differential Representation for spectinomycin bioproduction with glucose feed concentration of 125 $g l^{-1}$. ◆ spectinomycin data; — predicted spectinomycin concentration

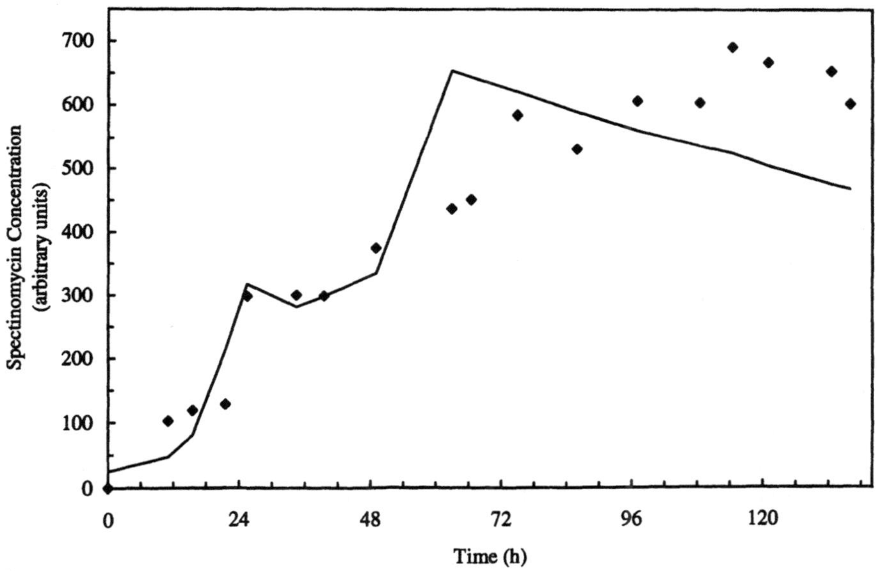

Fig. 22. Prediction of spectinomycin concentration from air flow rate data using the External Differential Representation for spectinomycin bioproduction with glucose feed concentration of 150 $g l^{-1}$. ◆ spectinomycin data; — predicted spectinomycin concentration

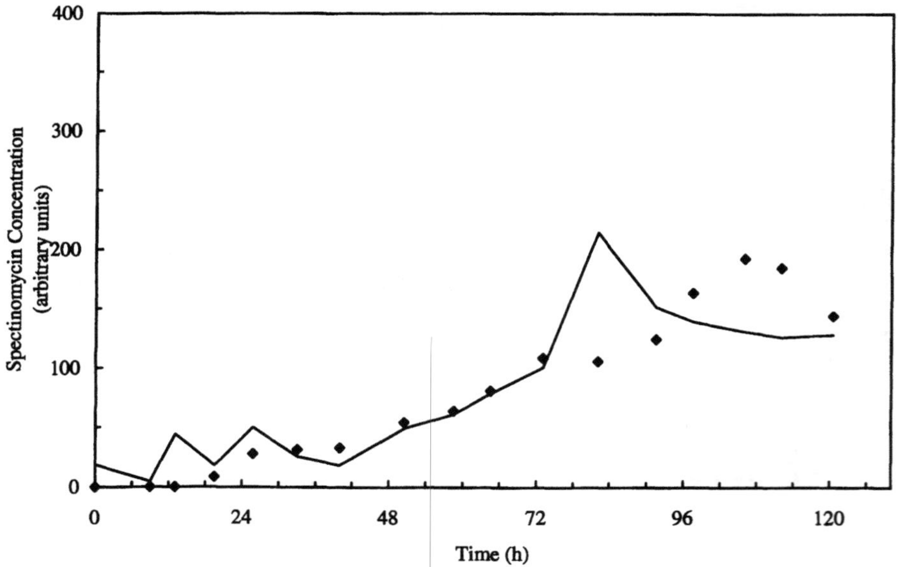

Fig. 23. Prediction of spectinomycin concentration from air flow rate data using the External Differential Representation for spectinomycin bioproduction with glucose feed concentration of 175 $g l^{-1}$. ◆ spectinomycin data; — predicted spectinomycin concentration

biosynthesis and towards growth for various levels of residual glucose concentrations. Figure 24 gives a diagrammatic representation of the metabolic state of the microorganism during spectinomycin biosynthesis. This figure shows how the metabolism is distributed between four characteristic zones: (1) synthesis of product 1, (2) synthesis of product 2 and growth, (3) exponential growth, and (4) decay and inhibition. The optimization should be aimed at maintaining the operational conditions so that the metabolism remains in zone (1). A constant dissolved oxygen condition and constant rate of feeding glucose will not achieve this objective since it is impossible to meet changing demands of the metabolism with constant rates. For example, even for a glucose feed concentration of 150 $g l^{-1}$, which produced the highest amount of spectinomycin, gradual drift from production to growth occurs towards the end of the bioproduction. The larger the deviation from the optimal conditions, the more noticeable are the drifts in the glucose uptake and antibiotic production profiles.

Low glucose feeding rates resulted in the early exhaustion of glucose and hence low yield of spectinomycin. As the glucose feed concentration increased the microorganism performed better and produced higher spectinomycin titers. However, as long as the glucose feed is insufficient a change in metabolism occurs in the later stages of the bioproduction. On the other side of the optimum, glucose is supplied at a slightly faster rate than required for maintaining the metabolism in zone (1). Consequently, most of the bioproduction occurs in zone (2). For example, when the glucose feed concentration is 200 $g l^{-1}$, the

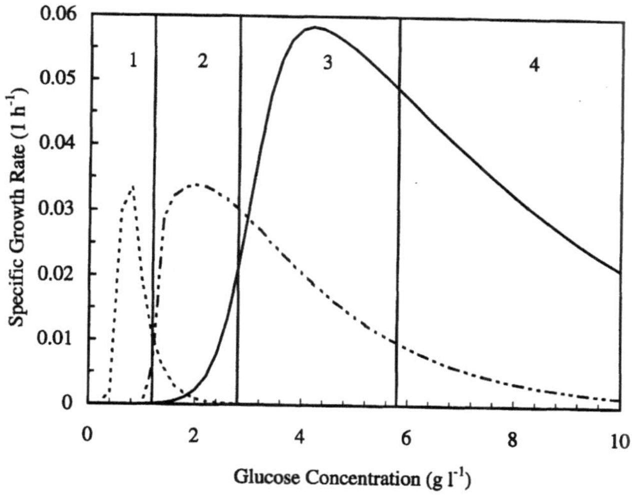

Fig. 24. Metabolic zones in spectinomycin bioproduction

preferred metabolism corresponds to zone (2) and zone (3). Finally for the glucose feed concentration of 400 $g\,l^{-1}$ inhibition and decay dominated the bioproduction corresponding to zone (4).

It is evident that the optimum operating range is narrow. Hence, controlling the bioproduction in this range requires the implementation of an advanced controller. Furthermore, the choice of the control variable plays an important role in the performance of the controller. The glucose concentration takes several hours to respond completely to an input or disturbance. Hence this variable is unsuitable for implementing a direct control with stringent requirements, especially since the process is sensitive to changes. In comparison, the oxygen response requires only a few minutes. Since the energetics of spectinomycin biosynthesis are coupled with the carbon regulation, spectinomycin productivity can be controlled by controlling the air flow rate (dissolved oxygen) along a predetermined optimal profile. The best set of optimum operating conditions for spectinomycin bioproduction will involve both an optimal glucose feeding profile and a corresponding regulation of the dissolved oxygen along an optimal profile. The derivation of the geometric control algorithm, which has been implemented to control the dissolved oxygen and pH for optimizing glucose feed as shown here, is presented elsewhere [49].

9 Conclusions and Future Prospects

Spectinomycin is an aminoglycoside antibiotic. Its synthesis begins from glucose-6-phosphate and proceeds along two branches to give the final product.

The overall yield of spectinomycin depends on two key intermediates, 1,3-myo-inosadiamine or actinamine. The dual maxima of spectinomycin with changing glucose feed concentrations and the two level saturations in the spectinomycin production profiles support this observation. Results suggest that the residual glucose concentration determines the flux of the carbon source through the two branches. Since one of these branches is energetically more favorable, there is a distinct residual glucose concentration for which the spectinomycin yield is maximum.

To describe the sharp changes from growth to production phase and the rapid switch from one level of saturation to another, we proposed an *Activation-Reaction-Inhibition* structure model. The classical Monod structure could not account for the special characteristics of spectinomycin bioproduction. The model comprised exponential kinetic structures in contrast to the hyperbolic structures of the Monod type kinetics. It was demonstrated that the model exhibited higher geometric and parameter sensitivity in comparison to the Monod type model.

Using this *Activation-Reaction-Inhibition* model it is possible to correlate the air flow rate to the antibiotic concentration. The EDR was developed for air flow and spectinomycin measurements. For this prediction the cell mass concentration values were not known because they could not be measured. Hence, the required cell mass concentration values for the prediction were evaluated with the EDR. Air flow rate profiles reconstructed from the identified model compare well with the raw data. Similarly, the spectinomycin concentrations were predicted from the air flow rate. This result is significant because if the air flow rate profile which gives highest spectinomycin yield is known, the process may be controlled along this trajectory every time to consistently achieve maximum productivity.

However, there are several aspects of spectinomycin bioproduction which need further investigation. The most important among them is to determine which branch of the spectinomycin biosynthetic pathway is the energetically favorable one. It would be critical also to determine the residual glucose concentration necessary to keep this biosynthetic pathway active during the entire production phase and study its variation at different scales of operation. An idea of the actual fluxes in the pathway, bottle necks and correlation with the EMP pathway will positively contribute to further enhancing the spectinomycin yield.

Since methionine and cobalamine affect the amination reactions, their influence on the overall spectinomycin productivity may also be investigated. Other important aspects relate to the effect of changing the glucose feeding rate during the production phase and simultaneously implementing the corresponding optimal air flow rate. Indeed a substantial amount of work remains with regard to the contribution of external parameters on the yield of spectinomycin bioproduction.

Acknowledgements. This work was conducted at the Department of Chemical Engineering. Tulane University, New Orleans, USA, and financially supported by the Abbott Laboratories and in part by Sigma Xi, Grants in Aid of Research. The authors also acknowledge the facilities and opportunity provided by IIT, Delhi, India in completing this manuscript. The critical comments of Dr. P.K. Roychoudhury and Dr. U.S. Agarwal which helped in shaping the final form of this manuscript are sincerely appreciated.

10 Appendix

Mathematical Basis for Exponential Structure

The Monod equation is a phenomenological model for representing the growth of microorganisms. If we expand the Monod equation in a polynomial form by performing the division we obtain the following solution

$$\frac{\mu_m S}{K_m + S} = \mu_m - \mu_m \frac{K_m}{S} + \mu_m \frac{K_m^2}{S^2} - \mu_m \frac{K_m^3}{S^3} \cdots \quad \cdots \tag{A.1}$$

Where μ_m is the maximum specific growth rate, K_m is the Monod constant and S is the substrate concentration. Now, comparing this series form of the Monod equation with the exponential series

$$\sum_0^\infty (-1)^n \frac{K_e^n}{n!S^n} = 1 - \frac{K_e}{S} + \frac{K_e^2}{2!S^2} - \frac{K_e^3}{3!S^3} + \cdots \quad \cdots = e^{-K_e/S} \tag{A.2}$$

we immediately notice that the factorial components present in Eq. (A.1) are absent in Eq. (A.2). Hence the exponential equivalent of the Monod equation is

$$\mu(S) = \mu_m \exp\left(-\frac{K_e}{S}\right) \tag{A.3}$$

Where K_e is the equivalent of the Monod constant K_m. The factorial components influence the shape of the generated curve in the lower range of substrate S values. At higher S values the difference between the Monod and the exponential kinetics becomes negligible. Figure A.1 presents the difference between the exponential and Monod structures with $K_e = K_m$. The exponential kinetic structure exhibits a delay in the initial slope resembling the prolonged lag phase observed in spectinomycin biosynthesis. Furthermore, the change of slope at $S = K_e = K_m$ is more sensitive to K_e (exponential structure) than K_m (Monod structure). Notice that both the Monod and the exponential structures saturate at μ_m. However, the constants K_m and K_e exhibit some difference. When $K_m = S$ we obtain the ratio $\mu/\mu_m = 2$ for the Monod equation. In the case of the exponential kinetics $\mu/\mu_m = e$.

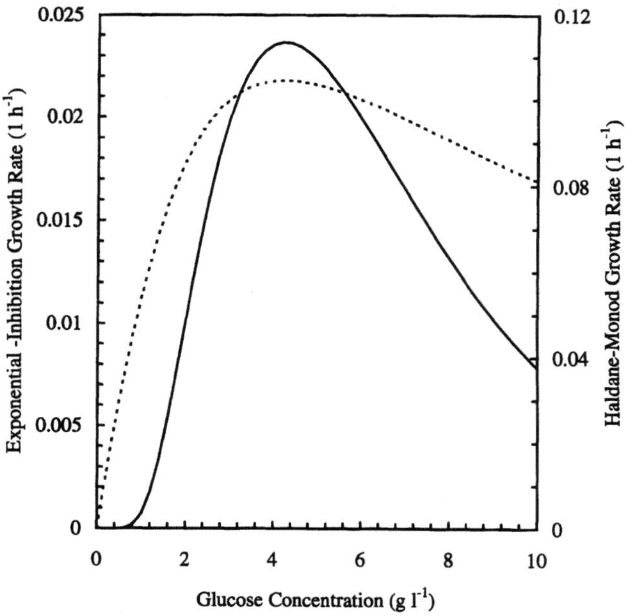

Fig. A.1. Variation of specific growth rate for Monod and exponential structures.

The exponential structure displays significant advantages over the Monod structure in evaluating parameters from experimental data. Obtaining kinetic parameters from experimental data with Monod equation accompanies error. For example, points least accurately measured and furthest from the origin dominate the slope of the Lineweaver-Burke plot ($1/\mu$ versus $1/S$). Accurately measured values are clustered near the origin and contribute less towards the slope of the curve. The exponential structure removes this problem in parameter evaluation. Here a semi-logarithmic plot of ln μ versus $1/S$ gives the value of K_e directly from the slope ($-K_e$) and the maximum growth rate from the intercept (ln μ_m). In the semi-logarithm plot, the logarithm of the values dampens the effect of noise in the data. Whereas in the Lineweaver-Burke plot the reciprocal magnifies this effect. The logarithm eliminates problems with order of magnitude differences in experimental values as well. Above all, this structure allows easy evaluation of the kinetic parameters. These properties of the exponential structure advantageous for describing spectinomycin biosynthesis and identifying the process parameters where the observed trends are quite complex.

Coincidentally, the maximum of this function with respect to the substrate concentration occurs at the same values as that for Haldane-Monod kinetics, namely, $S = \sqrt{(K_m/K_i)} = \sqrt{(K_e/K_i)}$. Figure A.2 presents a comparison of the Haldane-Monod and exponential inhibition kinetics for the same K_m and K_i values. The magnitude of the function at the maximum is smaller for the exponential structure because of the stronger influence of the inhibition

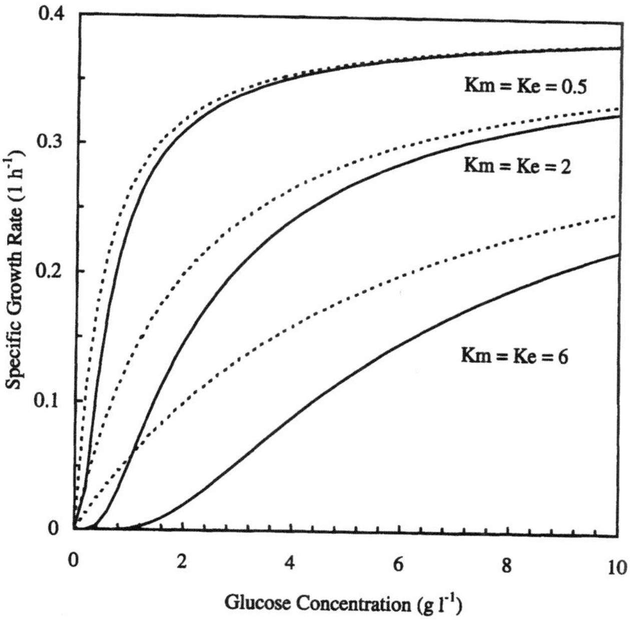

Fig. A.2. Position of specific growth rate for Haldane-Monod and exponential-inhibition structures.

coefficient K_i. Obviously, the much higher inhibition effect experienced from the exponential term results in this decrease. Hence, although the maximum occurs at the same point for identical values of K_m ($= K_e$) and K_i, the true significance of these constants depend on the kinetic structure. Due to the nature of the exponential function, the shape of the profiles are many times more sensitive than in the Haldane-Monod kinetics. Therefore, the exponential structure describes satisfactorily the sharp changes in slope observed in the spectinomycin concentration, the glucose concentration, and the air flow rate profile.

Although the Monod structure is the most widely used to describe cell growth in bioprocesses, it has two important drawbacks. First, it has a hyperbolic structure and hence the parameters of the denominator are relatively insensitive. Second, being similar to the Michaelis-Menten kinetics, which assumes the quasi-steady state condition of intermediates, restricts the validity of the model to steady state operations. The Haldane-Monod kinetics which is an extension of the Monod kinetics to circumstances where inhibition is observed, shows insensitivity in both the Monod constant K_m and the inhibition constant K_i. Simulations demonstrate that the exponential structure represents the low glucose range far better than the Monod structure. In addition, the derivative of the exponential structure is zero at $S = 0$ and is the correct physical representation of the process when substrate is absent whereas for the Monod kinetics this value is μ_m/K_m. Hence, the exponential structure is more suitable for describing the true mechanism of spectinomycin biosynthesis accurately.

The exponential form is excellent for parameter estimation. When converted into its logarithmic equivalent the resulting algebraic equation contains all the parameters independent of each other and in their native forms. This characteristic increases the sensitivity and accuracy in the estimation of its parameters. The exponential kinetics,

$$\mu(S) = \mu_m \exp\left(-\frac{K_e}{S}\right)\exp\left(-\frac{S}{K_i}\right) \tag{A.4}$$

where K_e is the exponential model constant equivalent to the Monod constant K_m and K_i is the inhibition constant, reduces to the form

$$\ln \mu = -\frac{S}{K_i} - \frac{K_e}{S} + \ln \mu_m \tag{A.5}$$

suitable for parameter estimation. The Haldane-Monod structure under similar manipulation gives

$$\mu = -\frac{S\mu}{K_i} + \mu_m - \frac{\mu K_m}{S} \tag{A.6}$$

The states of Eq. (A.6) cannot be decoupled. Consequently, parameter estimation from the Haldane-Monod kinetics suffers from problems associated with interaction and insensitivity. In the case of the exponential structure, a simple regression on the substrate data gives the estimates of the parameters. Whereas, for the Haldane-Monod kinetics, the coupled parameters make it difficult to estimate their value. The procedure is not straight forward because by coupling, such as, μ/S and μS, the states lose their individual properties and the associated errors amplify by propagation. This results in inaccuracies in the estimates of all three parameters. Therefore, from the discussions above we conclude that the exponential structure contains more characteristics for describing spectinomycin bioproduction. Furthermore, data analysis is easier in the exponential form. Above all, the exponential kinetics reduces to the Monod-type structure under special conditions.

Mechanistic Model

We have established in Section 5 that there are two alternate routes for the biosynthesis of spectinomycin. In these alternate routes, two different metabolites, P_1 and P_2, contribute to the final concentration of spectinomycin. Since the final concentration of spectinomycin depends on the concentrations of P_1 and P_2, we assume that the observed concentration of spectinomycin is the sum of the concentrations of P_1 and P_2. The other variables describing the fermentation are the substrate (glucose) concentration S, cell mass concentration C and the dissolved oxygen (from air) concentration A. We assume that both products P_1 and P_2 degrade at a constant rate K_d. Then the equations for the

Activation-Reaction-Inhibition model for the fed-batch production of spectinomycin are,

$$\frac{dC}{dt} = \alpha\,\mu_m \exp\left(-\frac{K_e}{S} - \frac{S}{K_i}\right) C - \frac{C}{V}\frac{dV}{dt}$$

$$\frac{dP_1}{dt} = \alpha_1\,\mu_{P_1} \exp\left(-\frac{K_{e1}}{S} - \frac{S}{K_{i1}}\right) C - K_d P_1 - \frac{P_1}{V}\frac{dV}{dt}$$

$$\frac{dP_2}{dt} = \alpha_2\,\mu_{P_2} \exp\left(-\frac{K_{e2}}{S} - \frac{S}{K_{i2}}\right) C - K_d P_2 - \frac{P_2}{V}\frac{dV}{dt}$$

$$\frac{dS}{dt} = -\frac{1}{Y_{CS}}\alpha\,\mu_m \exp\left(-\frac{K_e}{S} - \frac{S}{K_i}\right) C - \frac{1}{Y_{P1S}}\alpha_1\,\mu_{P_1} \exp\left(-\frac{K_{e1}}{S} - \frac{S}{K_{i1}}\right) C$$
$$- \frac{1}{Y_{P2S}}\alpha_2\,\mu_{P_2} \exp\left(-\frac{K_{e2}}{S} - \frac{S}{K_{i2}}\right) C + \frac{S_F}{V} - \frac{S}{V}\frac{dV}{dt}$$

$$\frac{dV}{dt} = F \qquad \text{(A.7)}$$

Here α, α_1, and α_2 are the activation components. The maximum specific growth rate is μ_m, and μ_{P_1} and μ_{P_2} the maximum specific product formation rates for P_1 and P_2. Also K_e and K_i are the exponential model and the inhibition constants for cell growth; whereas K_{e1}, K_{i1} and K_{e2}, K_{i2} are exponential model and inhibition constants for P_1 and P_2, respectively.

The *activation* terms are constant for the process and act as a trigger to describe the change from lag phase to growth phase. They have the form $\exp(a(S-b)/(1+\exp(a(S-b))$. The two parameters a and b need to be evaluated only once, so that they match with the start of spectinomycin synthesis. In subsequent estimation of the model parameters, α is simply treated as a constant premultiplier. In this sense, it does not belong to the set of model parameters evaluated by nonlinear regression. The estimation model parameters are presented in Table 2.

11 References

1. Martin JF, Demain AL (1980) Microbiol Rev 44: 230
2. Lilley G, Rowley BI, Bull AT (1974) J Appl Chem Biotechnol 24: 677
3. Gray PP, Bhuwapathanapun S (1980) Biotechnol Bioeng 22: 1785
4. Hirsch CF (1981) In: Schlessinger D (ed) Regulation of secondary metabolism in microorganisms, Am Soc Microbiol, Washington DC
5. Aharonowitz Y (1980) Ann Rev Microbiol 34: 209
6. Aharonowitz Y, Demain AL (1979) Can J Microbiol 25: 61
7. Demain AL, Kennel YM, Aharanowitz Y (1979) Symp Soc Gen Microbiol 29: 163
8. Basak K, Majumdar SK (1973) Antimicrob Agents, Chemother 4: 6

9. Weinberg ED (1977) In: Weinberg ED (ed) Mineral element control of microbial secondary metabolism, Microorganisms and minerals, Marcell Dekker, New York
10. Kominek LA (1972) Antimicrob Agents, Chemother 7: 856
11. Pirt SJ, Righelato RC (1967) Appl Microbil 15: 1284
12. Righelato RC, Trinci APJ, Pirt SJ, Peak A (1968) J Gen Microbiol 50: 399
13. Pla LC (1971) Biochim Biophys Acta 242: 541
14. Martin JF (1977) Adv Biochem Eng 6: 105
15. Martin JF, Demain AL (1977) J Bacteriol 132: 590
16. Atkinson DE, Walton GM (1967) J Biol Chem 242: 3239
17. Chapman AG, Fall L, Atkinson DE (1971) J Bacteriol 108: 1072
18. Fynn GH, Davison JA (1976) J Gen Microbiol 94: 68
19. Purich DL, Fromm HJ (1973) J Biol Chem 25: 461
20. Martin JF, Liras P, Demain AL (1978) Biochem Biophys Res Commun 83: 822
21. Mitscher LA, Martin LL, Feller DR, Martin JR, Goldstein AW (1971) Chem Commun 1541
22. Otsuka H, Mascaretti OA, Hurley LH, Floss HG (1980) J Am Chem Soc 102: 6817
23. Oliver TJ, Goldstein A, Bower RR, Holper JC, Otto RH (1961) Antimicrob Agents Chemother 495
24. Mason DJ, Dietz A, Smith RM (1961) Antibiot Chemother 11: 118
25. Wallace BJ, Tai PC, Davis BD (1974) Proc Nat Acad Sci USA 71: 1634
26. Wiley PF (1962) J Am Chem Soc 84: 1514
27. Hoeksema H, Argoudelis AD, Wiley PF (1962) J Am Chem Soc 84: 5212
28. Wiley PF, Argoudelis AD, Hoeksema H (1963) J Am Chem Soc 85: 2652
29. Mansouri S, Schultz JS (1984) Bio/Technology 2: 885
30. Phelps MR, Hobbs JB, Kilburn DG, Turner RFB (1995) Biotechnol Bioeng 46: 514
31. Xie X, Suleiman AA, Guilbault GG (1992) Biotechnol Bioeng 39: 147
32. Park JK, Shin MC, Lee SG, Kim HS (1995) Biotechnol Prog 11: 58
33. Karube I (1984) Biotechnol Genet Eng Rev 2: 313
34. Watanabe E, Takagi M, Takei S (1991) Biotechnol Bioeng 38: 99
35. Riedel K, Renneberg R, Wollenberger U, Kaiser G, Scheller FW (1989) J Chem Tech 44: 85
36. Leech D (1994) Chem Soc Rev 23: 205
37. Royce PN (1993) Crit Rev Biotechnol 13: 117
38. Heinzle E (1992) J Biotechnol 25: 81
39. Christensen LH, Schulze U, Nielsen J, Villadsen J (1995) Chem Eng Sci 50: 2601
40. Sipior J, Eichhorn LR, Lakowicz JR, Carter GM, Rao G (1996) Biotechnol Prog 12: 266
41. Brown LW, Bowman PB (1974) J Chromatogr Sci 12: 373
42. Wills PJ, Wise R (1978) J Antimicrob Chemother 4: 279
43. Borowiecka B, Chojnowski W, Pajchel G (1984) Acta Polon Pharm 61: 195
44. Myers HN, Rindler JV (1979) J Chromatogr 176: 103
45. Tsuji K, Jenkins KM (1985) J Chromatogr 333: 365
46. Hanka LJ, Mason DJ, Sokolski WT (1961) Antibiot Chemother 11: 123
47. Hwang K, Coen L, Johnson HE, Hunter WW, Cugier P (1961) Antimicrob Agents Chemother 507
48. Menawat AS (1988) Abbott internal report
49. Gomes J, Menawat AS (submitted for publication)
50. Gomes J (1993) Ph.D. Thesis, Tulane University, New Orleans
51. Gomes J, Menawat AS (1992) Biotechnol Prog 8: 118
52. van der Schaft AJ (1989) Systems Control Lett 12: 151
53. Boothby WM (1975) An introduction to differentiable manifolds and riemannian geometry, Academic Press, New York
54. Casti JL (1985) Nonlinear system theory, Academic Press, New York
55. Warner FW (1983) Foundations of differentiable manifolds and lie groups, Springer-Verlag, New York
56. Isidori A (1989) Nonlinear control systems – An Introduction, 2nd edn. Springer-Verlag, Berlin

Received October 1996

Role of Hydrodynamic Shear on Activity and Structure of Proteins

C.B. Elias and J.B. Joshi
Department of Chemical Technology, University of Bombay, Matunga, Mumbai-400 019, India

Proteins are important products used in industry. They may be enzymes which catalyze different reactions or they may be required for their biological activities as hormones, growth factors or therapeutics. During production and recovery, proteins are subjected to fluid forces which arise due to operations such as stirring, pumping and centrifugation. The resulting hydrodynamic shear forces may cause damage to the large molecular weight proteins, resulting in denaturation and inactivation of the protein. This is a major concern as it affects the overall efficiency of protein recovery and final yield of the product. A considerable amount of research has been devoted to studying the effects of hydrodynamic shear stress on proteins, especially with respect to the enzymes. Enzymes are subjected to shear stresses during their production in fermentors, during isolation and purification steps in downstream operations and also during their use in enzyme reactors, especially if stirred reactors are employed to perform enzyme catalysed reactions.

The present review discusses the effects of fluid shear stress on proteins including enzymes. A brief description on deactivation has been included in order to understand the effect of shear on the deactivation kinetics of proteins. The model systems used to subject proteins to shear and some unit operations during protein processing or use wherein they are exposed to shear stresses have also been presented. The significance of shear effects in designing bioprocesses involving shear sensitive biocatalysts as well as suggestions for future work have also been given.

Advances in Biochemical Engineering/
Biotechnology, Vol. 59
Managing Editor: Th. Scheper

List of Symbols and Abbreviations

a	enzymatic activity
a_0	initial enzyme activity
ADH	alcohol dehydrogenase
ABR	attrition bioreactor
CBH	cellobiohydrolase
CMC	carboxy methyl cellulose
Cx	endo-β-1-4-glucanase
Di	impeller diameter (m)
D	particle diameter (μm)
D_{max}, D_e	maximum and equilibrium particle diameter (μm)
E_0, E_1, E_2 ... E_n	specific activities of initial and active intermediate enzyme conformations ($g\,cm^{-3}$)
G	rms velocity gradient (s^{-1})
Hb S	sickle cell haemoglobin
k_1	first order deactivation rate constant (s^{-1})
K_d	protein aggregate breakup rate constant (t^{-1})
L	length of tube (m)
MCC	microcrystalline cellulose
N	impeller speed (s^{-1})
N_{Re}	Reynolds number
δP	pressure drop between two points ($Pa\,m^{-1}$)
Q	flow rate ($l\,min^{-1}$)
r	radius (m)
t	time (s)
T	torque (N m)
V	volume (m^3)
v	velocity

Greek Letters

$\alpha_1, \alpha_2, \ldots \alpha_n$	ratio of specific activities
γ	shear rate (s^{-1})
γ_r	shear rate at distance r from center of tube (s^{-1})
κ	ratio between radii of bob and cup
$\lambda_1, \lambda_2, \ldots \lambda_n$	function of deactivation rate constant
μ	dynamic viscosity ($N\,s\,m^{-2}$)
θ_r	angle between cone and plate surfaces
τ	shear stress ($N\,m^{-2}$)
ω	angular velocity of the rotating cup ($rad\,s^{-1}$)
$\gamma\theta$	product of shear rate and time of exposure

1 Introduction

Proteins constitute one of the largest and most important class of products of the biotechnological industry. Any parameter which affects either the yield or recovery of protein thus becomes the focus of vast research. Proteins which are composed of amino acids have a characteristic three dimensional configuration due to their specific tertiary or quarternary structure. The activity of the protein is closely related to its proper configuration. Any change in the physical structure of the protein may lead to its denaturation and deactivation. Proteins are produced by cells either intracellularly or they may be extracellular and are secreted into the medium, they may also occur as complexes associated with the cell membrane.

During the process of production and recovery proteins are subjected to fluid forces which arise due to operations such as stirring, pumping and centrifugation. The hydrodynamic shear forces resulting due to fluid motion cause damage to the large molecular weight proteins, resulting in denaturation and inactivation of the protein [1]. This is a major concern as it affects the overall efficiency of protein recovery and final yield of the product. A considerable amount of research has been devoted to studying the effects of hydrodynamic shear stress on proteins, especially with respect to the enzymes [2]. Enzymes are subjected to shear stresses during their production in fermentors, during isolation and purification steps in downstream operations and also during their use in enzyme reactors, especially if stirred reactors are employed to perform enzyme catalysed reactions.

Although a large amount of literature exists pertaining to enzyme deactivation which deals with different parameters such as pH and temperature, relatively little is known about the inactivation of enzymes due to hydrodynamic shear stress. It is likely that enzymes are inactivated when exposed to an air-liquid interface and that this effect is enhanced under conditions where agitation is also present. A variety of proteins have been studied with respect to their sensitivity to shear. Table 1 summarizes the different proteins and systems used for subjecting proteins to shear. From a review of the literature it can be seen that different types of proteins such as enzymes, antibodies, plasma proteins, milk and soya protein have been examined using diverse systems ranging from simple shaking flasks, viscometers, capillary devices to stirred tank reactors. Furthermore it is clear that enzymes are the proteins of choice when conducting experiments with shear stress, as the loss in catalytic activity resulting due to any changes in protein structure can be easily monitored. The other important area studied has been the precipitation of proteins. Soya protein precipitation has been the most commonly employed as it is easily available.

The present review discusses the effects of fluid shear stress on proteins including enzymes. A brief description on deactivation has been included in order to understand the effect of shear on the deactivation kinetics of proteins. The model systems used to subject proteins to shear and some unit operations

Table 1. Shear studies on proteins

Protein studied	System used	Ref.
Cellulase enzyme	shear due to shaking	[44]
Catalase and rennet	passing through capillary tubes	[27]
Invertase	recirculation through thin channel ultrafiltration module	[73]
Urease	urea hydrolysis in a coaxial cylinder viscometer	[38]
Horse liver alcohol dehydrogenase	recirculated through a hollow fiber reactor	[74]
Cellulase	recirculating enzyme solution through peristaltic pump	[30]
Yeast alcohol dehydrogenase	sealed coaxial cylindrical viscometer and agitated in a reactor	[86]
Bovine serum albumin monoclonal antibodies	capillary shear devices	[29]
Soya protein	precipitation in a continuous flow tubular reactor and stirred reactor	[66]
Soya protein precipitates	coaxial cylindrical viscometer	[65]
Penicillinase lactate dehydrogenase	immobilized in tubes with recirculation	[87]
Cellulase	attrition bioreactor	[45]
Cellulase	low shear stress rotating disk fermenter and conventional stirred tank reactor	[46]
Lipase	stirred tank reactor	[47]
Protease	turbine agitated reactor	[49]

during protein processing or use wherein they are exposed to shear stresses have also been presented. The significance of shear effects in designing bioprocesses involving shear sensitive biocatalysts as well as suggestions for future work have also been given.

2 Deactivation Kinetics

The optimum design of a biocatalytic process requires a proper understanding of not only the reaction kinetics but also of the deactivation of the biocatalyst. Since the most commonly used biocatalysts are enzymes the deactivation of enzymes and various models describing these kinetics have been described here. Deactivation may occur due to a number of reversible and irreversible processes. These include dissociation and denaturation which result due to changes in the tertiary structure of the protein molecule and are reversible with changes in physical conditions such as pH and salt concentration. The irreversible reactions involve decomposition, aggregation and coagulation which occur due to the changes in the secondary structure or due to chemical modification of the amino acid residues of the native protein molecule.

Enzyme deactivation may be represented by a simple first order model as

$$-dE/dt = k_1 E \tag{1}$$

where E is the specific activity of the enzyme, k_1 is the deactivation rate coefficient and the fractional rate of decrease of enzyme activity $[(-1/E)dE/dt]$ are both constant. This type of first order deactivation may be due to disruption of either a single bond or sensitive structure [3, 4]. Enzyme inactivation is characterized by fitting the enzyme activity versus time results to the equation

$$ln(a/a_0) = -k_{obs}t \tag{2}$$

where a_0 is the initial enzyme activity at time $t = 0$ and k_{obs} is the observed inactivation rate constant [5]. If this semilog plot is linear, it is assumed that the enzyme inactivation is first order. It might be difficult to visualize that enzyme molecules which are very complex structures exhibit simple inactivation kinetics. This has been explained on the basis that the disruption of a single bond or structure might eliminate enzyme activity [4]. However, there are several reports in the literature to show that enzyme inactivation follows a complex process and the simple first order model may not be sufficient to describe the enzyme inactivation process [6–9]. Enzymes are highly defined structures and although certain variations may not affect the catalytic activity of the enzyme, deviations from the native form are likely to alter the specific activity. Henley and Sadana [10] have reviewed the literature on enzyme deactivation curves for various enzymes and used a series type enzyme deactivation [6] scheme to model complex enzyme deactivation. Deactivation was represented as a series of reactions where $E \rightarrow E_1 \rightarrow E_d$. The enzymatic activity

$$a = (E + E_1)/E_0 \tag{3}$$

$$= 1/k_2 - k_1\,[k_2\,exp(-k_1t) - k_1\,exp(-k_2t)] \tag{4}$$

and can be expressed as

$$a = (E + \alpha_1 E_1)/E_0 \tag{5}$$

where $\alpha_1 = E_1/E$, the ratio of the specific activities of the intermediate and the initial enzyme states. This model visualizes an active enzyme precursor and a final enzyme state with possible non-zero activity. The enzyme activity here is a weighted function of the active enzyme states [10]. The advantage of this model is that it also allows for a rise in the activity, if α_1 is greater than 1. This may occur in cases where enzymes are stabilized or protected against deactivation [11–15]. This mechanism may be extended to include a number of steps and intermediate forms.

The series deactivation model assumes a single native state of the active enzyme molecule. However, the possibility of different states of the enzyme molecule being present initially cannot be ruled out, as in the case of isozymes which arise due to substitution of amino acid residues and are thus not interconvertible forms. Each of these may further deactivate by different pathways. The deactivation of these isozymes is described by a parallel mechanism [16]. The deactivation of each of these active states may follow a different pathway which in itself could be complex leading to a combination of parallel and series deactivation mechanisms. Henley and Sadana [17] proposed a unified general model which provides an insight into the enzyme structure and function and helps in determining the deactivation pattern which follows.

The model considers that deactivation follows a multistep process, occurring as a series, in parallel or both, by which the native enzyme, E, is altered to some intermediate forms, E_1, $E_2 \ldots, E_n$. Each of these may be deactivated to a final dead form E_d. Considering the general scheme of deactivation including a series and parallel deactivation, the total enzymatic activity was expressed as

$$a = C_1 e^{-\lambda_1 t} + C_2 e^{-\lambda_2 t} + \cdots + C_n e^{-\lambda_n t} \tag{6}$$

where C_i are the constants of integration and λ_i is a function of the forward and reverse first order rate and the first order deactivation rate constants. This general equation may be used to determine the number of steps required to represent the mechanism of a given deactivation process [17]. Each exponential term here represents an independent, potentially active form of the enzyme. If there is a rapid equilibrium between any two forms then only one of them is independent, as the concentration of all forms involved is governed by the equilibrium relationship. This has been used to explain deactivation of rabbit skeletal muscle AMP deaminase due to deprotonation of the active site [18]. Some exceptions to this model are deactivation due to higher order processes such as autolysis especially of proteolytic enzymes [19] and biological contamination.

Complex non-first order kinetics are displayed by various enzymes. These have been classified into two types.

1) a biphasic behaviour where a rapid inactivation is followed by a decelerated decay finally resulting in an activity plateau
2) a "grace period behaviour" where an initial stable phase of little activity loss is followed by periods of accelerating and decelerating inactivation [20].

Enzymes such as luciferase and acid phosphatase show biphasic behaviour [21–22] whereas galactosyl transferase and β-D-fructofuranosidase show the latter type of behaviour [23, 24]. Lencki et al. [20] described the effects of subunit dissociation, denaturation, aggregation, coagulation and decomposition on enzyme inactivation kinetics, and distinguished the reversible and irreversible mechanisms of deactivation.

The native to denatured structural transition of globular monomeric proteins is considered to be a reversible process. The presence of a stable intermediate during the transition process could lead to the observed grace period behaviour or to two first-order slopes i.e. biphasic behaviour, and has been suggested as the source of non-first order kinetics of enzyme inactivation kinetics [25]. However, these intermediates are generally observed only at very low concentrations relative to the native and denatured protein species and the time frame of these transitions is very short relative to the protein deactivation time and hence these may be considered to be in equilibrium and thus do not directly affect the enzyme kinetics. Many enzymes are olimeric comprising associated monomers. The interactions of these monomers have a strong influence on enzymatic activity. Irreversible mechanisms occurring at higher protein concentrations involve intermolecular interactions. These interactions occur via formation of covalent, hydrogen and ionic bonds. They may on one hand stabilize the structure of the active enzyme while their dissociation may expose the hydrophobic amino acid residues which subsequently interact irreversibly to form inactive aggregates [26].

A general model for enzyme deactivation would therefore include all possible inactivation reactions such as decomposition, dissociation, denaturation and aggregation. It is evident that this would result in an extremely complex situation. However, in most of the cases only one or two of the possible reactions dominate under the real conditions leading to simple behaviour (i.e. real or apparent first order kinetics). It should be noted that the slope of a first order plot can have a wide variety of phenomenological meanings [20].

3 Deactivation in Model Contactors

Proteins are subjected to shear during different steps of their production and recovery. These may occur in process equipment of varied geometric configuration. However, for carrying out studies on the effects of shear, proteins have been treated in model systems. These include cylindrical tubes and coaxial cylinder viscometers. Flow through cylindrical tubes is often encountered in protein isolation and purification operations, such as flow through pumps and ultrafiltration modules. The coaxial cylinder viscometer has been widely used to study the effect of shear on proteins as well as cells because it is possible to generate defined shear stresses in the laminar or turbulent region. These model contactors will be discussed in detail below.

3.1 Cylindrical Tubes

Fully developed laminar flow in a cylindrical tube is one dimensional and has a parabolic velocity profile. The shear stress in the tube varies linearly with the radial distance r. The shear stress is zero at the center of the tube and is maximum along the tube wall and is given as:

$$\tau = r\delta P/2L \tag{7}$$

where r is the radial distance from the tube axis, δP is the pressure drop between two points situated at distance L. The flow in the tube is turbulent above the critical Reynolds number of about 2100.

Protein solutions have been subjected to shear by passing them through capillary tubes of varying diameter under different conditions. Charm and Wong [27] investigated shear damage in the case of three different enzymes viz.

catalase, carboxypeptidase and rennet, by passing enzyme solutions through capillary tubes. They observed that the degree of inactivation of these enzymes depended on the shear rate and time of exposure. The combined effect of these parameters was found to be represented by the product. Thus the average value of the product of shear rate and the time of exposure was calculated from the experimental conditions using the flow rate, velocity of the stream and the radius of the tube as follows:

$$(\gamma\theta)_{ave} = \frac{2\pi \int_0^{r_w} r v \gamma_r \theta_r dr}{Q} \tag{8}$$

where v is the velocity of the streamline, r is the distance from the center, r_w is the tube radius, and Q is the flow rate. The effect of pumping alone was delineated by removing the capillary test section and recirculating enzyme solution through the pump for a number of passes at different flow rates. The range of shear stress used in their experiments varied between 0.02–2.6 $N m^{-2}$. They found that when the value of the product ($\gamma\theta$) exceeded a certain value, inactivation of the enzyme occurred. For all the three enzymes, deactivation was observed at shear stresses greater than 1.5 $N m^{-2}$.

In all these studies the velocity profile in the tube is considered to be parabolic implying a linear variation of shear stress from a maximum at the wall to zero at the center. The calculations of shear stress to which proteins were exposed were based on the average of this distribution [27]. It may be noted that the shear stress levels in the entrance region are higher than those obtained in the fully developed flow. The flow at the entrance region has been described by Schlicting [28] as a developing flow with a constant velocity across an interior portion of the tube, with boundary layers going out from the wall until they meet in the middle at a distance of the order of 40 and 80 diameters downstream. The velocity profile in the boundary layers has been approximated as a parabola, developed at some distance from the entrance. Therefore, in most of the experiments in tubes, where the proteins undergo several passes through the capillary device the protein will be exposed to considerably higher shear stresses than those calculated on the basis of a fully developed velocity profile [29].

The inactivation of the enzyme cellulase by shear is a major concern during the enzymatic hydrolysis of cellulose. Reese and Ryu [30] used a crude preparation of cellulase enzyme and subjected it to shearing conditions by recirculating it using a peristaltic pump at different flow rates. The shear stress generated in this system ranged between 1.0–2.3 $N m^{-2}$ and the Reynold's number N_{Re}, under conditions of maximum shear stress was 730 indicating that the flow was laminar over the entire range. They reported that the deactivation constant increased with an increase in the shear stress and the increase was especially significant at shear stresses higher than 1.5 $N m^{-2}$. Interestingly the effect of shear was primarily on the exo-glucanase cellobiohydrolase (CBH) component and not on the endoglucanase component of the cellulase complex. Thus the deactivation of cellulase by shear becomes evident when digestion of the more crystalline forms of cellulose is performed and this might be one of the reasons for the decrease in reaction rate during the enzymatic hydrolysis of cellulose.

In a further study on cellulase enzyme, Kim et al. [31] recirculated the enzyme through a capillary tubing in the presence of different stabilizers. They also investigated the effect of air-liquid interfacial forces by passing the enzyme solution through tubing from which all air bubbles had been removed. Under these conditions cellulase was not inactivated even after 4 h at a shear rate of $850\,s^{-1}$. However, in the presence of air in the tubing, a situation similar to those under normal operating conditions was observed and there was a 60% loss in activity. The deactivation of the enzyme also showed dependence on the protein concentration present. Various stabilization agents were used in the presence of air-liquid interface and it was found that Zonyl, Triton X-100 and bovine serum albumin were able to decrease the deactivation significantly whereas sodium dodecyl sulphate and lysozyme were not effective. The effect of these additives is mainly due to their ability to minimize the concentration of enzyme present at the air-liquid interface. Studies carried out with an extremely shear sensitive enzyme *viz.* ribulose-1,5-biphosphate carboxylase/oxygenase showed that bovine serum albumin was able to minimise shear damage in the presence of an air-liquid interface. This enzyme undergoes deactivation even under very mild conditions such as pipetting or vortexing in the laboratory [32]. The enzyme also shows spontaneous deactivation on standing. These effects were examined by subjecting the enzyme to shear under conditions similar to those used by Charm and Wong [27]. The enzyme showed deactivation at a relatively weak shear stress corresponding to an average shear rate x time of 10^5 and this deactivation was significantly decreased by the addition of bovine serum albumin ($1\,mg\,ml^{-1}$) [33].

From this discussion it becomes clear that any unit operation which requires a protein solution to be passed through a tubing has the potential to cause deactivation of the protein. The denaturation of proteins under these conditions is pronounced in the presence of an air-liquid interface. The air-liquid interface may exist due to the presence of small amounts of air in the form of bubbles being entrapped within the equipment or the solution itself. Addition of certain surface active agents such as Zonyl, Triton X-100, and serum albumin can alleviate the problem of shear induced denaturation of proteins especially in the presence of entrapped air. However, systematic work is needed for understanding the mechanism of action of stabilization of these additives. Furthermore, a systematic procedure is also required for the selection of the additive and its concentration for a given application.

3.2 Concentric Cylinder Viscometer

Another device used for generating defined shear stresses is a viscometer consisting of two cylinders placed concentrically. The flow in this device is of couette type and hence this is also referred to as couette viscometer. This comprises a stationary inner cylindrical bob and an outer rotating cylindrical cup. The sample to be subjected to shear is placed between the bob and the cup. The shear stress acting on this fluid is given by:

$$\tau = -2\mu\omega r^2(1/r^2)(\kappa^2/1-\kappa^2) \quad (9)$$

where ω is the angular velocity of the rotating cup; r is the cup radius, μ is the dynamic viscosity and κ is the ratio between radii of bob and cup.

This device is commonly used to study the bulk properties of cell suspension in particular the rheological properties of blood cell suspensions [34–36]. The effect of shear on protein solutions can be conveniently studied in these devices as shear stresses in both the laminar and turbulent regions can be generated using this system. Enzymatic solutions of catalase, rennet and carboxypeptidase which were subjected to conditions of shear in a narrow gap viscometer showed deactivation as a function of shear rate and time [37]. Tirrel and Middleman [38], studied the effect of shear on enzyme kinetics by carrying out the hydrolysis of urea by urease in a coaxial cylinder viscometer. The studies were carried out keeping the flow in the laminar region and maintaining the temperature at 23 °C. The results showed a continuous decrease in the rate of urea hydrolysis as a function of shear time at a given shear rate. There was a greater loss of enzyme activity with increasing shear rates. The rate of hydrolysis of urea by urease enzyme after it is subjected to shear showed that both reversible and irreversible inactivation of enzyme occurred. Recovery of enzyme activity on standing after being subjected to shear has been observed for rennet enzyme [39]. However, the catalase enzyme did not recover its activity upon standing. These enzyme solutions were tested for a wide range of shear rates between 0.01–1200 s^{-1}.

Virkar et al. [40] treated an alcohol dehydrogenase enzyme solution to a high shear rate of 26 000 s^{-1} generated in a concentric cylinder viscometer. In the case of this two cylinder viscometer, the outer cylinder was rotated while the inner one was held stationary in order to prevent the formation of Taylor vortices. Air entrapment was avoided by filling the apparatus completely with fluid, and closing the inlet and outlet ports. Alcohol dehydrogenase enzyme was not deactivated at this high shear rate and retained above 90% of its activity after treatment for a period of 1 h. However, when an enzyme solution containing ammonium sulfate was used, precipitation and loss of protein and activity was seen. This shows that deactivation of enzymes is minimized in the absence of an air-liquid interface. The deactivation of enzyme seen when a suspension of ADH in ammonium sulfate was used could be due to the existence of partially inactive forms of enzyme molecules present during long periods of storage as ammonium sulfate suspensions, due to the formation of intrasubunit disulfide bonds by oxidation of free sulfhydryl groups. Such partially inactive tetramers are more susceptible to shear than the native active enzyme molecules. These results indicate that enzymes in their native globular forms are not overly susceptible to shear damage and undergo shear associated damage only in the presence of an air-liquid interface.

4 Deactivation in Unit Operations

In the above section the study of shear effects on proteins had been carried out by treating the proteins to defined shear conditions in model contractors. However, these systems do not completely represent the conditions which proteins encounter in real operations. In routine use, proteins are subjected to conditions of shear in different unit operations. These include fermentation such as either production of protein in stirred, agitated reactors or also during their use in enzyme catalysed reactions. Proteins also undergo conditions of shear in other stages such as filtration and cell disruption. Some typical unit operations have been considered below.

4.1 Fermentation

It has been widely reported that the production of proteins can be correlated to the operating conditions existing in the bioreactor which includes the shear rate. In this context the term fermentation has been used in its widest possible meaning to include bioprocesses involving catalysts of biological origin such as cells or enzymes. Specifically the use of proteins is being considered here. The effects of shear on microbial, plant and animal cells was considered by the authors in a preceding review [41]. In the case of filamentous organisms such as actinomycetes and fungi it is known that the shear rate affects protein production indirectly by affecting the morphology. The formation of diffuse mycelia or spherical pellets is controlled by the agitation rate. The protein production is in turn influenced by the morphology of the organisms [42, 43].

Early studies by Basu and Pal [44] showed that fungal cellulase were deactivated due to shaking in flasks during the enzymatic hydrolysis of cellulose carried out in shaker flasks. This implied that the production or use of enzymes in stirred reactors would also be potentially damaging to the proteins. Virkar et al. [40] investigated the deactivation of the enzyme alcohol dehydrogenase (ADH) in a rotating disk reactor. This was a cylindrical reactor which had a 66 mm diameter disk placed at approximately half the vessel depth. They ensured complete filling of the vessel in order to eliminate entrapped air. The disk was rotated at a constant speed of 3600 rpm. At this speed the flow conditions in the reactor were turbulent and the average shear stress was estimated at about 70 $N m^{-2}$. Under these conditions no loss of activity was reported over a period of 5 h. In the cases, wherein the vessel was deliberately incompletely filled there was a linear decrease in the activity of the enzyme with time. This loss was accompanied by the formation of a very fine precipitate. After a period of 5 h the activity decreased to about 60% of the initial activity.

In another study, Jones and Lee [45] carried out a kinetic analysis of enzymatic hydrolysis of cellulose using a dried commercial preparation of

Trichoderma reesei cellulase in a bioreactor which combined the wet milling step with the hydrolysis. This reactor they termed as an attrition bioreactor (ABR). Two types of impellers were used, a marine type propeller and a curved blade turbine. Stainless steel balls of 4.69 mm diameter were used as milling media. The effect of an air-liquid interface was seen by sealing the reactor after completely filling it. It was shown that the enzyme solution was 50% deactivated in the presence of air whereas only 5% deactivation took place after 20 h in its absence at an agitation speed of 700 rpm. Under normal operating conditions a speed of 200 rpm was used at which there was no significant deactivation irrespective of the presence of air.

The above experiments deal with the effects of shear on cellulase enzyme during its use as a catalyst of cellulose hydrolysis. The enzyme however, is also exposed to shearing conditions during its production in reactors as it is secreted extracellularly. The production of cellulase by immobilized *Trichoderma reesei* has been studied in two types of fermenters, the conventional stirred tank reactor representative of high shear rates and a low shear stress rotating disk fermenter [46]. The enzymes produced from the two types of reactors were compared by subjecting them to conditions of shear by stirring at room temperature for varying time periods and estimating the activity of the component enzymes of the complex viz. cellobiohydrolase I (CHB I) and endoglucanase using standard assay methods for the respective enzymes. Their results showed that the productivity of enzyme per se was reduced in the low shear stress rotating disk fermenter. The important result that they obtained however, was that the enzyme produced under conditions of low shear stress exhibited a higher specific microcrystalline cellulose (MCC) hydrolyzing activity. This difference in the activity of the enzymes was not distinguishable in early stages of cellulose hydrolysis but was seen in the later stages when the action of enzymes is on the more crystalline domains of cellulose.

Furthermore, it was observed that not only was the enzymatic activity of the cellulase altered by shear forces but also the ability of the enzyme molecules to adsorb onto crystalline cellulose was affected. This may be a result of structural changes in the enzyme molecule. These effects are discussed in a separate section below.

Microbial lipases have widespread industrial applications as catalysts for hydrolysis, synthesis and transesterification of triglycerols and optical resolution. Very few studies exist on the production and stability of enzymes. Lee and Choo [47] described the deactivation of lipase due to shear forces. However, in a later study Gordillo et al. [48], examined the effects of shear on lipase from *Candida rugosa*. They subjected the enzyme to shear stress by stirring 500 ml culture broth free from biomass in a stirred bioreactor 1 l at speeds of 250, 500 and 750 rpm at 30 °C. The activity was greater than 90% after 25 h of stirring. In another set of experiments they studied the effect of surface forces generated due to gas-liquid interface by bubbling helium and air through a cell free culture broth in 250 ml flasks at different flow rates. At all the flow rates studied the loss in activity was less than 20%. A lipase containing broth when subjected to stirring at 500 rpm in the presence of air flow of 1 vvm also showed no deactivation after a time period of 30 h. From these results it may be deduced that the enzyme lipase is not affected by shear stress in conditions which exist either during its production or further use in enzyme catalysed reactions.

Proteases are another class of industrially important enzymes which find applications in different areas such as foods, detergents and the leather industry. Proteases are commonly produced using aerobic microorganisms which secrete the enzyme extracellularly. As stirred tank fermenters are most commonly used for the production of these enzymes, a need to examine the effect of fluid forces existing in these reactors on enzyme production and activity assumes significance. A study carried out by Gusek et al. [49] dealt with the production of a serine protease by a filamentous aerobic bacterium *Thermomonospora fusca*. The enzyme produced in conventional shake flask cultures as well in a stirred tank reactor (baffled with two Rushton disk impellers of diameter 75 mm) under agitated and aerated conditions was compared. The enzyme production in the shake flask culture was on an average 16 times higher than that in the stirred fermenter. Coating the impeller blades with tight fitting tygon tubing lead to improved biomass and enzyme yields. Another feature of this study was that in experiments conducted in shake flask and fermenter systems although the cell biomass obtained was comparable in both the culture systems the protease production in the shake flasks was significantly higher at 30 μg protease ml^{-1} as compared to 4.6 μg protease ml^{-1}.

The role of shear stress with respect to biosynthesis of enzymes, proteins and even other metabolites such as antibiotics has been well documented. In most of these cases hydrodynamic shear forces cause changes in the morphological forms of the organism especially filamentous forms such as actinomycetes and fungi [50]. As the production of metabolites is tightly coupled to the morphological form of the organism, the shear forces regulate metabolite formation [51, 52]. Several reports describing these effects exist in the literature, however since they primarily deal with the effect of shear on the microbes per se they are not considered here. These effects have already been discussed in an earlier review by the authors [41].

4.2 Cell Homogenization

Several microbial proteins are produced intracellularly. For their further application it is necessary to disrupt the cells and release product. There are several methods used for disruption of cells which include mechanical, enzymatic and chemical methods. Enzymatic treatment of cell disruption is mainly carried out only on a laboratory scale as on an industrial scale the cost of the enzymes used becomes prohibitive. Chemical methods of treatment are followed in certain cases where mainly solvents and detergents are used for cell disruption. These methods may be used when the cell walls have a greater content of lipopolysaccharides as in the case of Gram negative bacteria. However, when protein or enzyme production is carried out using organisms which have strong or rigid cell walls, as in the case of Gram positive bacteria such as yeast chemical methods are normally inadequate for cell disruption. The other disadvantage of this method is that the agents used for cell disruption such as solvents or detergents are denaturing to proteins and hence cannot be used. The scale up of these methods both enzymatically and chemically also pose difficulties in terms of equipment and further recovery of enzyme and solvent.

Commonly used methods of mechanical disruption include high pressure homogenisers, ultrasonic devices and bead milling. These systems have the advantage that they can be scaled up using known principles and can be used for almost any kind of microorganism. These features make mechanical disruption a popular technique in industrial scale applications although one of the problems associated with these systems is the excessive amount of hydrodynamic shear forces generated during the operation. This problem is mainly encountered during homogenisation and bead milling process. Cell disruption by ultrasound is dependent upon the mechanical effects of a gaseous and/or vaporous cavitation in the suspending medium. Ultrasonic vibrations cover a range of frequencies extending upwards from 20 kHz. At high acoustic power inputs microbubbles form at various nucleation sites in the fluid which grow during the rarefaction phase of the sound wave. During the compression phase the bubble contents are compressed to a minimum radius where the bubble collapses releasing a violent shock wave which is propagated through the fluid. In the collapse phase a large quantity of sonic energy is converted to mechanical energy in the form of elastic waves which disintegrate into eddies. Cell disruption occurs due to the mechanical stresses arising from the shock waves of cavity collapse as well as from flows and turbulence generated by the pulsating cavities. Inactivation of enzymes has been observed during ultrasonication [53, 54]. However, excessive heat generated during conversion of sonic energy to mechanical energy and chemical effects such as ionisation and subsequent free radical generation are reported as the main cause of enzyme inactivation [55]. Besides these known effects damage due to the hydrodynamic flow field may also occur [56].

The problem of denaturation of protein during cell disruption has serious implications especially in the case of membrane associated proteins because the shear forces which require to be applied to disrupt the rather strong microbial cell walls can potentially disrupt the membrane-enzyme complex. These effects have been studied in the progesterone 11 α hydroxylase complex. This enzyme is intracellular and involves a cytochrome P450 enzyme and a NADPH cytochrome P450 reductase and is membrane associated. The effect of shear on this enzyme complex during disruption has been reported for *Rhizopus nigricans* [57, 58]. The cell grown in shake flasks and in stirred fermenter was carried out in a concentric cylinder viscometer in order to disrupt cells under conditions of defined shear. The shear rates used in the experiment ranged from 2870–6460 s^{-1} for time periods up to 5 min. Both the magnitude of shear and its duration have an effect on the activity of the enzyme system. Disruption of cells at a shear rate of 5020 s^{-1} resulted in less activity than at 4300 s^{-1}. Further it was observed that at all the shear rates studied there was a reduction in the activity even after 3 min.

Experiments carried out using the cell free extract of the enzyme complex showed that treating the enzyme complex to shear rates between 720 and 2870 s^{-1} for a short time of 4 min led to an increase in the enzyme activity. After that there was a decline in the activity over a further exposure period of 40 min. Glycerol was added to the medium to increase the shear stress at a given shear rate, and showed that the enzymatic activity decreased with an increase in the shear stress. In these studies experiments were also carried out under conditions where air was excluded from the viscometer by sealing. The comparison of activity results in the systems wherein air was present and absent showed distinctly that air enhanced the subsequent decline of enzyme activity after the initial increase in activity seen after 4 min. These results are similar to those obtained with soluble enzymes wherein the presence of an air-liquid interface causes a marked increase in enzyme denaturation due to shear. However, the work with the progesterone 11 α hydroxylase complex which is membrane associated makes it clear that such enzymes are not as stable as soluble enzymes when exposed to shear forces. This should be kept in mind when carrying out cell disruption for release of these products.

4.3 Precipitation

Precipitation is an important unit operation used for the primary separation of proteins from culture fluids. It is practised widely for the purification of various proteins from blood [59], recovery of soya proteins [60], casein from milk and in the production of several microbial enzymes [61, 62].

Precipitation may be employed to reduce processing volumes and to solidify products for finishing. Differential precipitation is also carried out to fractionate protein mixtures [63]. Combining affinity ligands with precipitation provides a very versatile protein fractionation method [64].

Subsequent to precipitation, centrifugation is generally carried out to achieve separation of the precipitated protein. The efficient recovery of these protein aggregates is controlled by the size and mechanical strength of precipitates. These characteristics of precipitates are dictated by the mixing conditions in the reactor. Thus it is necessary to characterize the effect of hydrodynamics on the various aspects of protein precipitation.

Protein isolation is commonly done using isoelectric precipitation with acid to obtain a suitable pH. This method is commonly used in the precipitation of soya protein as well as in casein precipitation from milk for cheese manufacture. The advantage of this method is that the precipitated protein can be easily reconstituted by simple pH adjustment. The other commonly used methods for precipitation include salting out using ammonium salts or solvents such as acetone and ethanol.

Hoare [65], studied the precipitation of casein by ammonium sulfate salting out, with the view to optimization of its subsequent recovery in industrial centrifuges. He observed that the precipitate particle size distribution varied with the conditions of precipitation and ageing. Virkar et al. [66] examined the acid precipitation of soya proteins. This is an important step in the entire process of protein production from soya extract where separation of soluble protein from carbohydrates is carried out. A study of the effect of precipitation conditions was done by carrying out experiments in two systems i.e. in a stirred vessel in batch mode and in a continuous flow tubular reactor. The batch reactor consisted of a 250 ml glass beaker stirred with a magnetic stirrer. The acid addition was done either dropwise or at one time to see the effect of the contacting procedure on the precipitation. In the tubular reactor the protein and acid solutions were fed through two separate streams at controlled flow rates. The fluid in the reactor was turbulent (N_{Re} greater than 20000). The acid flow was always maintained two orders of magnitude lower than the protein feed rate, thus it contributed little to the overall flow in the reactor.

In the batch reactor, a small increase in the particle size was seen with an increase in the protein concentration. It was also seen that poor mixing resulted in an increase in the particle size with a broadening of the distribution curve. The rate of growth of the mean particle size was higher at high protein concentration. In the tubular reactor macromixing was complete at a distance of less than 0.3 m from the point of acid injection. The precipitation was studied at two flow rates and conditions were turbulent at both the flow rates. The initial growth rates were extremely rapid. The final mean particle sizes showed that the mean particle diameter was smaller in the high flow rates. This is due to an increase in the shear stress resulting in greater breakup rates. Thus the final particle size distribution is due to an equilibrium between the formation of particles and the breakup of particles by the shear stress.

An improvement in the aggregate strength making it more resistance to breakup by shear has been found to occur due to aging in the stirred vessel. Bell and Dunnill [67] studied this aspect of protein precipitates by examining the shear disruption of soya protein precipitate particles. This is important in the centrifugal recovery of protein precipitates as these are subjected to high shear fields in the entry zones of centrifugal feed streams. The protein precipitates were treated to capillary shear by passing them through capillary tubes at shear rates between 10^4 and 10^5 s^{-1} with exposure times between 0.004–0.2 s. These conditions are similar to those in the entry zones of centrifuges. The influence of the conditions of aging on the resistance of aggregates to shear breakup was evaluated by aging the dispersion in the reactor at a constant speed and then subjecting an aliquot to capillary shear. The changes in particle size were monitored using a Coulter counter. The stirrer speed used for agitation and precipitation were the same at 150–650 rpm. At these speeds the mean particle diameter was 53.5–8.8 μm, respectively.

Aging the precipitated protein solution by stirring leads to a state of dynamic equilibrium where the rate of formation of aggregate equals the rate of breakup.

A significant reduction in the number of small particles occurs during the aging period. This result has been explained based on Smoluchowshi's collision frequency theory which states that the probability of collision between small and large particles is greater than the probability of collision between small particles, hence large particles behave as nuclei around which small particles collect. Aging for a short duration resulted in a wide size distribution of particles after subsequent shearing. Shearing after prolonged aging times showed a bimodal distribution of particles as a result of fragmentation, a peak between 4–8 μm associated with fragmentation and an increase in the number of particles less than 3.5 μm produced due to erosion. The median size of the fragments (5.8 μm) was not affected by the time of aging. The results obtained suggest that aging leads to an improvement in aggregate strength. Denser aggregate may be formed by infilling of small, primary particles and by aggregate rearrangement. This rearrangement is due to aggregate breakup followed by more stable reaggregation and by the deformation of aggregate in the shear field in the stirred tank. This process results in an increase in the interparticle contacts leading to increased strength. Rearrangement of shear deformation is assisted by the existence of a distribution of shear rates within the stirred vessel. The impeller motion also causes the aggregate to pass from regions of low shear to regions of relatively high shear, imposing a dynamic stress on the aggregate and hence causing a shear induced rearrangement.

These mechanisms of improvement of aggregate strength depend upon the effective pumping rate, the level of shear and the time of exposure in the tank. These variables have been incorporated in an aging parameter Gt, where G is the rms velocity gradient and t is the time of aging, expressed for a given impeller and tank geometry.

$$G = (2\pi NT/V\mu)^{1/2} \tag{10}$$

where torque (T) is evaluated by an empirical correlation:

$$T = 7.21 \times 10^2 N^2 Di^5. \tag{11}$$

The parameter Gt was used to correlate the ratio of final to initial mean aggregate diameters resulting from capillary shear exposure and it was found that an approximate value of $Gt = 10^5$ gave the maximum aggregate strength. The capillary shear experiments showed that particle breakup is a function of both shear rate and time of exposure.

Bell and Dunnill [68] further studied the effects of precipitation reactor configuration on the strength and further recovery of protein aggregates of soya protein. Precipitation of protein was carried out in two different reactor configurations, viz. batch and continuous tubular reactor. The batch precipitation was carried out in a single baffled stainless steel vessel (i.d. = 670 mm; h = 670 mm) with an off centered three bladed pitched blade paddle (paddle diameter = 190 mm; speed = 230 rpm). The tubular reactor consisted of an acid extract mixing zone and a 19 m length of 15 mm diameter tubing. The recovery of precipitates in two models of industrial centrifuges was investigated. The precipitates obtained in the two different modes of precipitation were further characterized. Sedimentation of the batch tank prepared material with a large mean size resulted in a hazy supernatant. In contrast the material from the tubular reactor resulted in a clear supernatant. The particles obtained from the tubular reactor were found to be easily disrupted in subsequent capillary shear experiments and prior aging in a stirred reactor resulted in improved aggregate strength and recovery. Microscopic observation of the particles prepared by the two systems showed that the continuously prepared aggregate was irregular in shape and hence more susceptible to disruption as compared with that prepared in the stirred tank. The very wide range of velocity gradients existing in the stirred vessel coupled with increased residence time resulted in improved aggregate strength. This is supported by the observations of Virkar et al. [66] who found that use of grid turbulence promoters installed at regular intervals improved the particle strength in a continuous tubular reactor.

The centrifugal separation of the precipitates was further examined by Bell and Dunnill [68] using disk and scroll discharge centrifuges. For the disk centrifuge no significant differences on the separation efficiency were seen for the precipitate prepared by the different methods. However, in the scroll discharge or decanter centrifuge, the batch prepared precipitate gave 30% greater throughput before efficiency dropped below 98%, while the solid content for the tubular reactor material was approximately 30% higher than the decanter centrifuge material. It is possible that the particles produced by shear breakup in the equipment and resulting in fines are not recovered, i.e. when the throughput is increased the clarification efficiency is reduced due to non recovery of fines which are probably produced in the feed zone. The stirred tank prepared material showed considerable resistance to long exposure times of turbulent shear, suggesting that disruption in the scroll centrifuges occurs due to the intense shear in the feed zone and not due to turbulence existing in the bowl of the centrifuge. This was supported by the results of the short duration high shear capillary experiments where the sizes of particles generated by breakup were similar to the unrecovered particles in the scroll centrifuges. This work clearly points out the need of design of centrifugal separation processes keeping in mind the nature of precipitates and integrating the precipitation and centrifugal recovery operations.

Aggregate stability may be described by equating the size dependent hydrodynamic forces to the size dependent strength of the aggregate. The dependence on shear rate takes the form

$$D_{max} \propto G^{-\gamma} \tag{12}$$

where D_{max} is the largest possible size which can avoid disruption and γ is indicative of the mechanism of breakup [69, 70]. The disadvantage of this model is that there is no information on the kinetics of breakup or on the size of distribution below the maximum size. Twineham et al. [71] proposed a displacement model describing the aggregation process, including breakup in terms of the rate of change of aggregate diameter at a fixed cumulative volume percent.

Including terms for aggregate growth via collision and breakup via hydrodynamic shear they obtained an empirical expression for the rate of change of D, the size of the aggregate, as it approached the equilibrium size D_e, where growth and breakup are in balance as

$$dD/dt = .- K_d(D - D_e) \tag{13}$$

where $-K_d$ is the protein aggregate breakup rate constant. Here again although the kinetics of breakup have been included, the distribution of sizes has not been considered. Brown and Glatz [72], reviewed the above models as well as the model elucidated by Glasgow and Luecke [73] for a change in aggregate numbers at a particular size as a result of breakup via both hydrodynamic and collisional forces. They carried out experiments with isoelectric precipitates of soya protein and studied its breakup in an agitated vessel driven by a six-bladed turbine impeller in order to evaluate these models. The mean shear rates used in their study ranged from 1030–1340 s^{-1}. They concluded that the breakup of protein aggregates smaller than the Kolmogoroff microscale, occurs predominantly due to collision fragmentation. Although the mechanism of aggregate growth cannot be completely described for all situations, it can be generalized that the size of the final precipitate depends upon a balance between growth and fragmentation. Fragmentation by different mechanisms includes deformation and rupture of particles due to fluctuations in hydrodynamic pressures, erosion of the aggregate and rupture of the primary particles by shear or collision. As further recovery of precipitates is strongly dependent on their size and strength it is imperative to describe the behaviour of these systems under different flow conditions comprehensively, incorporating particle strength, aging parameters and kinetics of the precipitation process besides the mean velocity gradient.

4.4 Membrane Filtration

Membrane filtration processes are being increasingly used for concentration, separation or purification of biological material in laboratory as well as industrial scale applications. Ultrafiltration is an especially popular method for the concentration and purification as it is versatile and selection of membrane allows fractionation of proteins based on their molecular weights with relative ease as compared to other traditional purification techniques. This method involves recirculation of enzymes through ultrafiltration modules which may be in tube like configuration or they may be used in flat membrane type devices. In either case the protein solution is subjected to fluid forces and it is important to characterize the performance of these devices not only in terms of their separation efficiencies but also to examine whether this operation causes any damage to the protein molecules.

One of the problems with membrane filtration systems is renewing of the surface to prevent reduction in the flux due to concentration polarization phenomenon. This renewing is achieved by different methods including generating high shear at the surface by rapid flow or by vibration of a porous plate above the membrane. These methods however may also damage the proteins. Charm and Lai [39] have compared four different types of membranes used for the filtration of suspensions and solutions. Suspension of protein micelles (casein), cell debris as well as enzyme solutions have been studied. The ultrafiltration equipment included in the study included a vibrating porous plate with plate openings (0.3 cm at the bottom and 0.9 cm at the top) accommodating a 47 mm membrane filter which could be emptied by periodically flushing out the contents with air pressure. The second system was a mounted ultrafiltration module with a channelled membrane support, which allowed a high shear rate ($10\,000\,s^{-1}$) at the membrane surface and laminar flow in the rectangular channels with a 150 mm diameter membrane. A rectangular ultrafiltration module with turbulent flow, 0.035 m^2 membrane area and hold up volume of 7 l and a vibrating plate filter accomodating 90 mm membrane with hold up of 0.0112 l were the other systems included in their study. The study also involved different types of membranes viz. 0.22 μm and 0.1 μm diameter isotropic microporous membranes, anisotropic diffusive membranes with molecular weight no greater than 35 000 and anisotropic microporous membranes with molecular weight no greater than 30 000. The proteins studied were also subjected to shear stress in a viscometer at comparable levels. They observed that for casein suspension, the filtration rates were highest in the vibrating plate filter. The vibration increased the filtration rate by 1.5-fold and prevented rapid clogging of the membrane. Catalase enzyme showed a 7–40% loss in activity in the different filtration systems whereas the rennet enzyme showed no loss in activity. However, the rennet enzyme showed a decrease in activity when treated to shear in a viscometer but recovered its activity on keeping post shear treatment, indicating that the loss in activity was reversible. The loss of activity in the catalase enzyme was dependent on the shear stress and time of exposure.

Bowski and Ryu [74] showed that the recirculation of invertase enzyme through a thin channel ultrafiltration module did not cause any loss in activity. Fink and Rodwell [75] observed loss in horse liver alcohol dehydrogenase when recirculated through a hollow fiber reactor. The apparently different susceptibility of enzymes to shear may be due to differences in their molecular weight and tertiary structures. Narendranathan and Dunnill [76] investigated loss in activity of yeast alcohol dehydrogenase when recirculated through ultrafiltration modules consisting of bundles of hollow fibres. Two types of hollow fiber capillaries were studied, porous and nonporous capillaries. The results with the non-porous capillaries showed that there was an initial loss of about 20% protein but subsequently the protein concentration remained constant. The specific activity of the enzyme however, was close to 100% of the initial activity throughout the experiment. For the different types of porous capillary membranes tested there was a similar initial loss of protein ranging from 24–37% but there was no loss in the final specific activity of the enzyme. The maximum shear stress at the capillary wall ranged from 9–31 N/m^2 for Reynolds number < 2300 for the different modules tested and 34 and 60 N/m^2 were used for Reynolds number > 2300. The above results indicate that the globular proteins are not necessarily damaged by shear forces during ultrafiltration.

5 Role of Hydrodynamic Shear on the Structure of Proteins

The effect of fluid shear stress on the catalytic activity of proteins has been described in several literature reports. Although the loss of activity in proteins is related to changes in structure there are very few studies where this has been directly reported. This may be due to the difficulties in detecting and monitoring minor changes in protein structure whereas catalytic activity is relatively easy to monitor. It is important however, to examine the changes in the structure of proteins when subjected to shear especially in the case of therapeutic proteins where the small changes in the tertiary structure may result in changes in bioactivity. These effects also play an important role when the protein under consideration is a multienzyme complex where the different enzyme molecules associated in the complex are differentially affected by the shear forces and hence the overall reaction process needs to be suitably controlled. Minor changes in protein structure also result in altered sensitivity to shear forces. This has been demonstrated for sickle cell haemoglobin (Hb S) [77]. The oxy-form of Hb S is extremely susceptible to fluid shear forces and results in turbidity and precipitation when exposed to stirring. The normal heomoglobin does not show similar shear sensitivity. This feature has clinical importance as the shear denaturation of Hb S results in increased red cell rigidity and change in red cell shape.

Reese and Mandels [78] have studied the shear deactivation of the cellulase enzyme complex. The stability of this enzyme is an important aspect of cellulose saccharisation processes. Agitation is needed during enzymatic digestion to keep the solid substrate suspended and for ensuring homogeneity in the reaction vessel. The cellulase enzyme is a complex of three components endo-β-1,4-glucanase (Cx), CBH and β-glucosidase (cellobiase). Cx activity can be measured using carboxy methyl cellulose (CMC) as the substrate. CBH activity cannot be directly measured as there is no substrate which is not also a substrate for Cx activity. However, microcrystalline cellulose (Avicel) which due to its crystalline nature is a poor substrate for either pure CBH or Cx is synergistically acted upon by both the enzyme components. After acid pretreatment, Avicel makes a good substrate for CBH. An enzyme mixture which shows a good Cx activity but no activity against Avicel may be considered to be inactive due to loss of CBH activity. Deeble and Lee [79] studied the enzymatic hydrolysis of newsprint in an attrition bioreactor and found that the CBH component was relatively more sensitive to shear deactivation. This was reflected in a rapid deactivation of cellulase initially after which the deactivation rates became constant. This shearing effect has been attributed to breakup or disturbance in the tertiary structure of the cellulase enzyme [80, 81]. These results were further supported by the studies of Sachse et al. [46] who investigated shear effects during the production of cellulase using immobilized cells of *Trichoderma reesei.* They observed that production of cellulase in a stirred bioreactor which generated higher shear rates led to a decrease in the enzyme produced and this was mainly due to damage of the CBH enzyme of the cellulase complex. Changes in enzyme structure was also deduced by its reduced adsorbability onto microcrystalline cellulose under higher conditions of shear. van Tilbeburgh et al. [82], demonstrated that the decrease in the ability of CBH to adsorb onto insoluble cellulose occurred when the enzyme molecule was damaged by enzymatic cleavage of the carbohydrate-rich moiety and this was concurrent with the decrease in enzyme activity towards crystalline cellulose.

Changes in protein structure have also been observed in structural proteins in cells. The cytoskeletal proteins show depolymerization and changes in polymer structure when treated to shear stress. This is important when cells of higher eukaryotes such as plant and animal cells are considered, where proteins play an important role in maintaining cell shape and hence integrity and function [83] and has been reported in endothelial and other animal cells [84, 85]. A detailed understanding of how shear forces modulate protein structure and hence their stability would be extremely beneficial in designing and optimizing bioprocesses.

6 Conclusions and Perspectives

Proteins are exposed to an environment of hydrodynamic stress during fermentation, cell homogenisation, membrane filtration, chromatography and during pipeline transport. On a commercial basis flow is usually turbulent in all these equipments. Since there is no published work on quantitative measurements of Reynolds stresses, a very systematic investigation is needed in the future. Studies relating to shear effects on proteins have been mainly carried out with enzymes and furthermore, these studies have been restricted only to some industrially important enzymes such as cellulase, catalase, rennet and alcohol dehydrogenase. Besides these, there are several other enzymes which are being increasingly used in commercial applications such as proteases and lipases, where studies on shear stability are not comprehensive. From the literature it can also be seen that very few studies have been carried out on enzymes under fermentation conditions. This is especially important in the cases where the enzyme secretion is extracellular and hence the protein product will be exposed to the turbulent conditions existing in the fermenter. Apart from enzymes, there are also several biologically active proteins which are produced via bioprocesses including vaccines, monoclonal antibodies and interferons. These proteins have high molecular weights and complex three dimensional structures. It is therefore important to know the effect of hydrodynamic forces on these proteins as the native conformation of these structures is important for their bioactivity. Considering the importance of these biomolecules in the future generation of therapeutics it is important to establish a direct relationship between hydrodynamic shear and structure of proteins. Hence, future work should include simultaneous measurement of stresses and protein activity and monitoring the changes in the structure.

From a review of the literature it can be deduced that although proteins appear to be more robust than cells (microbial, plant or animal) with respect to shear damage due to fluid forces, in the presence of air they are rapidly denatured. This effect can be greatly enhanced under conditions where a gas-liquid interface occurs along with agitation. This condition is commonly observed in the aerobic fermentations for the production of extracellular enzymes and proteins. In the case of gas-liquid dispersions the turbulence structure is even more complex. Fortunately techniques are now available (laser Doppler anemometry and its modified versions) for the measurement of stresses in such complex systems. Future studies in this area should include a detailed investigation of stress in gas-liquid dispersions. Earlier workers [31] have shown that the addition of stabilization agents such as Zonyl, Triton X-100 and bovine serum albumin can prevent protein deactivation in the presence of a gas-liquid interface. Measurements of shear stresses are needed to investigate these systems together with their biological effects to enable us to develop a rationale for the selection of additives (and their concentration) for given applications.

A complete understanding of the shear stresses and their effects on the structure and function of proteins will help develop new designs of equipment (such as fermentors, cell homogenisers, solid-liquid separation devices and pumps) which generate stresses within the permissible levels without compromising on the efficiency for the purpose for which they are to be used. For instance in the case of fermenters, it will be possible to evolve new impeller designs which generate low levels of stresses and still give fairly high rates of oxygen dissolution. Similarly, in the case of separation operations, such as precipitation, there are no models which completely describe the process. The influence of shear stress, especially during the aging of precipitates, on the final particle size of precipitate and its subsequent separation by centrifugation also needs to be studied in detail. Such rational designs of equipment for bioprocesses assume greater significance in view of the increased number of protein products which are being produced commercially with applications in several areas including food, pharmaceuticals, agriculture and health care.

7 References

1. Prokop A, Bajpai RK (1992) Adv Appl Microbiol 37: 165
2. Thomas CR (1990) In: Winkler MA (ed) Chemical Engineering Problems in Biotechnology, Elsevier Applied Science, London and New York, p 23
3. Atwood KC, Norman A (1949) Proc Natl Acad Sci 35: 695
4. Reiner JM (1969) Behavior of Enzyme Systems, 2nd ed, Van Nostrand, New York
5. Laidler KJ, Bunting PS (1973) The Chemical Kinetics of Enzyme Action, 2nd ed, Clarendon Press, Oxford
6. Henley JP, Sadana A (1984) Biotechnol Bioeng 26: 959
7. Wada A. Saito Y, Okogushi M (1983) Biopolymers 22: 93
8. Cardosa JP, Emery AN (1978) Biotechnol Bioeng 20: 1471
9. Sadana A (1981) Ind Eng Chem Fundam 20: 336
10. Henley JP, Sadana A (1985) Enzyme Microb Technol 7: 50
11. Sadana A (1981) Enzyme Microb Technol 3: 357
12. Mozhaev VV, Martinek K (1982) Enzyme Microb Technol 4: 299
13. Sela M, White FH, Anfinsen CB (1957) Science 125: 691
14. Wiseman A (1978) In: Wiseman A (ed) Topics in Enzyme and Fermentation Technology Vol 2, Ellis Horwood, Chichester, p 280
15. Mozhaev VV, Martinek E (1984) Ann NY Acad Sci 434: 50
16. Saheki S, Saheki K, Tanaka T (1982) Biochim Biophys Acta 704: 484
17. Henley JP, Sadana A (1986) Biotechnol Bioeng 28: 1277
18. Ranieri-Raggi M, Raggi A (1983) Biochim Biophys Acta 742: 623
19. Kawamura Y, Nakanishi K, Matsuno R, Kamikubo T (1981) Biotechnol Bioeng 23: 1219
20. Lencki RW, Arul J, Neufeld DJ (1992) Biotechnol Bioeng 40: 1421
21. Chase AM (1950) J Gen Physiol 33: 533
22. Gianfreda L, Marrucci G, Grizzuti N, Grace Jr. G (1985) Biotechnol Bioeng 27: 877
23. Baldwin RL (1975) Annu Rev Biochem 44: 453
24. Sadana A (1988) Biotech Adv 6: 446
25. Schmid RD (1979) Adv Biochem Eng 12: 42
26. Contaxis CC, Reithel FJ (1971) J Biol Chem 246: 677
27. Charm SE, Wong BL (1970) Biotechnol Bioeng XII: 1103
28. Schlicting H (1979) Boundary Layer Theory 7th ed, McGraw Hill, New York
29. Carlisle RG, Wood HG, Benjamin DC (1988) Bioproc Eng Symp 117
30. Reese ET, Ryu DY (1980) Enzyme Microb Technol 2: 239
31. Kim MH, Lee SB, Ryu DDY, Reese ET (1982) Enzyme Microb Technol 4: 99
32. Gibbons GC (1978) Carlsberg Res Commun 43: 195
33. Penarrubia L, Moreno J (1987) Biochim Biophys Acta 916: 227
34. Chien S, King RG, Kaperonis AA, Usami S (1982) Blood Cells 8: 53
35. Tran-Soy-Tay R, Nash GB, Meiselman HJ (1986) J Rheol 30: 231
36. Drasler WJ, Smith CM, Keller KH (1987) Biophys J 52: 357
37. Van Wazar JR, Lyons JW, Kim KY, Colwell RE (1963) Viscosity and Flow Measurement, Interscience Publishers,
38. Tirrel M, Middleman S (1975) Biotechnol Bioeng XVII: 299
39. Charm SE, Lai CJ (1971) Biotechnol Bioeng 13: 185
40. Virkar PD, Narendranathan TJ, Hoare M, Dunnill P (1981) Biotechnol Bioeng 23: 425
41. Joshi JB, Elias CB, Patole MS (1996) Chem Eng J Biochem Eng J 62: 121
42. Mukhopadhyay S, Ghose TK (1992) Biochem Eng 2001: 162
43. Morimura S, Kida K, Sonada Y (1992) J Ferment Bioeng 74: 129
44. Basu SN, Pal PN (1956) Nature 178: 312
45. Jones EO, Lee JM (1988) Biotechnol Bioeng 31: 35
46. Sachse H, Kude J, Kerns G, Berger R (1990) Acta Biotechnol 10 (6): 523
47. Lee YK, Choo CHL (1989) Biotechnol Bioeng 33: 183
48. Gordillo MA, Obradors N, Montesinos JL, Valero F, Lafuente J, Sola C (1995) Appl Microbiol Biotechnol 43: 38
49. Gusek TW, Johnson RD, Tyn MT, Kinsella JE (1991) Biotechnol Bioeng 37: 371
50. Rau U, Brandt C (1994) Bioproc Eng 11: 161

51. Bronnenmeier R, Mark H (1982) Biotechnol Bioeng 24: 553
52. Markl H, Bronnenmeier R, Wittek B (1991) Int Chem Eng 31: 185
53. Lilly MD, Dunnill P (1969) In: Perlman D (ed) Fermentation Advances, Academic Press, London, pp 163–197
54. Wiseman A (1969) Proc Biochem 4: 63
55. Brookman JSG (1974) Biotechnol Bioeng 16: 371
56. Chisti Y, Moo-Young M (1986) Enz Microb Technol 8: 194
57. Talboys BL, Dunnill P (1985) Biotechnol Bioeng 27: 1726
58. Talboys BL, Dunnill P (1985) Biotechnol Bioeng 27: 1730
59. Cohn EJ, Gurd FRN, Surgenor DM, Barnes BA, Brown RK, Derouaux G, Gillespie JM, Kahnt FW, Lever WF, Liu CH, Mittelman D, Mouton RF, Schmid K, Uroma E (1950) J Am Chem Soc 72: 465
60. Bell DJ, Hoare M, Dunnill P (1983) Adv Biochem Eng 26: 1
61. Foster PR, Dunnill P, Lilly MD (1976) Biotechnol Bioeng 28: 1056
62. Suominen IA, Ford C, Stachon D, Heimo H, Niederauer M, Nurmela H, Glatz C (1993) Enzyme Microb Technol 15: 593
63. Rothstein F (1994) In: Harrison R (ed) Protein Purification Process Engineering, Marcel Dekker, New York, p 115
64. Iyer H, Przybycien TM (1995) Biotechnol Bioeng 48: 324
65. Hoare M (1982) Trans Inst Chem Eng 60: 79
66. Virkar PD, Hoare M, Chan MYY, Dunnill P (1982) Biotechnol Bioeng 24: 871
67. Bell DJ, Dunnill P (1982) Biotechnol Bioeng 24: 1271
68. Bell DJ, Dunnill P (1982) Biotechnol Bioeng 24: 2439
69. Tomi DT, Bagster DF (1978) Trans Inst Chem Eng 56: 1
70. Tambo N, Hozumi H (1979) Water Res 13: 421
71. Twineham M, Hoare M, Bell DJ (1984) Chem Eng Sci 39: 509
72. Brown DL, Glatz CE (1987) Chem Eng Sci 42: 1831
73. Glasgow LA, Luecke RH (1980) Ind Eng Chem Fundam 19: 148
74. Bowski L, Ryu DDY (1974) Biotechnol Bioeng XVI: 697
75. Fink DJ, Rodwell VW (1975) Biotechnol Bioeng XVII: 1029
76. Narendranathan TJ, Dunnill P (1982) Biotechnol Bioeng 24: 2103
77. Asakura T, Agarwal PL, Relman DA, McCray JA, Chance B (1973) Nature 244: 437
78. Reese ET, Mandels M (1980) Biotechnol Bioeng 22: 323
79. Deeble MF, Lee JM (1985) Biotechnol Bioeng Symp 15: 277
80. Nielson MJ, Kelsey RG, Shafizadeh F (1982) Biotechnol Bioeng 24: 293
81. Reese ET, Ryu DDY (1982) Enz Microb Technol 4: 99
82. Van Tilbeburgh H, Tomme P, Claeyssens M, Bhikabhai R, Pettersson G (1986) FEBS Lett 204: 223
83. Morita T, Kurihara H, Maemura K, Yoshizumi M, Nagai R, Yazaki Y (1994) Circ Res 75: 630
84. Levesque MJ, Sprague EA, Schwartz CJ, Nerem RM (1989) Biotechnol Prog 5: 1
85. McIntire LV, Papoutsaki ET, Petersen JF (1991) Bioproc Eng Symp ASME 29
86. Thomas CR, Nienow AW, Dunnill P (1979) Biotechnol Bioeng 21: 2263
87. Harrington TJ, Gainer JL, Kirwan DJ (1991) Enzyme Microb Technol 13: 610

Received July 1996

Novel Screen Methodologies for Identification of New Microbial Metabolites with Pharmacological Activity

D.C. Hill, S.K. Wrigley and L.J. Nisbet
Xenova Ltd, 240 Bath Road, Slough, Berkshire SL1 4EF, UK

Advances in Biochemical Engineering/
Biotechnology, Vol. 59
Managing Editor: T. Scheper

Micro-organisms continue to provide an important source of chemical diversity for the discovery of compounds with new biological activities. Microbial metabolites discovered recently using assays to detect compounds with potential pharmacological utility are surveyed and found to represent an extensive range of structural types produced by a wide variety of organisms. Assays used for screening samples produced by microbial processes must be robust, sensitive and specific and able to operate above a background of potential interferences from a number of sources. Discovery assays curently in use fall into three main categories: cell-based, receptor-ligand interaction and enzyme inhibition assays. Trends in the use of these assays and new developments in assay technology applicable to the screening of microbial samples are examined with particular reference to the high throughput screening environment. For microbial screening to be a competitive route to new drug leads, the disciplines involved must be engineered into a seamlessly integrated process to deliver novel compounds with the required biological properties rapidly.

1 Introduction

1.1 The Value of Microbial Chemical Diversity

Microorganisms produce an impressive range of secondary metabolites representing a rich diversity of chemical types. Some of these compounds have proved to be of great economic importance, not only to the pharmaceutical industry, but also as sources of new leads for the agrochemical sector. Organised exploitation of microorganisms as a source of chemical diversity began in the 1940s and 1950s as the significance of the discovery of penicillin was realised [1, 2]. During this period, screening programmes for antimicrobial activity were successful in identifying many new antibiotics such as the cephalosporins, tetracyclines and aminoglycosides. The search for microbial metabolites with other biological activities began in the late 1960s with the development of assays for the discovery of enzyme inhibitors of microbial origin [3]. This led to the discovery of many new compounds including two currently successful clinical agents, clavulanic acid and mevinolin. The former is a β-lactamase inhibitor [4] which is used in combination with β-lactam antibiotics in antibacterial therapy whilst the latter is the first of a series of related hydroxymethylglutaryl – CoA reductase inhibitors for treatment of hypercholesterolemia [5]. Further phamacologically significant natural products include the immunosuppressants cycloporin A [6] and FK506 [7].

The number of reported microbial metabolites with non-antibiotic biological activities has increased steadily since the early 1970s, and by 1990 it exceeded the number of antibiotics reported [8]. Advances in biotechnology and rapid progress with the human genome project are providing access to a wide range of new molecular targets implicated in various human disease states [9], and this explosion in target identification will yield many new opportunities for the exploitation of the molecular diversity generated from natural product sources. The following is a survey of assays used in the search for phamacologically active microbial metabolites during the period 1988–1995, together with some observations on current trends in assay technology for new drug discovery.

1.2 Demands of Assays Designed for Testing Microbial Samples

A diverse array of assays has been, and is being, used in microbial screening programmes to detect novel lead compounds with pharmacological activities. Such assays must fulfil a number of criteria if this search for potential new drugs is to be successful.

Firstly, they must be designed to function in the presence of samples with a range of physico-chemical properties. Various sample preparation techniques are used to generate libraries for screening, resulting in test mixtures which may have been extracted in a range of solvents, or simply filtered to remove high molecular weight material. In the latter case, samples will contain a range of medium components, may be highly coloured and may have a low or high pH or ionic strength. Microbial screens must therefore be able to detect active metabolites above a background of potential interference.

Secondly, they must be sensitive. Assuming that active metabolites are produced in fermentations at a concentration of 1–10 μg ml^{-1}, have an average molecular weight of 500 Da, and are diluted in the assay 20- to 100-fold, the required detection limit of the screen is in the 20-200 nM range.

Thirdly, to enhance metabolite detection, the assay should be highly specific for the molecular or cellular target. To aid the determination of selectivity, data generated from the screening programme can be compared with that from similar and unrelated assays before progressing samples to metabolite isolation and structure elucidation.

Finally, the throughput of the microbial screen should be carefully considered before commencing. Most assays can be converted to high throughput screens, but cell-based assays are inherently more complex and can take significantly more resources and time to operate. Miniaturisation of assay formats and the use of robotics can have a considerable impact on resource management for screening.

Effective management of the data generated from advanced screening programmes is critical for success. Microbial screening cannot progress without complete integration of input from scientists from the disciplines of microbiology, biochemistry and natural products chemistry to facilitate effective analysis of their data. Information from all three disciplines needs to be interrogated thoroughly when progressing samples through to metabolite isolation. Microbiological information such as organism source, description and fermentation data should be recorded in detail. New taxonomic techniques should be introduced where possible and their outputs incorporated into the data analysis process. In the screening group, use of advanced data analysis packages should be exploited to identify true hits and determine selectivity. This should include use of suitable controls to check data validity and assess assay performance continuously. Finally, in natural products chemistry, the use of chromatographic techniques to separate and purify the active compounds, and sensitive spectroscopic techniques to characterise them, should be coupled with the searching of a range of spectral libraries and natural products databases to recognise and eliminate known compounds. Valuable resources and expertise can then be focused on the isolation of metabolites with potential as lead templates for development into new drugs.

Development of a fully integrated bioinformatics system is highly desirable to facilitate rapid detection and identification of lead compounds. Such a system combines new methods, existing tools, hardware and developed networks, with basic computer technology such that data from all disciplines is linked together in an apparently seamless unit. This allows rapid and detailed analysis of data from all disciplines prior to selection of hits for progression to and during bioassay guided purification and structure elucidation of biologically active compounds.

1.3 Overview of General Assay Categories

There are many assay types which have been exploited in natural products screening programmes. For the purposes of this review, assays have been broadly characterised into three types: cell-based assays, receptor binding assays and enzyme assays.

A range of cell-based screens have been used in the detection of novel microbial metabolites. These are often employed in the search for molecules which inhibit a cellular function, but where a specific molecular target has not been defined. This means that active compounds may interact with the cells at a variety of targets from ligand-receptor binding, through intracelllar signal transduction, to effector protein synthesis and secretion. As there is a complex series of available intervention sites, compounds identified in cell-based screens need to be subjected to a series of selectivity and secondary assays to determine their mechanism of action. Cytotoxicity assays are often essential, and simultaneous operation of similar screen types can generate valuable data which helps to eliminate non-specific activities.

Receptor binding assays have provided a major focus for drug discovery programmes. Preparations from animal tissues have been used as sources of membrane receptor in a wide variety of targets. In addition, the use of cell lines expressing target receptors has given the opportunity for screening against whole cells and membrane receptors which can be expressed in vitro. Binding assays have most commonly measured the binding of radioactive ligand to receptors in the presence of test samples. The bound ligand is then separated from the free ligand, either by centrifugation or filtration, and a percentage inhibition calculated in comparison with controls.

The use of enzymes to discover pharmacologically active enzyme inhibitors by in vitro screening was pioneered by Umezawa in 1966 [10]. Advances in molecular biology have provided a huge array of enzyme systems as targets for selective cytotoxicity or pharmacological intervention. However, screening against isolated enzymes can result in too many false positive inhibitors which are active in vitro and not in vivo. This is because compounds detected may affect enzymes or their environment non-specifically. In addition, whilst compounds may be very active against the target enzyme, they may not be able to reach their target when tested in whole cells or in vivo. Despite these limitations, enzymes remain a popular choice when selecting targets to begin a primary screening programme, and continue to be a rich source of novel inhibitors from natural sources.

2 Cell-Based Screens

2.1 *Inflammatory Disease*

The cellular nature and complexity of interactions involved in the immune response and inflamatory disease has often driven the search for new active metabolites into cell-based screen formats.

2.1.1 Immunomodulation

Compounds with immunomodulatory activity, such as kifunesine (FR-900494) and conagenin, have been isolated and characterised using relatively simple cell-based assays to monitor the incorporation of [^{3}H]-thymidine into mouse spleen cells and T cells. Kifunesine is a cyclic oxamide derivative of 1-amino-substituted mannojirimycin which was isolated from the actinomycete *Kitasatosporia kifunense* [11, 12]. Conagenin is a 2-methylserine derivative isolated from *Streptomyces reseosporus* [13].

Both compounds demonstrated activity in more complex functional assay systems. Kifunesine concentrations of 0.1–1 $\mu g\,ml^{-1}$ restored the ability of immuno-deficient mice to produce antibody to sheep red blood cells (SRBC), and conagenin was found to modulate T cell activity in vivo [14, 15]. Kifunesine is also a weak inhibitor of α-mannosidase.

2.1.2 Immunosuppression

Assays based on the mixed lymphocyte reaction (MLR) have been used in screening programmes for novel compounds with potential immunosuppressive activity, as well as activity in the suppression of allograft rejection in transplantations, therapy of autoimmune disease and hypersensitivity.

A productive programme in terms of compounds reported has identified ISP-1 (myriocin), a number of synthetic myriocin derivatives including 14-deoxymyriocin, and the related fungal metabolites, the mycestericins, as immunosuppressive agents. In the assay used to discover these compounds, responder cells (mice splenocytes) and stimulator cells (co-irradiated splenocytes treated with mitomycin C) were mixed in 96-well plates and cell proliferation was measured using [^{3}H]-thymidine incorporation. Screening was focused on *Trichoderma* sp. and *Isaria* sp. as these fungi are known producers of cyclosporins and immunosuppressive cyclic depsipeptides. It is interesting that the compounds detected were all related and form a new class of immunosuppressive agents.

Myriocin (Fig. 1), an unusual lipophilic amino acid, was found to be five to ten times more potent than cyclosporin A (CsA) in inhibiting the MLR [16].

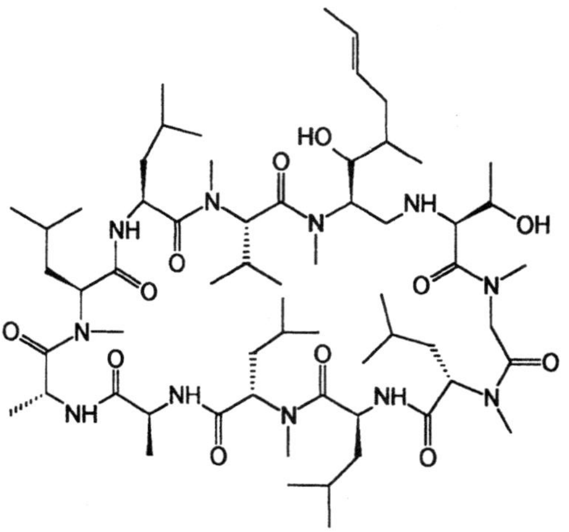

1

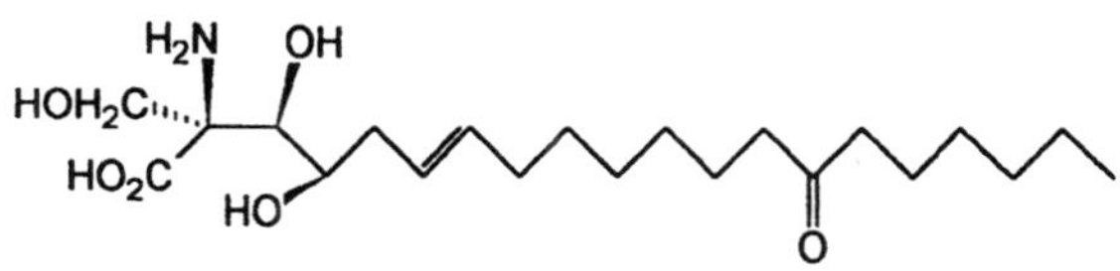

2

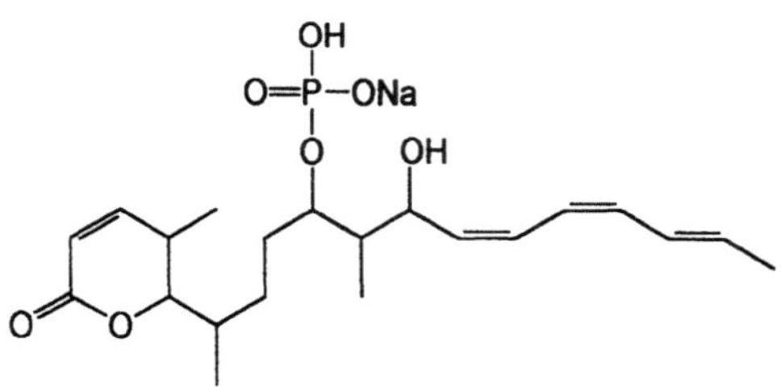

3

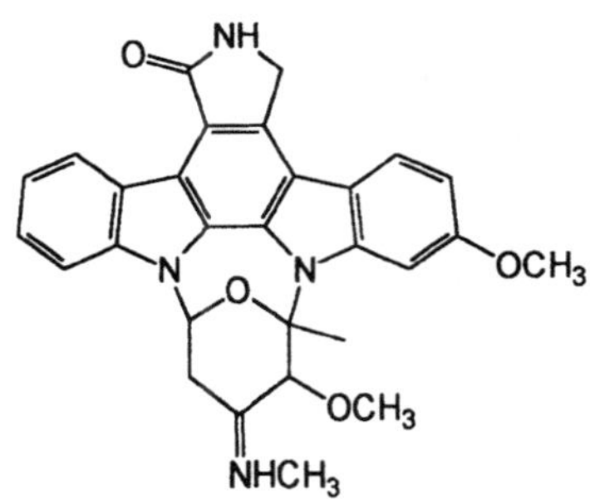

4

Fig. 1. Structures of the immunosuppressants FR901459, **1**, and myriocin, **2** (Sect. 2.1.2), the cell adhesion inhibitor cytostatin, **3** (Sect. 2.1.3), and the macrophage activation promoter TAN-999, **4** (Sect. 2.1.4)

Myriocin had no effect on the growth of human leukaemia cell lines demonstrating that, like CsA, its anti-proliferative activity was not due to non-specific growth-inhibitory activity. Myriocin was ten times more potent than CsA in inhibiting T cell dependent antibody production in an in vivo delayed type hypersensitivity (DTH) response to SRBC antigen. The compound was also 100 times more potent than CsA in suppression of allograft rejection. As a result of this promising activity, 14 myriocin derivatives were synthesised and evaluated [17]. Of these, 14-deoxomyriocin was found to have the most potent immunosuppressive activity on the mouse allogenic MLR and was ten times more potent in allograft rejection in vivo. Mycestericins A, B, D and E had similar potency to myriocin in the mouse allogenic MLR [18].

Other programmes using MLR assays for microbial product discovery have identified FR901459 (Fig. 1), a novel cyclosporin analogue from *Stachybotrys chartarum* [19] and MLR-52, a novel staurosporine analogue from a *Streptomyces* sp. [20], which was found to inhibit protein kinase C (PKC). FR901459 inhibited lymphocyte proliferation in a dose-dependent fashion with approximately one third of the potency of CsA. As CsA effects are primarily mediated by inhibition of lymphokine mRNA transcription, the effect of FR901459 on IL-2 production was examined. FR901459 was found to suppress IL-2 mRNA transcription, and cell proliferation. In addition, FR901459 inhibited a DTH reaction to the T cell dependent antigen on SRBC in Balb/c mice, and increased graft survival time in orally treated mice. MLR-52 had similar potency to staurosporine in inhibition of the MLR, and was demonstrated to inhibit PKC-catalysed phosphorylation of a small tyrosine-containing peptide.

2.1.3 Inhibition of Cell Adhesion

Recent studies have shown that cell adhesion plays an important role in inflammatory responses and tumour metastasis. IC101, a novel, cyclic depsipeptide, was isolated from *Streptomyces albulus* and characterised as an inhibitor of cell adhesion using concanavalin A (con-A) activated EL4 cells and B16 cells [21]. Cells were added to microtitre plates coated with the extracellular matrix (ECM) proteins fibronectin, laminin and type IV collagen. The assays were performed in the presence of cycloheximide to minimise the production of endogenous adhesion proteins.

IC101 strongly inhibited binding of EL4 and B16 cells to each ECM component. IC101 also inhibited the MLR as determined by [^{3}H]-thymidine incorporation into rat spleen cells, did not suppress antibody formation on inoculation of mice with SRBC, and strongly suppressed the DTH response in vivo. IC101 showed some toxicity in vitro and in vivo.

Further extracellular matrix receptor antagonists, the delaminomycins, were also isolated from spent media of the IC101-producing organism [22]. The delaminomycins have complex, non-peptidic structures based on substituted

N-acyltetramic acids [23]. Delaminomycins A, B and C inhibited cell adhesion to fibronectin, laminin and type IV collagen. In addition A and B inhibited adhesion of B16 melanoma cells to all ECM components but C did not inhibit adhesion to fibronectin. Delaminomycin A had the most potent activity. All three compounds inhibited the MLR, as determined by [^{3}H]-thymidine incorporation into rat spleen cells, in a dose-dependent manner. Delaminomycin A did not suppress SRBC induced antibody formation and strongly suppressed the DTH response in mice. The compound had low toxicity but did not inhibit tumour metastasis in vivo.

Cytostatin (Fig. 1), a novel phosphorylated α-pyrone derivative produced by a *Streptomyces* sp., was identified as an inhibitor of cell adhesion to ECM components during the screening programme in which the delaminomycins were discovered [24, 25]. It did inhibit tumour metastasis in vivo, exhibiting antimetastatic activity on B16 melanoma cells in mice.

2.1.4 Macrophage Activation

Macrophages act as regulatory and effector cells in the host defence system and are potentially destructive to invading micro-organisms and neoplastic cells. Macrophages reach target sites and act on target cells by a variety of mechanisms including production of oxygen radicals and release of lysosomal enzymes.

Two novel staurosporine analogues were characterised as microbial promoters of macrophage activation using an assay which involved observation of the induction of spreading and elongation of Mml cells by scanning electron microscopy [26, 27].

TAN-999 (Fig. 1) and TAN-1030, new related indolocarbozole alkaloids isolated from *Nocardiopsis dassonvillei* and a *Streptomyces* sp. respectively, induced Mml cell elongation similar to that seen in activated macrophages when tested in this system.

Using fluorescein isothiocynate (FITC)-labelled latex particles, these metabolites were shown to augment phagocytic activity of Mml and a highly phagocytic murine macrophage cell line J774A.1. Fcγ receptors on the cell lines were upregulated, β-glucuronidase (and hence lysosomal) activity was increased, and phagocytosis-dependent respiratory burst in mouse peritoneal macrophages was greatly enhanced both in vitro and in vivo. It was concluded that both TAN-999 and TAN-1030A were capable of activating macrophages. In all cases TAN-999 was the most active metabolite.

Because of their similarity to staurosporine, a potent protein kinase inhibitor which is active particularly against PKC, TAN-999 and TAN-1030 were evaluated further. TAN-999 was also found to inhibit PKC, and genistein, a tyrosine kinase inhibitor, induced Mml cell spreading and enhanced peritoneal mouse macrophage respiratory burst. This implied that macrophage activation is protein kinase regulated, confirming that protein kinases are an important target for development of therapeutic agents in inflammatory diseases and cancer.

2.1.5 Other Screen Methodologies and Targets

A spectrophotometric assay for prevention of myeloperoxidase release from polymorphonuclear leucocytes was used to discover the aseanostatins [28]. Inducers of the glycoprotein hormones known as colony stimulating factors (CSF) which regulate blood cell formation and function have also been sought. Leustroducsins A, B and C were discovered in one such programme using an assay which monitored the effect of test samples on CSF production by KM-102 cells [29–31].

2.2 Cancer

Cellular physiology, proliferation and cell-cell interactions also play an important part in the establishment and development of cancer. Indeed several of the targets for inflammatory disease and cancer overlap as described in the macrophage activation screening programme in Sect. 2.1.4.

2.2.1 Multidrug Resistance (MDR)

A major cause of failure of cancer chemotherapy is multidrug resistance, which is the development of resistance to several structurally unrelated anticancer agents. A specific glycoprotein, P-170, which apparently actively exports cytotoxic compounds, is generally associated with multi-drug resistant (MDR) cells. MDR modulators are therefore potentially important adjuncts to chemotherapeutic agents for the treatment of cancer.

From *Aspergillus fischeri* var. *brasiliensis*, 5-N-acetylardeemin was isolated as an MDR modulator using assays on human tumour KB3-1 (normal) and KBV-1 (multidrug resistant) cell lines [32, 33]. Compounds were tested against KBV-1 cells and serially diluted vinblastine, in microtitre plates. After incubation, the cells were stained with methylene blue and assessed in comparison to a control well containing no cytotoxic compound. Resistance and reversal were calculated by comparing the concentration of the cytotoxic compound required to reduce cells numbers to $< 10\%$ of the control well in each cell line. KBV-1 cells were 1600 times more resistant to vinblastine compared with KB3-1 cells. In the presence of 10 $\mu mol\, l^{-1}$ 5-N-acetylardeemin, however, KBV-1 cells were significantly sensitised, with their comparative resistance to vinblastine being reduced from 1600 to 6 times. Similarly, 5-N-acetylardeemin was shown to be ten times more effective in chemosensitising KBV-1 cells than verapamil, a known MDR modulator.

2.2.2 Cell Cycle Inhibitors

Events in the eukaryotic cell cycle are also potential targets for discovery of new anti-cancer drugs. The eukaryotic cell cycle consists of two major events: DNA replication in S phase and mitotic cell division in M phase. These are separated by two intervening (gap) phases, G1 and G2, during which many proto-oncogene functions are in operation. Cancer can be characterised as a failure of cell cycle regulation resulting in uncontrolled proliferation. Hence inhibition of the cell cycle offers opportunities to combat cancer.

A screening programme has been based on monitoring effects of microbial metabolites on an early S phase synchronous culture of rat normal 3Y1 fibroblasts by flow cytometry. The distribution of DNA content in naked nuclei of 3Y1 cells was determined using propidium iodide staining. The stage of cell arrest was estimated by comparison with data generated using known compounds (vinblastine-M phase, trichostatin A-G2 phase, leptomycin B-G1 and G2 phases). This screen was used to identify leptolstatin, a new gap phase specific inhibitor [34, 35]. Leptolstatin is an unsaturated α-pyrone derivative which is closely related to the leptomycins produced by a *Streptomyces* sp.

Leptolstatin had a strong reversible inhibitory effect on 3Y1 cells. Removal of leptolstatin from the arrested culture induced additional DNA synthesis in G2-arrested cells. This suggested that diploid G2 cells were converted to tetraploid G1 cells during G2 arrest. Compound removal induced formation of proliferative tetraploid cells.

2.2.3 Other Screen Methodologies and Targets

In cancer cells various steps of such signal transduction pathways are activated, or short-circuited, by oncogenes and this leads to less regulated growth. Therefore these events are a possible target for discovery of selective anti-tumour agents.

Differences in cell phenotypes resulting from aberrant expression of the signal transduction molecule *ras* have been used to detect compounds active in cancer targets. A screen using transformation of *ras*$^+$ cell morphology to the *ras* phenotype led to the discovery of reductoleptomycin A from a *Streptomyces* sp. [36]. Reductoleptomycin A is structurally very similar to leptolstatin (Sect. 2.2.2). It is noteworthy that the dorrigocins, a series of glutarimide metabolites from *Streptomyces platensis* subsp. *rosaceus*, recently reported to cause a change in the morphology of *ras*$^+$ NIH/3T3 cells to that of normal cells by a process involving inhibition of carboxyl methylation, appear to have been discovered during the course of screening for antifungal antibiotics [37–39].

Angelmicins A and B from a *Microbispora* sp. were isolated as selective inhibitors of growth of *src*-transformed cells in a separate screen which used a difference in cell growth between transformed and normal cells as its detection system [40]. Angelmicin B had no effect on *ras* transformed cells but specifically inhibited *abl* and *src* transformed cells via selective inhibition of tyrosine kinases.

2.3 Cardiovascular Disease

Platelet aggregation and new blood vessel formation assays have been used in the discovery and characterisation of novel microbial metabolites potentially active in cardiovascular disease.

2.3.1 Inhibition of Platelet Aggregation

Platelet aggregation inhibitors may be useful in the treatment of vascular disorders. FK409, a novel inhibitor of platelet aggregation, was assayed in a turbidimetric assay for observing thrombin-induced rabbit platelet aggregation [41]. In addition, FK409 levels in the spent medium were quantified by measuring relaxation activity on noradrenaline-induced contraction of rat aorta.

FK409 (Fig. 2) is a semi-artificial bio-product of *Streptomyces griseosporus* [42]. It is a hexenamide derivative bearing nitro- and oxime functions generated from a precursor by the presence of sodium nitrite in the medium under acidic conditions.

FK409 exhibited a marked relaxation activity on rat aorta and showed a hypotensive effect equal to that of nitroglycerin, a drug used in the treatment of angina pectoris, when tested in vivo.

Further platelet aggregation inhibitors which have been found using similar assay systems include the naphthalenecarboxylic acid derivatives PI-200 and PI-201 from a *Streptomycyes* sp. [43], which are weak inhibitors of ADP-induced aggregation of rabbit platelets, and Sch 38519, a novel isochromanequinone produced by a *Thermomonospora* sp. [44, 45], which is reported to be a weak inhibitor of thrombin-induced aggregation of human platelets.

Another related approach resulted in the identification of a basidiomycete metabolite, which may inhibit platelet aggregation by acting at the thromboxane A_2 receptor. Thromboxane A_2 (TxA_2) is the main product of arachidonic acid metabolism in platelets, and a very potent inducer of platelet aggregation and vasoconstriction. Collagen-induced bovine platelet aggregation assays were performed in microtitre plates leading to the discovery of 2-methoxy-5-methyl-1,4-benzoquinone (MMB) produced by *Lentinus adhaerans* [46]. Further biological characterisation of MMB suggested that it is a competitive TxA_2 receptor antagonist.

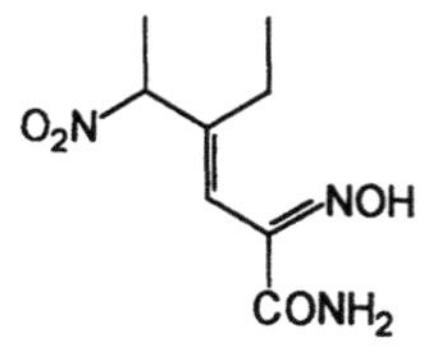

1

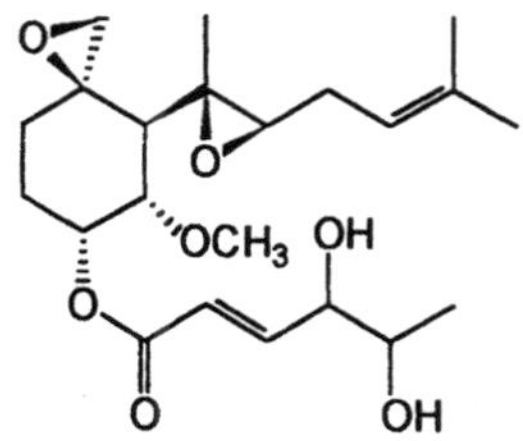

2

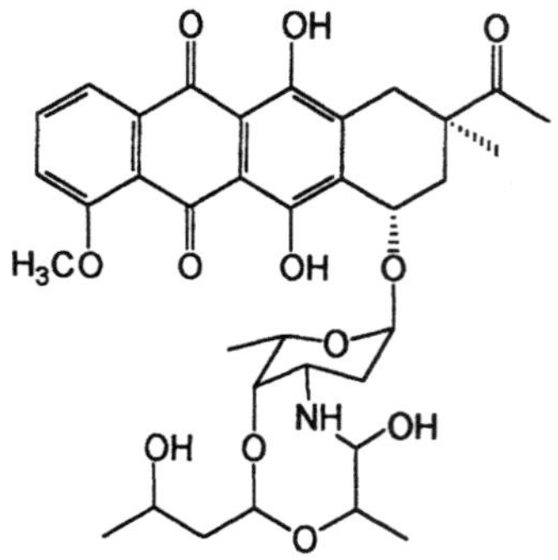

3

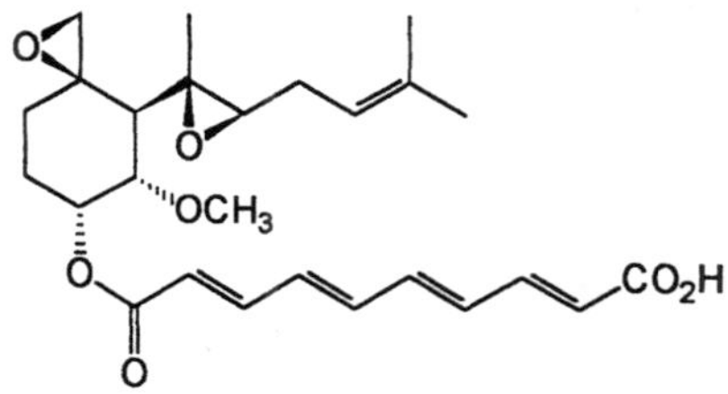

4

Fig. 2. Structure of the platelet aggregation inhibitor FK409, **1** (Sect. 2.3.1), and the angiogenesis inhibitors FR111142, **2**, TAN-1120, **3**, and fumagillin, **4** (Sect. 2.3.2)

2.3.2 Angiogenesis Inhibitors

Angiogenesis is the process of new blood vessel formation by endothelial cells and is essential for aggressive solid tumour growth. A novel metabolite of the fungus *Scolecobasidium arenarium*, FR-111142 (Fig. 2), was detected while screening for angiogenesis inhibitors from soil microorganisms [47]. Its biological activity was characterised by testing angiogenic effects in chick embryo chorioallantoic membrane (CAM). Test samples were dried onto Teflon rods and the resultant pellets placed on 5-day CAMs in fertilised chick eggs. After incubation, fat emulsion was injected and the antiangiogenic response measured by qualitative scoring for capillary density of an avascular zone around the test pellet.

FR-111142 had potent antiangiogenic activity in the chick egg CAM, but apparently caused no damage to embryo growth. Cytotoxicity was measured against human umbilical vein endothelial cells (HUVEC), murine leukaemia P388 cells, fibrosarcoma Meth A cells, lymphoma EL-4 cells, and baby hamster kidney (BHK) cells by incubation with FR-111142, followed by staining viable cells with neutral red. The compound had cytostatic effects against HUVEC and

EL-4 cells. FR-111142 suppressed Meth A fibrosarcoma and solid colon tumour growth in vivo in mice with no associated weight loss, but was not active against leukaemia P388. The compound had low toxicity in vivo.

FR-111142 is closely related to the *Aspergillus fumigatus* metabolite fumagillin (Fig. 2) [48], which has also been reported to have antiangiogenic effects and to have promising in vivo anti-tumour activities. Synthetic derivatives of FR-111142 were prepared, and one of these, FR-118487, was found to have superior properties and in vivo anti-tumour activity [49].

Other microbial metabolites discovered during screening programmes for angiogenesis inhibitors include the pyridine derivatives WF-16775 A_1 and A_2 from *Chaetasbolisia erysiophoides* [50] and the anthracycline TAN-1120 (Fig. 2) from *Streptomyces triangulatus* sbsp. *angiostaticus* [51]. TAN-1120 is reported to have particularly potent in vivo angiostatic activity.

2.4 CNS Disease

Cell-based assays involving growth and differentiation of neuroblastoma cell lines have also been used successfully in the detection and isolation of novel microbial metabolites.

2.4.1 Nerve Growth Factor (NGF) Mimics

Nerve growth factor is known to be important for the growth, survival and functional maintenance of neurons in the central and peripheral nervous system. Decrease in NGF availability has been implicated in various CNS disorders such as Alzheimer's disease. NGF treatment has been shown to ameliorate age-related impairment in memory tests.

A screening programme for NGF-like substances involved monitoring differentiation of the stable transformed mouse neuroblastoma cell line Neuro 2A. Among the culture broths of 6000 soil isolates tested, only one culture, a *Streptomyces* sp., was found to cause generation of a neurite-like structure from the Neuro 2A cell body. The metabolite responsible for this effect was purified and identified as lactacystin, which has a novel thioester structure containing γ-lactam and *N*-acetylcysteine moieties [52, 53].

Further examination of the effects of lactacystin showed that the compound induced formation of a microtubule-containing structure similar to that observed in established neurites. Neurite formation in Neuro 2A cells may be mediated by changes in intracellular cyclic AMP, protein kinase C and cAMP-dependent protein kinase, and a proteinase inhibitor is known to have a physiological role in neurite outgrowth. Lactacystin was assessed for activity across these areas but had no effect on any of them apart from causing a transient, dose-dependent increase in cAMP. This effect was coincident with the formation of bipolar projections from the cells but did not take place at the same time as effects seen on cAMP with adenylate cyclase modulators. It was therefore concluded that lactacystin is potentially a new type of NGF.

In a second study, a rat pheochromocytoma cell line PC12 was used to assess neurite outgrowth. In this case, cells were seeded onto collagen-coated 24-well plates and neurite growth scored microscopically after 48 h incubation with the sample. This study led to the discovery of two novel potentiators of NGF: NG-011 and NG-012 [54, 55]. These compounds are cyclic polylactones produced by *Penicillium verruculosum.* Optimum potentiation of NGF-induced neurite outgrowth was only achieved at relatively high concentrations ($100\ \mu g\ ml^{-1}$) of NG-011 and NG-012.

In a similar screening programme using PC12R cells, test samples were added to cells pretreated with NGF. After incubation, cells were fixed with gluteraldehyde and stained with Harris hematoxylin solution. Neurite outgrowth was monitored using a phase contrast microscope connected to a video camera CRT system. BU-4514N, a metabolite of a *Microtetraspora* sp., was discovered using this assay [56]. It has a complex structure similar to those of the delaminomycins (Sect. 2.1.3), containing *N*-acyltetramic acid and decalin moieties and an additional aminosugar portion. Low micromolar concentrations of BU-4514N significantly induced neurite outgrowths both in number and length.

3 Receptor Binding Assays

3.1 *Inflammatory Disease*

A number of ligand receptor interactions are known to be involved in mediation of inflammatory responses, and these have been targeted in the search for new pharmacologically active agents.

3.1.1 Leukotriene B4 (LTB4) Binding Inhibitors

Leukotriene B4 (LTB4) stimulates aggregation and degranulation of polymorphonuclear leucocytes (PMNLs), and promotes chemotaxis and chemokinesis. LTB4 has been detected at sites of inflammation in sufficient concentration to alter PMNL function, suggesting that it may serve as a mediator in inflammatory response. LTB4 antagonists may therefore be useful in treating diseases associated with inflammatory cell infiltration.

In this screen, [^{3}H]-LTB4 and a PMNL membrane suspension were incubated with test samples. After the incubation period, the bound [^{3}H]-LTB4 was separated from the free ligand using a glass fibre filter and a cell harvester. This led to the discovery of WF11605 (Fig. 3), a novel tetracyclic triterpene glucoside produced by an unidentified fungal strain [57, 58].

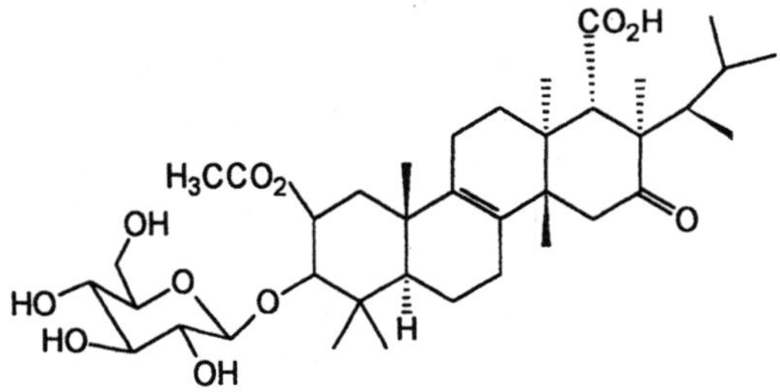

1

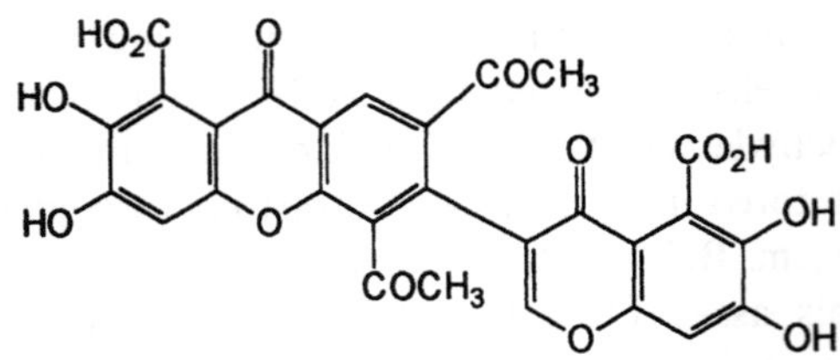

2

Fig. 3. Structures of the leukotriene B4 binding inhibitor WF11605, **1** (Sect. 3.1.1), and the CD4 binding inhibitor 411F, **2** (Sect. 3.1.2)

LTB4-induced chemotaxis in rabbit PMNLs in the presence of WF11605 was studied using a chamber containing a two compartment system. WF11605 significantly inhibited PMNL migration to LTB4. In addition, WF11605 inhibited LTB4-induced rabbit PMNL degranulation as quantified by mesuring β-glucuronidase activity, but had no effect on PAF and FMLP-induced degranulation, suggesting that it is a specific antagonist of LTB4.

3.1.2 CD4 Binding Inhibitors

CD4 is a glycoprotein expressed on the surface of mature helper/inducer T lymphocytes and has an essential role in many immune responses. It binds to class II molecules of the major histocompatibility complex (MHC II) and this stabilises the interaction between the T cell receptor and its ligand, the antigen-MHC II complex. Inhibition of this interaction can have profound suppressive effects on immune responses, making it an attractive target for the discovery of immunosuppressants. The CD4 molecule is also an important antiviral target as it is the cellular receptor for the human immunodeficiency virus (HIV). The HIV – gp120 protein binds to CD4 at a site distinct from that of MHC II. A screen for the detection of potential immunosuppressive and antiviral agents was undertaken at Xenova in a partnership programme with Roche. The assay used was based on the interaction between soluble recombinant CD4 (srCD4) and the monoclonal antibody anti-Leu 3a [59]. The binding site for anti-Leu 3a on CD4 is a critical domain for CD4 function and also overlaps the HIV – gp120 binding site.

This assay was used to screen microbial samples for CD4-binding activity, resulting in the discovery of a family of novel fungal metabolites produced by *Penicillium glabrum* (Wehmer) Westling [60]. The most potent member of this family, 411F (Fig. 3), inhibited anti-Leu 3a binding to srCD4 with an IC_{50} of 2 $\mu mol\,l^{-1}$. It only weakly inhibited the binding of HIV-gp120. 411F inhibited the binding of a representative MHC class II molecule, HLA-DRB*0401, with an IC_{50} of 1 $\mu mol\,l^{-1}$. In functional assays, 411F inhibited the antigen-induced, CD4-dependent, T cell proliferative responses of peripheral blood mononuclear cells. The compound also exhibited selectivity at the clonal level by inhibiting peptide-induced $CD4^+$ T cell proliferative responses, but not alloantigen-induced $CD8^+$ T cell proliferation.

411F is an acidic chromone-substituted xanthone which carries negative charge in aqueous solution at neutral pH. The compound was hypothesised to be acting by binding over a broad area of the positively charged amino terminal domains of CD4 which are implicated in the interaction with MHC Class II molecules. It has also been reported as a phospholipase C inhibitor (Sect. 4.2.1).

3.2 Cancer

Receptor binding assays have also been used to search for compounds active against targets for cancer therapy.

3.2.1 Hormone Binding Inhibitors

Assays based on hormone-receptor interactions have been used to search for new compounds which may be used for treatment of hormone responsive cancer and other diseases.

An assay for binding of [^{125}I]-17β-estradiol to estrogen receptors in the hog uterus cell cytosol was used to discover the BE-14348 family of substituted flavanones from *Streptomyces graminofaciens*, four of which were novel [61]. In this case the bound ligand was separated from free by addition of dextran-coated charcoal followed by centrifugation, and measurement of radioactivity in the supernatant.

As BE-14348B was the most active estrogen binding inhibitor from this series, its effects were compared with tamoxifen, an estrogen antagonist which has been used for breast cancer hormone therapy, and diethylstilbestrol, an estrogen agonist used in the treatment of prostate cancer. BE-14348B inhibited receptor binding with much more potency than tamoxifen, and a little more strongly than diethylstilbesterol. Measurement of the effect of the BE-14348 family on proliferation of estrogen-responsive human breast carcinoma MCF-7 cells showed that, like estrogen, compounds stimulated proliferation in a dose-dependent manner. It ws concluded that the compounds were estrogen agonists. In addition, BE-14348B only inhibited progesterone-receptor binding weakly, showed little cytotoxicity against MCF-7 and Hela cells, and had low toxicity in vivo.

Napyradiomycins A and B1 were isolated from a *Streptomyces* sp. during a screening programme for estrogen-receptor antagonists [62]. They are quinone antibiotics which had previously been isolated from *Chainia rubra* [63]. Napyradiomycins A and B1 inhibited estrogen-receptor binding in a dose dependent manner, and were less potent than tamoxifen. Napyradiomycin B1 was an estrogen-receptor antagonist, as determined by effects on growth of estrogen stimulated MCF-7 cells. It reduced estrogen-induced increase in rat uterus carcinoma slightly but significantly. Tamoxifen and napyradiomycin did not show dose dependent inhibition in vivo, suggesting that they exhibited partial estrogen-receptor agonistic activity.

R1128 A, B, C and D were discovered during the same screening programme. They are alkylated anthraquinones which were isolated from a *Streptomyces* sp. [64]. These compounds were less potent than tamoxifen in the receptor-binding assay, and had weak cytotoxicity against human lung adenocarcinoma A459 cells, human adenocarcinoma MCF-7 cells, mouse lymphocytic leukaemia P388 cells and mouse bone marrow cells. R1128B had very low toxicity in mice and rats. Based on experiments measuring the colony formation of MCF-7 cells in soft agar, the R1128 compounds were concluded to be estrogen-receptor antagonists.

Androgen is also known to play an important role in cancer, being active in benign prostatic hyperplasia and prostate cancer. A receptor binding assay based on the binding of [^{3}H]-mibolerone to rat prostate cytosol, and separating the bound from the free ligand by centrifugation with dextran coated charcoal, was used in the isolation and identification of WB2838 from a *Pseudomonas* sp. [65]. Triamicinolone acetonide was present in the reaction mixture to inhibit ligand binding to progesterone and glucocorticoid-receptors. WB2838 is a chlorinated phenol-substituted pyrrole which was discovered using this assay, and had been isolated and identified previously as an antifungal antibiotic [66]. WB2838 inhibited androgen binding to its receptor in a dose dependent manner, and was more potent than flutamide, but less potent than chlormadinone acetate (CMA), two known androgen-receptor antagonists. The compound was a competitive inhibitor of androgen-receptor binding, and only weakly inhibited estrogen-receptor binding. In addition, WB2838 had characteristics of an androgen-receptor antagonist, and a weak partial agonist, as determined by effects on androgen-responsive mouse mammary carcinoma SC3 cells in vitro. Oral administration of WB2838 reduced testosterone-induced weight increases in rat ventral prostate and seminal vesicle slightly but significantly.

WS9761A and B, anthracene derivatives produced by a *Streptomyces* sp., were identified as androgen receptor antagonists during the same screening programme [67]. WS9761A and B were competitive inhibitors of androgen-receptor binding. The compounds inhibited binding in a dose dependent manner, and showed only weak activity against estrogen-receptor binding. Both compounds were characterised as androgen-receptor antagonists by their inhibitory activities against the growth of SC-3 cells.

3.3 Cardiovascular Disease

Receptor-binding assays have also been used to screen for inhibitors of targets associated with cardiovascular disease.

3.3.1 Inhibition of Atrial Natriuretic Factor (ANF) Binding

Atrial atriuetic factor is a cardiac hormone which is released into plasma following atrial stretch. ANF binds with high affinity to membrane receptors resulting in the upregulation of guanylate cyclase, and subsequent accumulation of cyclic guanosine monophosphate (cGMP) in tissues and cells. This has many effects, including direct potent vasodilation of the peripheral blood vessels, and enhancement of various functions in the kidney. ANF is therefore actively involved in blood pressure homeostasis.

A binding assay has been used to screen for microbial metabolites which interfere with the binding of ANF to its receptors. This assay involved use of rat [^{125}I]-ANF and membranes from bovine adrenal cortex. The bound and free radioligand were separated by filtration through glass fibre filters. Anantin, a cyclic peptide produced by *Streptomyces coerulescens*, was discovered as a result of this screening programme [68, 69]. Further characterisation of anantin activity suggested it was a competitive inhibitor of ANF-binding. Assays using bovine aorta smooth muscle showed that the ANF-induced increase in intracellular cGMP was significantly inhibited by anantin, whereas the compound alone had no effect on intracellular cGMP levels. Anantin therefore behaved as a functional ANF-antagonist.

3.3.2 Endothelin (ET) Binding Inhibitors

The endothelin receptor has been a productive target in terms of microbial antagonists and binding inhibitors reported during recent years. Endothelin (ET) is a 21 amino acid polypeptide with several isoforms (endothelin-1, -2 and -3 : ET-1, ET-2, ET-3) which has potent, long-lasting vasoconstrictor activity in vitro and in vivo. ET is produced from endothelial cells along with TxA_2 and angiotensin II, and is an important regulator of vascular tone. Two subtypes of endothelin receptors, classified as the ET_A and ET_B receptors, have been cloned and characterised in mammalian systems, and both have been used as the basis of assays in screening programmes for microbial products.

Some of the first natural product ET binding inhibitors to be reported were two cyclic pentapeptides produced by *Streptomyces misakiensis*. A binding assay of the same type as that used to discover anantin, in this case using [^{125}I] ET-1 binding to ET_A receptors on porcine aortic smooth muscle, was used in the discovery of BE-18257 A and B [70, 71]. BE-18257 A and B inhibited [^{125}I]-ET-1 binding in a concentration-dependent manner, and are specific for the ET_A receptor. The compounds did not exhibit acute toxicity in vivo. The structure of BE-18257 B is shown in Fig. 4. The same compounds, together with two additional analogues, were discovered independently using a similar screening assay and named WS7338 A-D [72].

Further microbial endothelin binding inhibitors have been reported to be produced by actinomycetes, including the benzanthraquinones WS009A (Fig. 4) and B from a *Streptomyces* sp. [73], the depsipeptide cochinmicins from a *Microbispora* sp. [74], and the cyclic peptide RES-701-1 from a *Streptomyces* sp. [75, 76]. Endothelin binding inhibitors from fungi include asterric acid

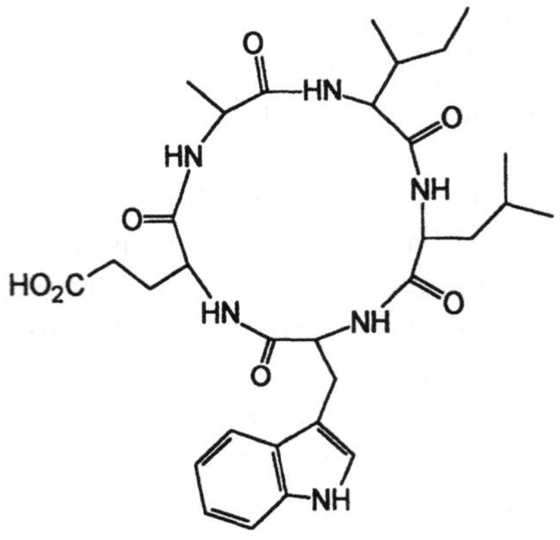

1

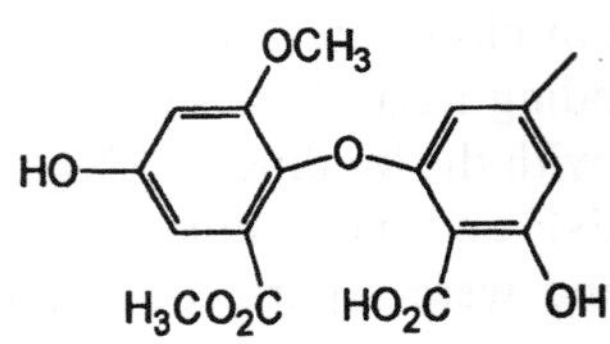

2

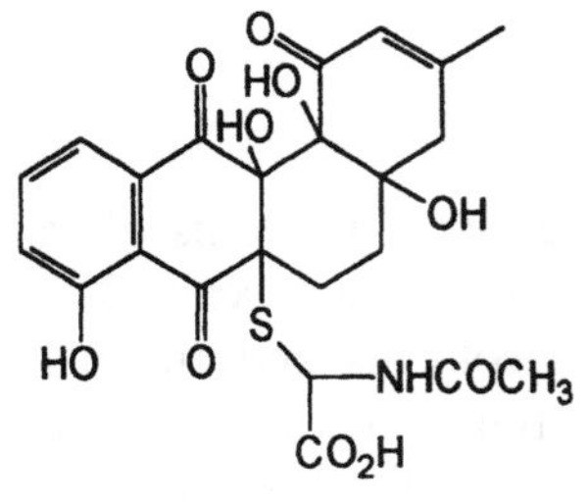

3

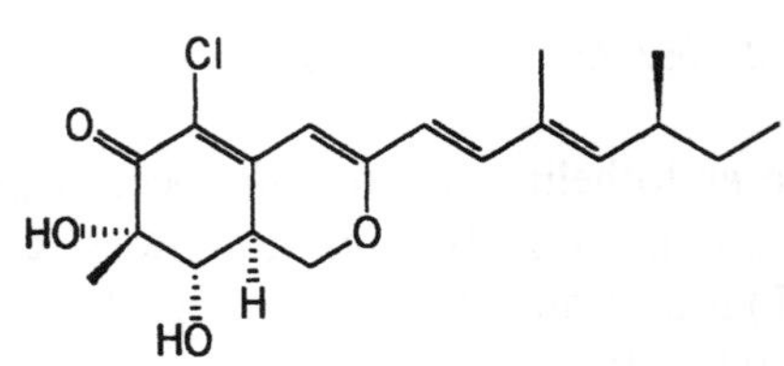

4

Fig. 4. Structures of the endothelin binding inhibitors BE-18257B, **1**, asterric acid, **2**, WS009A, **3**, and an azaphilone from *Penicillium sclerotiorum*, **4** (Sect. 3.3.2)

(Fig. 4) from an *Aspergillus* sp. [77], the aselacins from an *Acremonium* sp. [78, 79], the haloemodins from *Fusarium aquaeductum* [80], and new azaphilones from *Penicillium sclerotiorum* (the structure of one of which is shown in Fig. 4) which were discovered at Xenova in a partnership programme with Parke-Davis Pharmaceutical Research [81]. The ET_A/ET_B receptor selectivity of these compounds is varied. RES-701-1 and the haloemodins are reported to be ET_B-specific.

3.4 CNS Disease

Receptor binding assays have also been used to screen for inhibitors which may be useful in the treatment of CNS disease.

3.4.1 Inhibition of Tachykinin Binding

Tachykinins may play an important role in the pathophysiology of airway diseases, especially asthma. There are at least three related peptide tachykinins: substance P (SP), neurokinin A (NKA) and neurokinin B (NKB), and three different receptor types NK-1, NK-2 and NK-3. It is thought that the mechanical and chemical stimulation of the sensory airway nerve endings causes release of tachykinins, such as SP and NK, from axon collaterals leading to smooth muscle contraction, mucus secretion, vasodilation and protein extravasation.

Using [^{3}H]-SP binding to guinea-pig lung membranes and separating the bound from the free ligand by filtration, WS9326A, a novel acylated macrocyclic hexapeptide lactone, was discovered [82, 83]. The compound is a tachykinin receptor-active metabolite of *Streptomyces violaceusniger*, which was found to be a dual NK1, NK2 antagonist on further evaluation.

Anthrotainin (Fig. 5) was discovered using an assay based on the binding of [^{125}I]-SP to NK-1 receptor rich rat forebrain membranes [84]. Anthrotainin is a novel tetracyclic metabolite of *Gliocladium catenulatus* which is related to viridicatumtoxin. Treatment of anthrotainin with diazomethane produced a more stable but unusual adduct in which the amide moiety was methylated. Anthrotainin inhibited [^{125}I]-SP binds to rat forebrain membranes in a concentration dependent manner. However, the more stable methylated product was inactive. Studies on the dissociation rate of [^{125}I]-SP in the presence and absence of anthrotainin indicated that interaction of the compound with the receptor is not strictly competitive with SP binding.

Fiscalins A, B and C were discovered as new inhibitors of the binding of SP to human NK-1 receptors from U373 MG cells [85]. The structure of fiscalin A is shown in Fig. 5. In the binding assay, [^{125}I]-Bolton-Hunter-SP (BH-SP) and test samples were added to cells grown to confluence in 96-well plates. After incubation and washing, cells were detached from the plates with detergent and radioactivity was measured. The fiscalins are metabolites of the fungus

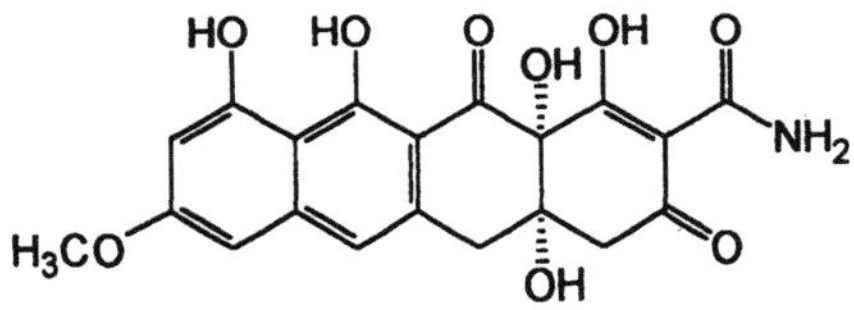

1

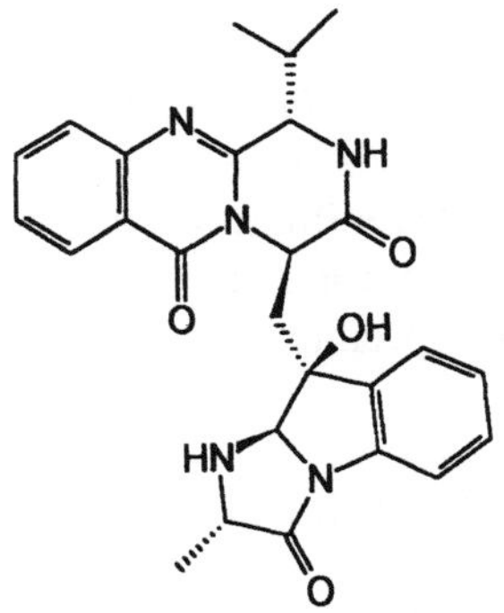

2

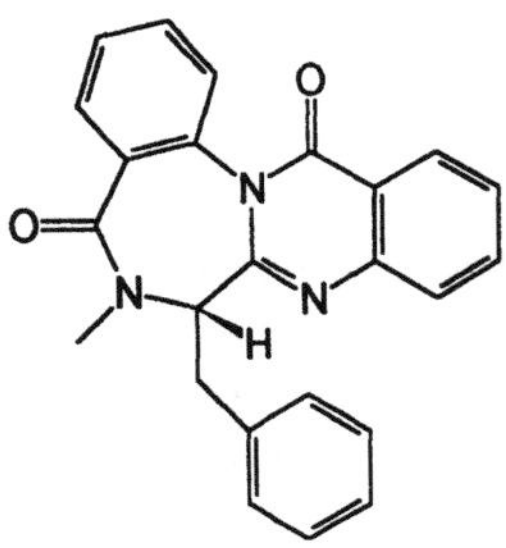

3

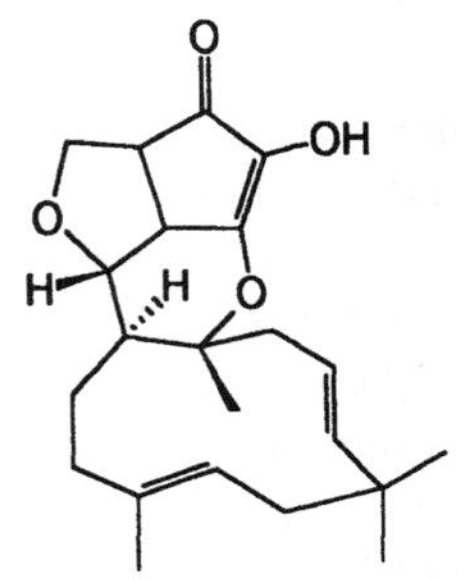

4

Fig. 5. Structures of the tachykinin binding inhibitors anthrotainin, **1**, fiscalin A, **2**, benzomalvin A, **3** (Sect. 3.4.1), and the GABA-benzodiazepine receptor binding inhibitor xenovulene A, **4** (Sect. 3.4.2)

Neosartorya fischeri and are related to the tremorgenic tryptoquivaline mycotoxins.

Using an assay involving BH-SP binding to intact human astrocytoma cells (HAC), WIN 64821 was discovered [86, 87]. In this assay cells were incubated with $[^{125}I]$-BHSP and test sample, followed by removal of unbound $[^{125}I]$-BH-SP by washing with ice cold buffer. Cells were detached from the plates by addition of detergent, and the level of bound ^{125}I measured. WIN 64821 is a diketopiperazine dimer produced by an *Aspergillus* sp. and is related to the known *A. flavus* metabolite ditryptophenaline [88]. Ditryptophenaline was markedly less active than WIN 64821. WIN 64821 was four times more potent in HAC than in human fetal brain membranes. The compound was a competitive inhibitor of $[^{3}H]$-SP binding. WIN 64821 also had a similar activity against human urinary bladder NK2 receptors, and was less potent against guinea pig brain NK-3 receptors.

On further evaluation, WIN 64821 was found to inhibit SP induced calcium release from astrocytoma cells, as measured by monitoring $^{45}Ca^{2+}$ efflux from pre-treated U-373 MG cells, but the compound had no effect in the absence of SP. In an NK-2 rat vas deferens assay, etedoisin-induced twitch potentiation was inhibited by WIN 64821 in a concentration-dependent manner. Specificity of WIN 64821 was demonstrated in a range of receptor binding and ion channel assays.

Further microbial tachykinin receptor antagonists reported recently include the NK-1/NK-2 receptor-active cyclic peptide WIN 66306 from an *Aspergillus* sp. [89], the benzodiazepine benzomalvins (the structure of one of which is shown in Fig. 5) from a *Penicillium* sp. [90], and the fiscalin-related spiroquinazoline from *Aspergillus flavipes* [91].

3.4.2 GABA$_A$-Benzodiazepine Receptor Binding Inhibitors

Benzodiazepines with tranquilising and anticonvulsant activity act by modulating the major inhibitory neuro-transmitter, gamma-amino butyric acid (GABA). Activation of the $GABA_A$ receptor by GABA agonists causes the chloride channel to open, and the resulting influx of chloride anions inhibits neuronal firing by generating hyperpolarisation. The effect of benzodiazepines is to increase the frequency of channel opening. Benzodiazepine therapy, although effective, can lead to the development of dependence and has a number of side effects. A search for alternatives to the benzodiazepines at Xenova utilized an assay based on the binding of flunitrazepam to the $GABA_A$ benzodiazepine receptor in an assay where the bound radioligand was separated from free radioligand using filtration.

The use of this assay to screen microbial fermentation samples led to the discovery of the xenovulenes, potent inhibitors of GABA-benzodiazepine binding produced by *Acremonium strictum* [92]. The xenovulenes contain oxygenated polyketide moieties fused to a sesquiterpenoid humulene ring. The most potent member of the series is xenovulene A (Fig. 5), which inhibits flunitrazepam binding to GABA-benzodiazepine receptors in a synaptosome membrane preparation from ox cerebral cortex with an IC_{50} of 40 $nmol\,l^{-1}$. Pharmacological evaluation of xenovulene A has shown that it acts selectively on a subset of receptors, giving the potential for a reduction in the side-effects associated with the benzodiazepines (unpublished data; Xenova).

4 Enzyme Assays

As discussed in Sect. 1.3, a very wide variety of enzyme targets have been used in the search for pharmacologically active compounds. Details of some of the compounds isolated over the last seven years are included in this section.

4.1 Inflammatory Disease

4.1.1 Sugar Metabolism Inhibitors

Glycoconjugates on mammalian cell surfaces have an important functional role, not least in inflammatory disease where cell adhesion is a key process in leukocyte rolling and, ultimately, extravasation. Cell adhesion is also important in cancer metastasis and viral infection. A plethora of enzymes are involved in the metabolism of sugars presented at cell surfaces. One such enzyme which has been used in a microbial screening programme is α-D-mannosidase.

Rat epididymis α-D-mannosidase activity was determined colorimetrically by measuring the amount of *p*-nitrophenol released from *p*-nitrophenyl mannose. A screening programme based on this assay resulted in the discovery of the α-D-mannosidase inhibitors, mannostatins A and B, from the culture broth of *Streptoverticillium verticillius* var. *quintum* [93]. These aminocyclopentane derivatives differ in that the methylthioether moiety in mannostatin A (Fig. 6) is oxidised to a methylsulphoxide moiety in mannostatin B. Further biological characterisation of the mannostatins revealed that they were competitive inhibitors of the enzyme and had low toxicity in vivo.

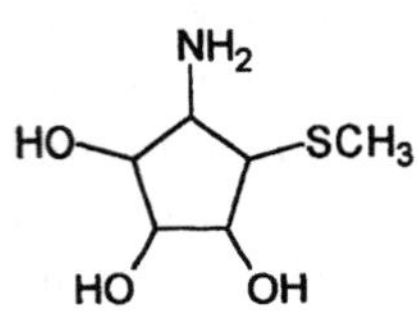

1

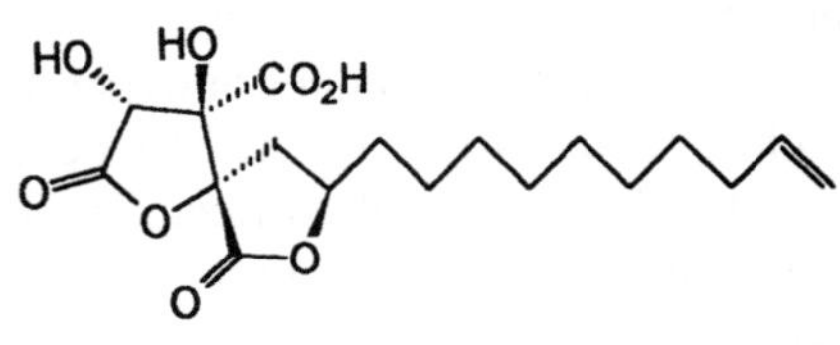

2

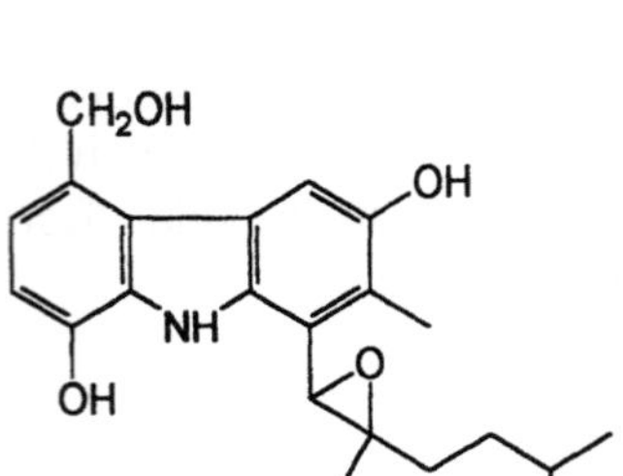

3

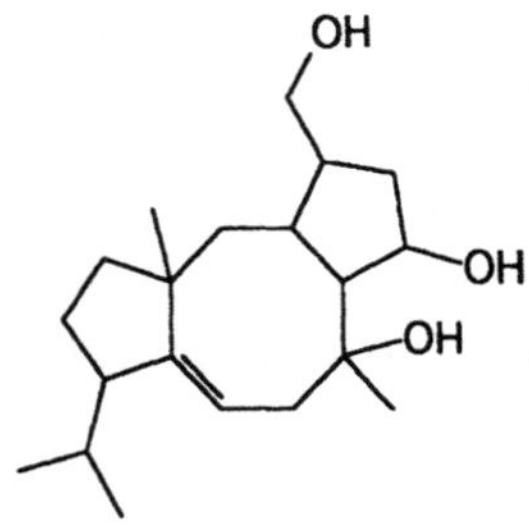

4

Fig. 6. Structures of the α-D-mannosidase inhibitor mannostatin A, **1** (Sect. 4.1.1), phospholipase A_2 inhibitor cinatrin A, **2**, the 5-lipoxygenase inhibitor epocarbazolin A, **3**, and the lysophospholipase inhibitor cycloocatatin, **4** (Sect. 4.1.2)

4.1.2 Inhibition of Lipid Metabolism

Phospholipase A_2 (PLA_2) is a lipolytic enzyme which plays a major role in the release of arachidonic acid from phospholipids in cell membranes. Further metabolism of arachidonic acid leads to the formation of prostaglandins and leukotrienes, which are potent mediators of inflammatory diseases.

Human PMNL PLA_2 was assayed by incubation with [^{14}C]-oleate labelled *Escherichia coli* as substrate. After terminating the reaction, released [1-^{14}C]-oleic acid was separated from the substrate on silicic acid columns and determined by radiometry. This assay was used in the characterisation of duramycins B and C, complex lanthionine containing peptidic antibiotics isolated from a *Streptoverticillium* strain and *Streptomyces griseoluteus*, respectively [94]. These compounds inhibited PLA_2 from synovial fluid as well as that from PMNLs.

Other reports of novel microbial PLA_2 inhibitors include the potent inhibitor thielocin A1β, a complex depside incorporating a xanthene moiety produced by *Thielavia terricola* [95, 96]; the cinatrins, spiro-γ-dilactones and γ-lactones from *Circinotrichum falcatisporum* [97]; and folipostatin, a depsidone from *Aspergillus unguis* [98]. The structure of cinatrin A is shown in Fig. 6.

The PLA_2 assay used in the characterisation of the thielocins involved the preparation of substrate by mixing 1-palmitoyl-2[1-^{14}C] linoleoyl phosphatidylethanolamine with L-α-phosphatidylethanolamine, drying and resuspension. After the enzyme reaction, released free fatty acid was extracted and measured radiometrically. Thielocin A1β strongly inhibited rat PLA_2 II, but weakly inhibited PLA_2 I from rat pancreas. The compound was also active against human PLA_2 II from rheumatoid synovial fluid, and not human pancreas PLA_2 I. Inhibition of rat PLA_2 II was Ca^{2+}-independent and independent of substrate concentration. Thielocin A1β showed similar PLA_2 inhibitory activity on a variety of substrates and was shown to be a non-competitive, reversible inhibitor.

Arachidonate 5-lipoxygenase is another enzyme involved in arachidonic acid metabolism. In this case the target enzyme was prepared from RBL-1 cells, and activity measured by determining the amount of 5-hydroxyeicosatetraenoic acid (5-HETE) produced from arachidonic acid. 5-HETE was measured by HPLC. This assay was used in the discovery and characterisation of the epocarbazolins, carbazole antibiotics bearing epoxide-containing side chains from *Streptomyces anulatus* [99]. The structure of epocarbazolin A is shown in Fig. 6. The activity of the epocarbazolins was stated probably to be due to their free radical scavenging activity.

The nitrosoxacins, a family of *N*-nitroso-*N*-alkylhydroxylamines [100], a series of 5-hydroxyanthranilic acid derivatives [101], and lagunamycin, an unusual diazo-tetraoxoquinoline [102], were all isolated from various *Streptomyces* strains during the same screening programme for 5-lipoxygenase inhibition. The activity of the nitrosoxacins was probably due to their chelating activity.

Eosinophilic leukocytes accumulate in blood and tissue in many allergic and inflammatory reactions. Sonicates of eosinophils express three to eight times as much lysophospholipase (Lyso PL) activity as comparable numbers of neutrophils or mononuclear leukocytes. Lyso PL catalyses the hydrolysis of fatty acid ester bonds of lysophospholipids, liberating the corresponding free fatty acid and glycerophosphate.

Partially purified Lyso PL from bovine liver was incubated with 1-palmitoyl lysophosophatidylcholine (Lyso PC) and Lyso PC-[palmitoyl-1-^{14}C]. After stopping the reaction, the product was separated with heptane and measured radiometrically. This assay was used in the detection, isolation and characterisation of cyclooctatin (Fig. 6), a diterpene metabolite of *Streptomyces melanosporofaciens* [103, 104]. Cyclooctatin is a competitive inhibitor of Lyso PL. This compound was also slightly active against PLA_2 and Phospholipase C (PLC). Cyclooctatin had low toxicity when tested in vivo.

4.1.3 Human Leukocyte Elastase Inhibitors

Human leukocyte elastase (HLE) is one of the most destructive enzymes known. This enzyme hydrolyses several connective tissue components, including elastin, proteoglycan and certain types of collagen. Release of HLE from PMNL by inflammatory stimuli may play a role in destructive processes associated with chronic inflammatory diseases. HLE activity was measured by monitoring the release of *p*-nitroaniline from methoxysuccinyl-$(Ala)_2$-Pro-Val-*p*-nitroanilide. The effects of inhibitors on other proteinases were measured using other *p*-nitroanilide substrates. This assay was used in the discovery of FR901277, a peptidic metabolite of *Streptomyces resistomycificus* [105], and FR901451, a tricyclic depsipeptide produced by a *Flexibacter* sp. [106, 107].

FR901277 inhibited HLE, porcine pancreatic elastase (PPE) and chymotrypsin but was only a weak inhibitor of trypsin. α1-Antitrypsin (α1AT), an endogenous serine proteinase inhibitor, inhibited all enzymes tested. FR901277 was a competitive HLE inhibitor, and moderated HLE-induced paw edema in mice. FR901451 was also shown to be a competitive HLE inhibitor, and to have promising activity in an in vivo model based on elastase induced changes of respiratory mechanics in hamsters.

4.2 Cancer

4.2.1 Inhibition of Cell Signalling Enzymes

Ras poteins, which are localized on the inner side of the plasma membrane, are important in signal transduction. The interaction of these proteins with the plasma membrane requires post-translational modification of their carboxy-terminus. All *ras* proteins share a common sequence, the CAAX box which consists of a conserved cysteine (C), two aliphatic amino acids (AA) and a carboxy-terminal residue (X), which is a pre-requisite for post-translational processing and subsequent malignant transformation. One of the key steps here is farnesylation of the conserved cys residue by farnesyl protein transferase (FPT), an enzyme which has been targeted in an inhibitor screening programme.

In the screening assay, FPT from human THP-1 cell cytosol was incubated with sample, recombinant p21 protein and [^{3}H]-farnesyl pyrophosphate, (FPP). After incubation, the reaction was stopped, the mixture filtered through a glass fibre filter, and the activity of the test samples determined with respect to controls. This assay was used to discover the pepticinnamins, a family of *N*-acylpeptides produced by a *Streptomyces* sp. [108, 109].

Protein tyrosine phosphatase (PTPase) is another enzyme which is thought to be important in the regulation of intracellular signal transduction. Tyrosine kinases associated with *src*-family protooncogene products are inactivated by phosphorylation of a tyrosine residue near their carboxyl terminus.

In the assay involved in the discovery of dephostatin, Jurkat cell membranes were used as a source of PTPase activity, and were assayed using *o*-phospho-L-tyrosine as a substrate and measuring liberated inorganic phosphate [110]. Dephostatin is 2-(*N*-methyl-*N*-nitroso)-hydroquinone, a novel metabolite of a *Streptomyces* sp. [111]. Dephostatin is a competitive PTPase inhibitor, has similar toxicity to oncogene-expressing cells as to normal parent cells, and is of low toxicity to mice.

A number of oncogene products are tyrosine kinases which have also been used in enzyme inhibitor screening programmes. Aberrant activation of such enzymes has been implicated in human clinical cancers.

The target in one such screening programme was the EGF receptor tyrosine kinase (EGF rTK). The incorporation of (γ^{32}P) from (γ^{32}P)-ATP was measured using human recombinant EGF as a substrate. The reaction was terminated with tricholoroacetic acid and incorporated radioactivity measured after separation through filter mats. Use of this assay resulted in the detection and discovery of BE-23372M, a substituted butenolide produced by *Rhizoctonia solani* [112, 113]. BE-23372M is a potent selective inhibitor of EGF rTK, having poor activity against mouse brain PKC and cAMP-dependent protein kinase. The compound also inhibited proliferation of A431 cells (which overexpress erb-B1, the EGF receptor) and MKN-7 cells (reported to express erb-B2 or HER-2, a truncated and mutated form of the EGF receptor that lacks the external EGF binding domain of the receptor and is activated in the absence of ligand binding, in human cells). BE-23372M did not block binding of EGF to its receptor.

Phospholipase C (PLC) is another enzyme which is important in signal transduction, having a key role in phosphoinositol metabolism, and thereby controlling various cellular responses to actions of hormones and peptide growth factors. Hispidospermidin, a novel cage compound with a trimethylspermidine side chain produced by *Chaetosphaeronema hispidulum*, was discovered using an assay involving rat brain PLC, and monitoring [^{3}H]-phosphatidyl inositol metabolism [114, 115].

Hispidospermidin inhibited rat brain PLC in a dose-dependent manner and was cytoxic to HeLa S3 cells. Another microbial metabolite discovered during the same screening programme was vinaxanthone, a chromone-substituted xanthone from *Penicillium vinaceum* [116], which has also been characterised as a CD4-binding inhibitor (Sect. 3.1.2).

4.2.2 Compounds Inhibiting DNA-Dependent Enzymes

Enzymes responsible for altering topological conformation of DNA have also been chosen as targets for dicovery of compounds which may be useful in cancer therapy.

DNA topoisomerases are nuclear enzymes which catalyse breaking and rejoining of DNA strands, and therefore control topological states of DNA. Anti-tumour drugs such as camptothecin and its derivatives have been shown to interfere with single strand breakage-rejoining of DNA catalysed by topoisomerase I (Topo I), by stabilising a tight Topo I-DNA complex termed a "cleavable complex" and preventing the final rejoining step of the reaction.

An assay based on Topo I mediated cleavage of plasmid pUL402 DNA by calf thymus Topo I was used to screen microbial samples and resulted in the discovery of UCE6 (Fig. 7), a naphthacene-5,12-dione produced by an actinomycete culture [117]. UCE6 activity was dose-dependent and comparable to that of camptothecin. The compound did not interfere with the breakage-rejoining reaction of calf thymus topoisomerase II (Topo II), and did not inhibit other DNA active enzymes such as DNA ligase. UCE6 did not induce changes in the supercoiled structure of pUL402 DNA in the absence of Topo I and had cytotoxic activity against HeLaS3 cells.

A second approach to the detection of Topo I inhibitors involved an assay for relaxation of supercoiled pBR322 plasmid DNA by calf thymus Topo I. Metabolites identified because of their inhibitory activity in this assay include naphthacenecarboxamides TAN-1518 A and B from a *Streptomyces* sp. [118], and the epioligothiadiketopiperazines TAN-1496-A, C and E from a *Microsphaeropsis* sp. [119]. The structure of TAN-1518 A is shown in Fig. 7. Both sets of metabolites strongly suppressed the growth of various murine and human tumour cells and induced apoptosis.

A comparable assay was used to discover the Topo II inhibitor UCT4B [120, 121]. The assay monitored the relaxation of supercoiled pUL406 DNA by calf thymus Topo II. UCT 4B (Fig. 7) is a diterpene closely related to the known antitumour antibiotic terpentecin, with which it was co-produced by a *Streptomyces* sp. UCT4B had no effect on pUL406 DNA in the absence of Topo II. The compound prolonged the life span of mice inoculated with P388 lymphocytic leukemia cells.

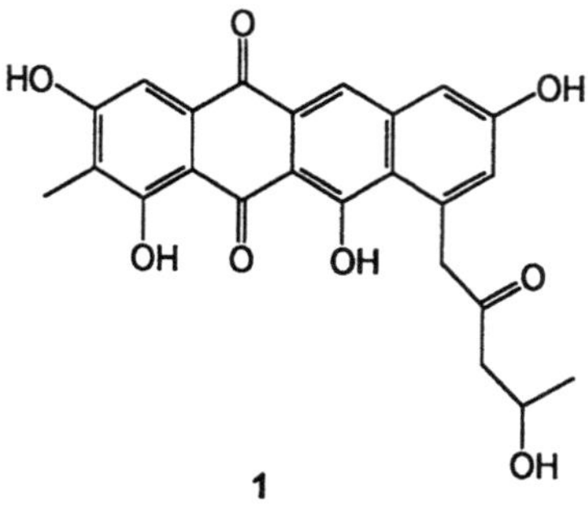

1

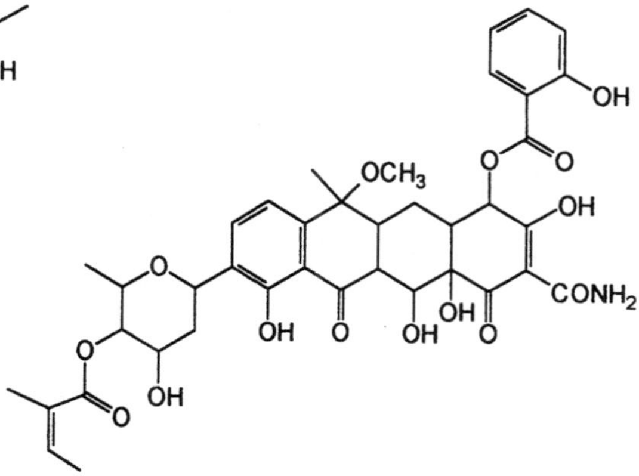

2

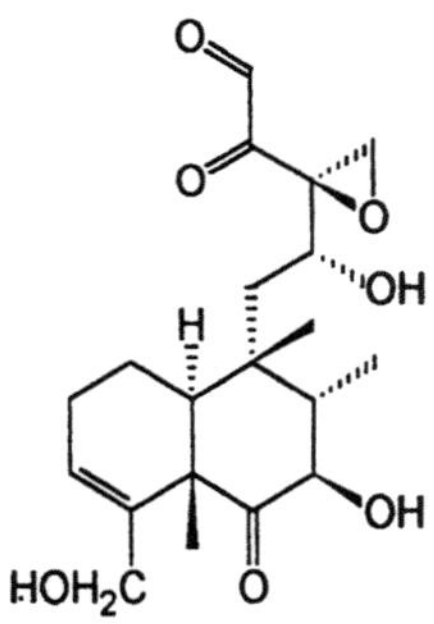

3

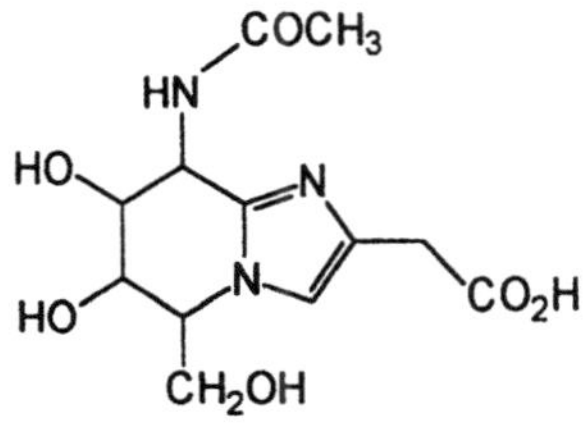

4

Fig. 7. Structures of the topoisomerase I-mediated DNA cleavage inducer UCE6, **1**, the topoisomerase I inhibitor TAN-1518 A, **2**, the topoisomerase II inhibitor UCT4B, **3** (Sect. 4.2.2) and the *N*-acety-β-D-glucosaminadase inhibitor nagstatin, **4** (Sect. 4.2.5)

BE-22179, an unusual cyclic depsipeptide containing two thioester linkages produced by a *Streptomyces* sp., was discovered in a screening programme for Topo II inhibitors using a similar assay [122]. BE-22179 inhibited murine leukaemia cell Topo II relaxation of supercoiled pBR322 plasmid DNA, but did not inhibit Topo I from the same source. The compound was active against P388, L1210 (murine leukaemia) and MKN-45 (human stomach adenocarcinoma) cell lines. BE-22179 prolonged survival time of mice transplanted with L1210 leukaemic cells at doses lower than those at which toxicity was observed.

4.2.3 Inhibition of Glandular Estrogen Biosynthesis

Aromatase is an NADPH-dependent monoxygenase which catalyses the conversion of androgen to estrogen, and is therefore a key enzyme in glandular estrogen biosynthesis. Aromatase is found in breast cancer tissue, and specific inhibitors may lower circulating estrogen levels without interfering with biosynthesis of other steroids.

An aromatase assay has been used to screen microbial metabolites. Human placental or rat ovarian aromatase was assayed by measuring the conversion of [1β, 2β-^{3}H]-androstenedione to ^3H_2O and unlabelled estrone. After incubating the enzyme with the substrate, the reaction mixture was extracted with chloroform, residual steroid in the aqueous fraction further adsorbed with activated charcoal, and the activity of the supernatant determined radiometrically. This work led to the identification of TAN-931, a substituted benzophenone derivative produced by *Penicillium funiculosum* [123, 124].

The inhibitory activity of TAN-931 was more potent against human than rat enzyme, and was less potent than 4-hydroxyandrostenedione (4-OHA) against either enzyme. The mechanism of aromatase inhibition by TAN-931 was different from that of 4-OHA. TAN-931 was shown to be more potent than 4-OHA in vivo where it suppressed ovarian growth, reduced plasma 17β-estradiol levels, and inhibited aromatase activity. TAN-931 did not displace 17β-estradiol from the high affinity binding sites on MCF-7 cells, indicating that it does not have affinity for estrogen receptors.

4.2.4 Testosterone Metabolism Inhibitors

Development of prostatic hyperplasia is mediated via androgen metabolism. One of the key steps in testosterone metabolism is formation of 5-α-dihydrotestosterone. A testosterone 5α-reductase inhibition assay was used to discover WS-9659A and B, novel phenazine derivatives with a monoterpenoid substituent produced by a *Streptomyces* sp. [125, 126]. WS-9659 A and B inhibited testosterone 5α-reductase in a dose-dependent manner, and were competitive inhibitors. WS-9659 A also inhibited testosterone 5α-reductase from dog and human prostates. The compound did not, however, inhibit aldose reductase and lactate dehydrogenase. WS-9659 A inhibited growth of ventral prostates induced by testosterone propionate in immature castrated rats.

4.2.5 Other Screen Methodologies and Targets

Increased glutathione *S*-transferase (GST) activity may be important in cellular resistance to anti-cancer drugs. Screening programmes for GST inhibitors have led to the discovery of a number of microbial metabolites including benastatins A-D [127–129] and bequinostatins A-D [130, 131], all of which possess a benzo[*a*]naphthacene skeleton and are produced by the same *Streptomyces* sp. This same *Streptomyces* sp. also produced an unrelated fluorene hydroquinone derivative, cysfluoretin, which also inhibits GST [132].

The complex hydroxamic acid-containing matlystatins from *Actinomadura atramentaria* were discoverd as type IV collagenase inhibitors [133, 134]. Type IV collagenase can break down collagen in basement membranes leading to tumour invasion.

Levels of *N*-acetyl-β-D-glucosaminidase (NAGase), an enzyme responsible for release of *N*-acetylglucosamine from glycoproteins and glycolipids, are known to increase in conditions such as leukaemia and cancer. Nagstatin (Fig. 7), a substrate-competitive inhibitor of NAGase which potentiates cellular immune response in normal mice and reactivates the depressed immune response in tumour bearing mice, was discovered as a metabolite of *Streptomyces amakusaensis* in a NAGase screening programme [135, 136].

4.3 Cardiovascular Disease

4.3.1 Inhibition of Cholesterol Metabolism

Acyl CoA: cholesterol acyltransferase (ACAT) plays an important role in atherogenesis and cholesterol adsorption from the intestines. ACAT inhibitors may therefore be useful in the treatment of atherosclerosis and hypercholesterolemia. ACAT has been a fruitful enzyme target in terms of novel microbial metabolites detected and identified during recent years.

ACAT activity can be assayed using cholesterol, rat liver microsomal protein and [^{14}C]-oleolyl-CoA. After incubation and termination of the reaction, total lipids are separated using TLC, and activity of the cholesteryl oleate region of the chromatogram determined radiometrically. Assays of this type have been used in the discovery of the purpactins, glisoprenins, pyripyropenes, terpendoles, AS-183, lateritin and gypsetin.

The purpactins are depsidones produced by *Penicillum purpurogenum* [137, 138]. Purpactin A was demonstrated to inhibit cholesterol formation in J774 macrophages. In this assay, the cells were dispensed in 24 well microtitre plates and incubated with cholesterol rich liposomes and [^{3}H]-oleate. After incubation, cholesteryl-[^{3}H]-oleate formation was measured using TLC. The compound was one hundred times more potent in this assay than in the in vitro enzyme assay. The purpactins had no antimicrobial activity and purpactin A was not toxic in vivo.

The glisoprenins, hydroxylated nonaprenol derivatives from a *Gilocladium* sp., inhibited ACAT activity in the microsomal enzyme assay and in J774 macrophages, and were not toxic in vivo [139, 140].

The pyripyropenes are produced by *Aspergillus fumigatus* and are the most potent ACAT inhibitors of microbial origin [141–143]. They possess steroid-like skeletons consisting of pyridine, α-pyrone and sesquiterpene moieties.

Pyripyropene A (the structure of which is shown in Fig. 8) inhibits ACAT activity in microsomes with an IC_{50} of 58 nmol l^{-1}. It has no effect against protein farnesyl transferase and diacylglycerol acyltransferase. The pyripyropenes were not cytotoxic in vitro and had low toxicity in vivo. Pyripyropene A showed significant activity in an in vivo model of cholesterol adsorption in hamsters at a dose of 25 mg kg^{-1}. ACAT inhibition by

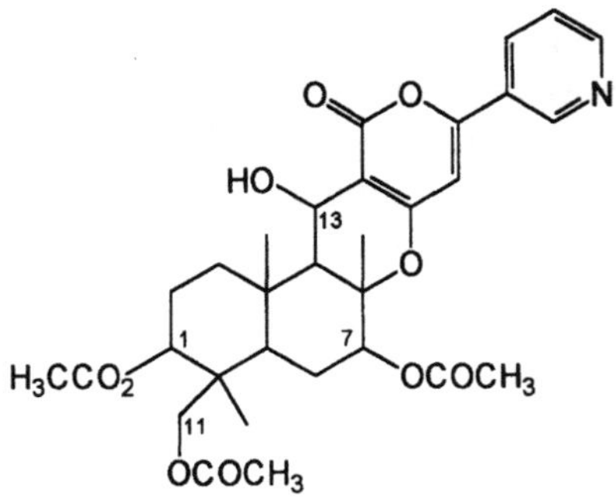

1

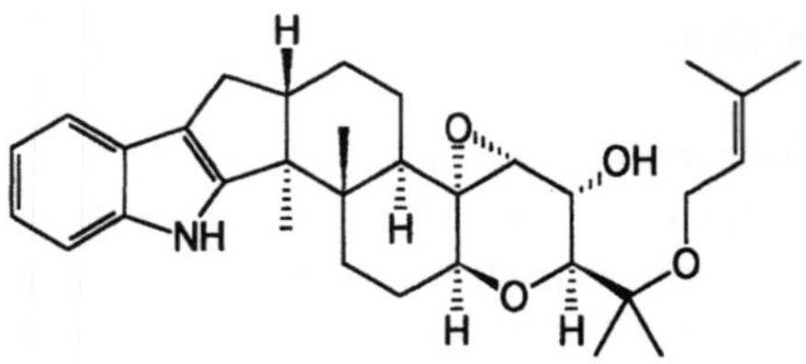

2

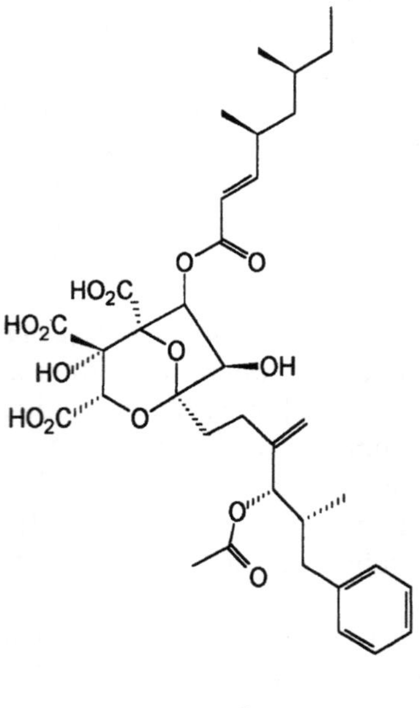

3

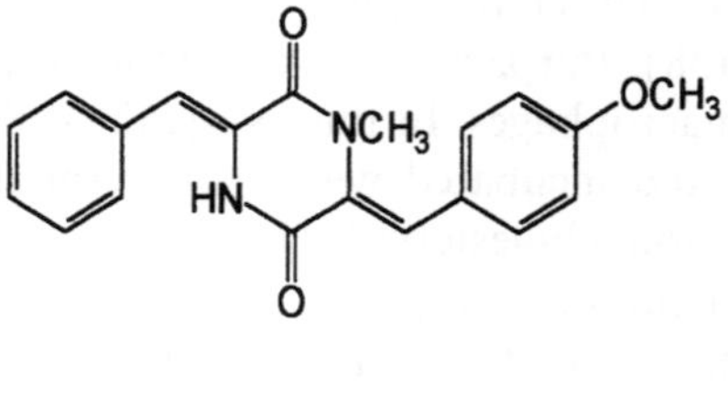

4

Fig. 8. Structures of the acyl CoA: cholesterol acyltransferase inhibitors pyripyropene A, **1**, and terpendole D, **2**, the squalene synthase inhibitor squalestatin SI, **3** (Sect. 4.3.1), and a diketopiperazine, **4**, which inhibits plasminogen activator inhibitor-1 (Sect. 4.3.3)

pyripyropene A was found to be reversible and non-competitive with the substrate oleoyl-CoA; it appeared to be competitive with cholesterol which pyripyropene A structurally resembles [143]. Around 150 derivatives of pyripyropene A have been synthesised to investigate the structural features of the molecule responsible for ACAT activity [143]. One of these, a 1,11-benzylidene acetal-7-O-*n*-valeryl derivative, PR109, was thirteen times more potent than pyripyropene A in vitro and more effective in vivo. The 13-hydroxy group and pyridine moiety were found to be very important for activity.

The terpendoles from *Albophoma yamanashiensis* constitute a series of epoxidated indoloditerpene derivatives [144, 145]. Terpendole D (structure shown in Fig. 8) was stated to be the most specific microbial ACAT inhibitory yet evaluated in the J774 macrophage assay, as judged by its ratio of ACAT inhibitory activity to cytotoxicity.

AS-183 is a novel furan-3-one derivative which exhibits potent inhibition in the in vitro ACAT enzyme assay [146]. It had no effect on acyl CoA: sn-glycerol-3-phosphate acyltransferase (AGAT), lecithin: cholesterol acyltransferase (LCAT), cholesterol 7α-hydroxylase, and 3-hydroxy-3-methylglutaryl coenzyme A (HMG-CoA) reductase activity. The compound did affect cholesterol ester formation in vitro in HepG2 (human liver), $CaCO_2$ (human small intestine) and THP-1 (human monocyte) cell lines.

Lateritin, a substituted 1,4-perhydrooxazine-2,5-dione derivative from *Gibberella lateritium* [147] and gypsetin, a complex diketopiperazine from *Nannizzia gypsea* var. *incurvata* [148, 149], inhibited cholesteryl ester formation in J774 macrophages and did not affect surface binding, uptake or degradation of oxidised [^{125}I]-LDL.

Squalene synthase catalyses the first pathway specific step in cholesterol biosynthesis, and inhibitors of this enzyme may have a cholesterol-lowering effect.

Screening of microbial samples using a squalene synthase assay involving rat liver homogenate protein, [1-^{14}C]-isopentenyl pyrophosphate (IPP) and *trans, trans*-[2-^{14}C]-farnesyl pyrophosphate led to the discovery of the squalestatins. This assay allows measurement of squalene production from either [1-^{14}C]-IPP as a radiolabelled precursor of FPP, or from [2-^{14}C]-FPP as a direct substrate for squalene synthase. Ascorbate and ascorbate oxidase were included in the reaction mixture to consume oxygen and prevent further conversion of IPP or FPP to metabolites beyond squalene. The squalestatins were discovered as potent inhibitors of squalene production from [1-^{14}C]-IPP or [2-^{14}C]-FPP [150, 151].

Squalestatins S1-3 are *Phoma sp.* metabolites containing a trihydroxylated, tricarboxylated 2,8-dioxobicyclo-[3.2.1]octane core. Squalestatin 1 (the structure of which is shown in Fig. 8) is identical to zaragozic acid A, which was isolated from an unidentified sterile fungal culture as a result of its squalene synthase inhibitory activity [152]. The squalestatins were shown to be potent inhibitors of rat liver squalene synthase activity, and squalestatin 1 was a potent inhibitor of *Candida albicans* microsomal squalene synthase. Zaragozic acid

A showed potent activity (50% inhibitory dose of 200 μg kg^{-1} body weight) in an acute in vivo model of hepatic cholesterol synthesis in mice [152].

4.3.2 Smooth Muscle Contraction Inhibitors

Myosin light chain kinase (MLCK) is a regulatory enzyme involved in smooth muscle contraction. MLCK catalyses transfer of the γ-phosphate of ATP to ser-9 of the 20 kDa myosin light chain. Phosphorylated myosin can then interact with myosin and generate contractile force. Since smooth muscle cells are distributed in arteries and bronchi, it is postulated that MLCK inhibitors would be potential vasodilators and bronchodilators. MLCK has been the target of a recent microbial screening programme which resulted in the identification of MS-347a, MS-282a and MS-282b. Chicken gizzard smooth muscle MLCK activity was monitored by assaying phosphorylation of a peptide substrate in the presence of calmodulin. After incubation, the reaction was terminated by addition of acid and product measured by HPLC analysis.

MS-374a, a novel xanthene epoxide, was isolated from an *Aspergillius* sp. together with MS-347b, the known fungal metabolite syndowinin B [153]. MS-347a inhibited Ca^{2+} and calmodulin-dependent and independent activity of chicken gizzard MLCK in a dose-dependent manner. MS-347b did not inhibit the enzyme. MS-347a also inhibited PKC, but did not inhibit cAMP-dependent protein kinase, cGMP-dependent protein kinase, and calmodulin-dependent cyclic nucleotidase phosphodiesterase (PDE). MLCK inhibition was dependent on the length of a pre-incubation period, suggesting irreversible inhibition.

MS-282a and MS-282b, macrocyclic dilactones from *Streptomyces tauricus*, inhibited the calmodulin-dependent, but not the calmodulin-independant, activity of MLCK, suggesting that these compounds block MLCK activity by antagonising calmodulin [154]. MS-282a also inhibited calmodulin-dependent PDE.

A screening programme targeted at calmodulin-dependent PDE inhibitors led to the discovery of SCH 45752 [155]. Calmodulin-dependent bovine heart PDE was assayed by hydrolysis of cyclic AMP, or cyclic GMP. SCH 4572 is a symmetrical bis-naphthopyrone from a culture of an unidentified fungus. It is identical to the known *Verticillium* sp. metabolite cephalochromin.

SCH 4572 was a potent inhibitor of bovine heart and bovine aorta PDE. The compound was much less active against calmodulin independent PDE, MLCK, cAMP-dependent and cGMP-dependent protein kinase, and protein kinase C. SCH 4572 stimulated carotid artery smooth muscle contraction

4.3.3 Inhibition of Plasminogen Activator Inhibitor-I

A key event in the regulation of thrombus formation and clearance is the generation of plasmin by tissue plasminogen activator (tPA). tPA activity is controlled by PAI-1, a fast acting inhibitor belonging to the serine protease inhibitor (serpin) superfamily of protease inhibitors. Elevated plasma concentrations of PAI-1 are thought to be important risk factors in thrombotic disease, and modulators of PAI-1 inhibition of tPA have therefore been sought as potential antithrombotic agents. An amidolytic assay of tissue plasminogen activator mediated plasmin generation has been used to screen micorbial samples at Xenova. The assay mixture contained a concentration of PAI-1 which was just sufficient to inhibit tPA activity. Inhibition of PAI-1 activity resulted in the tPA mediated clearance of plasminogen. The action of the released plasmin on a chromogenic substrate ws then measured spectrophotometrically.

Two diketopiperazines from an unindentified *Strepotomyces* sp. were characterised by their ability to inhibit PAI-1 activity in this assay [156]. The structure of one of these is shown in Fig. 8. Their structures were confirmed by chemical synthesis, and further synthetic studies generated analogues with superior activities in vitro and in vivo. Preclinical studies have confirmed that this series of compounds is effective in preventing clot formation in laboratory models of thrombosis. These compounds do not inhibit normal blood coagulation, suggesting that they will not cause excessive bleeding. They are the first compounds of a novel therapeutic class which, although discovered in a enzyme assay, inhibit an interaction between two large proteins.

4.4 CNS Disease

4.4.1 General Screen Methodologies and Targets

Screening programmes reported in the literature for enzyme inhibitors targeted at CNS disease have included: aminopeptidase-M (AP-M), an enkephalin inactivating enzyme key to opioid peptide metabolism, and prolylendopeptidase (PEP), which degrades a range of biologically active peptides and is found at elevated levels in the occipital lobe of Alzheimer patients. Probestin, and tetrapeptide from *Streptomyces azureus* [157] and leuhistin, an imidazole derivative produced by *Bacillus laterosporus* [158] were discovered in an AP-M screening programme. Poststatin, a pentapeptide metabolite of *Streptomyces viridochromogenes* [159, 160] and eurystatins A and B, cyclic tripeptides from *Streptomyces eurythermus* [161, 162] were discovered during a PEP screening programme.

5 Discussion and Conclusions

5.1 Review of Compounds Identified

The microbial metabolites described in the previous sections illustrate the continuing importance of micro-organisms as a source of chemical diversity for the discovery of compounds with new biological activities. They represent a wide range of structural types including simple and complex peptides, nitrogen and oxygen-based heterocyclic compounds, terpenoids, quinones, simple and complex aromatic systems, macrocyclic lactones and simple sugar analogues. Their producing organisms ranged from various *Streptomyces* spp. and representatives of some of the rarer antinomycete genera through a wide variety of fungi from simple *Aspergillus* and *Penicillium* spp. to some higher fungal species, highlighting the value of a broad microbial collection for screening.

None of the compounds described have yet progressed into a full development programme. Some, such as the myriocins (Sect. 2.1.2), FR-111142 (Sect. 2.3.2), the pyripyropenes (Sect. 4.3.1) and the squalestatins/zaragozic acids (Sect. 4.3.1), have been the subjects of preclinical research efforts which have resulted in analogue synthesis projects. At Xenova we have worked with great success on the lead optimisation of the diketopiperazine PA1-I inhibitors (Sect. 4.3.3), which are very amenable to synthetic modification. We have also embarked on a programme of semi-synthetic modification of xenovulene A as a selective benzodiazepine $GABA_A$ receptor antagonist (Sect. 3.4.1). It is difficult to gauge the extent of preclinical research projects based on microbial products in other companies as such information tends to be very closely guarded.

5.2 Review of Screen Methodologies

A wide range of assays yielded the compounds described in Sects. 2–4. For the cell based screens, a common assay choice was to monitor cell growth and morphology. Measurement of the behaviour of cells was commonly by microscopic observation. Although this approach can yield valuable information, throughput may be limited due to the need for rigorous manual scoring of the results. Other screens, such as those used to search for inhibitors of cell adhesion (Sect. 2.1.3), and myeloperoxidase release (Sect. 2.1.5), for example, have the advantage that results can be obtained directly from plate readers and analyzed using a suitable data management system. This allows increased sample throughput thereby broadening the opportunity for new drug discovery by accessing greater molecular diversity.

In the case of the receptor binding assays, all of the published compounds, except the CD4 binding inhibitors (Sect. 3.1.2), were discovered using assays to detect inhibition of binding of radiolabelled ligands to various cell membranes. Whilst this technique remains a popular choice for screening, there are increasing concerns over the use of radioactivity in large scale screening programmes and alternative technology for exploitation of recombinant receptors, as well as membranes and whole cells, is now being used more widely.

Finally, for the enzyme targets the screens used for discovery of new inhibitors included simple colorimetric assays with end-point read-outs, radiometric assays where the product was measured after a separation step, and electrophoretic assays for determination of enzyme activity on DNA conformation.

Again, in the latter two assay types, opportunities for automation of these techniques are currently limited.

5.3 Developments in Assay Technology for New Drug Discovery

Due both to the anticipated explosion in targets available for screening and the need to test a large number of samples quickly, a wealth of new assay formats are becoming available.

Cell-based assays which are now commonly used in screening programmes include those for inhibitors of extracellular protein secretion, adhesion assays, and reporter gene assays [163] which are shown in Fig. 9.

One recent innovation in cell-based systems is melanophore technology, which uses cells from the frog *Xenopus laevis.* The movement of pigment in this system is sensitive to intracellular concentrations of cAMP and diacylglycerol. If receptor stimulation activates adenyl cyclase and phospholipase C, darkening of cells is induced. Conversely, stimulation of a receptor which inhibits adenyl cyclase induces cell lightening. Recombinant receptors can be stably transfected into this system and pigment hue measured following receptor stimulation. Such changes are appreciable within minutes, making this technology a useful tool for high throughput screening [164], although success is dependent on the genuine coupling of the recombinant receptor with the secondary signalling pathways.

In addition, scintillation proximity assay (SPA) for cell-based assays has recently been developed. SPA is a homogeneous radioisotopic technique which relies on the limited path length of a β particle travelling through aqueous media. If the electron collides with a scintillant particle, energy is transferred and light is emitted. In SPA, scintillant is coated onto the surface of microtitre plate (MTP) wells, or incorporated into beads or the MTP itself, making it possible to attach reagents to the scintillating surface and to monitor the proximity of radiolabelled substances. In the case of cellular assays the scintillant is incorporated into the base of the MTP. The walls of the well are opaque to exclude optical cross talk, and the base is transparent to facilitate visualisation of cells during culture. Cells are grown in monolayers in the wells and radioisotope in solution is too distant from the scintillating base plate to generate a signal. Using this type of assay it is possible to monitor in situ receptor binding as well as cellular metabolism and motility [165].

For receptor binding assays wider availability of cloned receptors has led to changes in technology, including the development of "capture" assays, the use of SPA [166, 167] and introduction of fluorescence polarization as a screening tool, and examples of these are shown in Fig. 10.

In fluorescence polarization assays, the ligand is labelled with a fluorescent tag and used in a technique based on molecular rotation which measures the decrease in depolarization of fluorescence exhibited by small molecules when their rotation is slowed by binding to large molecules [168]. The quantification

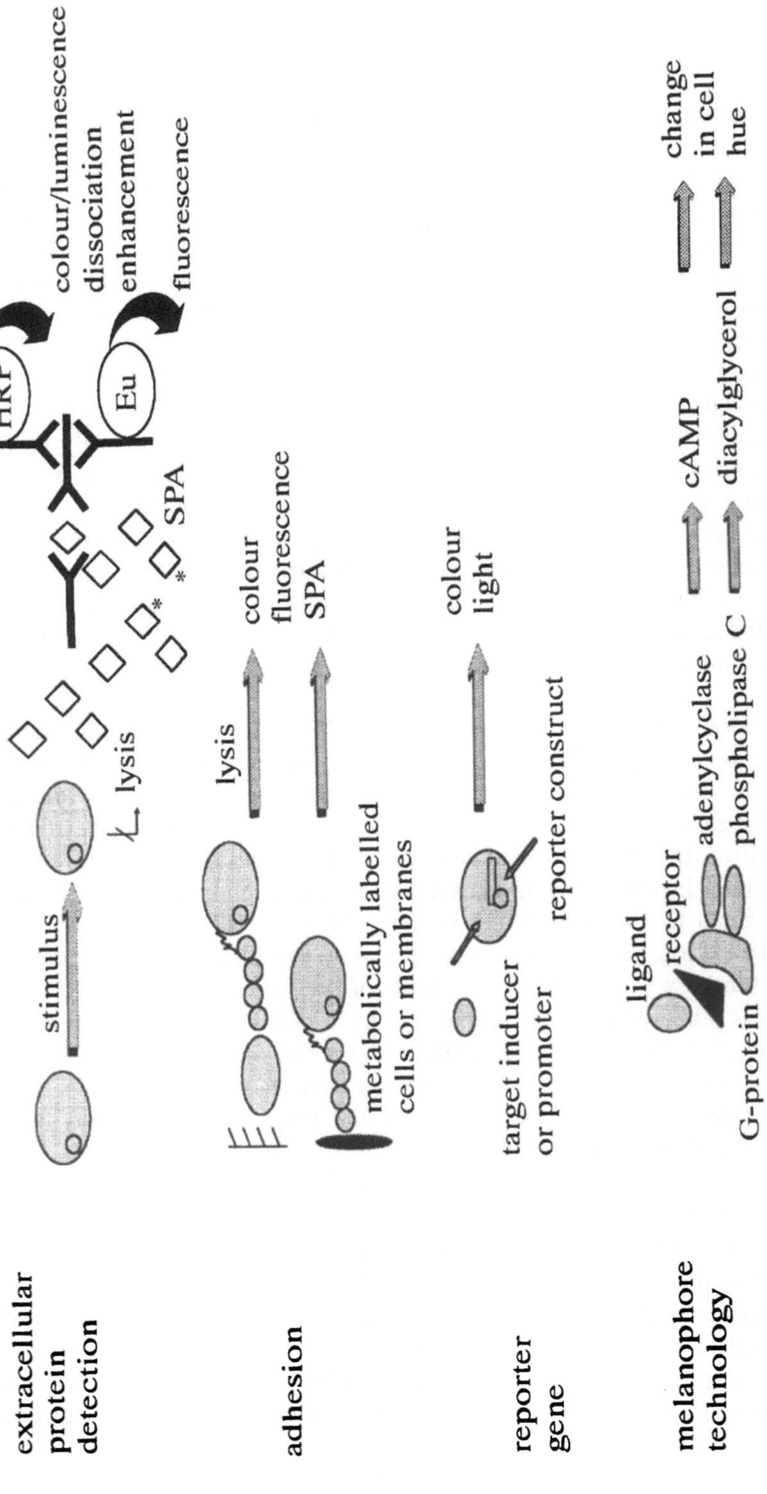

Fig. 9. Overview of a range of cell based assays used in screening programmes

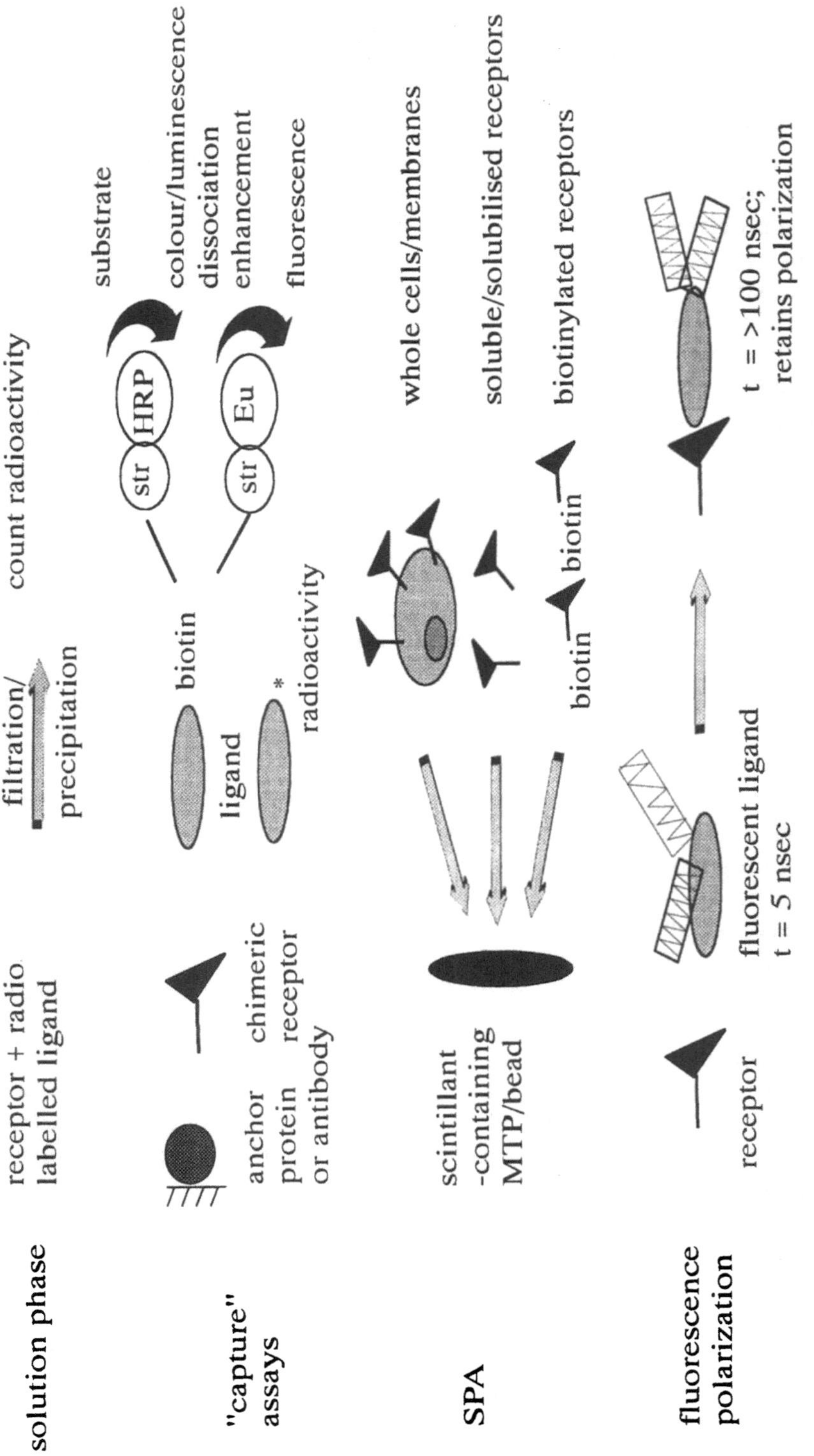

Fig. 10. Overview of a range of receptor-ligand based assays used in screening programmes

of biomolecular interactions can then be achieved using competition methods, making this technique useful as a high throughput screening method [169]. Fluorescence polarization and SPA assays both allow monitoring of 'real time' kinetics of association or dissociation of the receptor-ligand/protein-protein complex as they do not require a separation step.

One possible alternative as a detection system for heterogeneous assays is to use time resolved fluorescence (TRF). TRF utilizes lanthanide chemistry to overcome problems such as quenching and background flourescence which are often observed with standard fluorescence systems. Under appropriate conditions, lanthanides produce a high fluorescence intensity with a sharp emission peak and long decay time, and exhibit a large Stoke's shift (i.e. difference between the excitation and emission wavelength). In practice this means that fluorescence associated with the target interaction can be measured after background has decayed and that assays are highly sensitive and have a wide dynamic range [170]. In addition, a stable signal is generated making the technology ideal for automation [171]. Streptavidin, antibody and protein-lanthanide complexes are available commercially and most proteins can be readily labelled to take advantage of this versatile technology which is easily adaptable to natural products drug screening [172].

Time resolved fluorescence technology can also be used in delayed fluorescence energy transfer DEFRET techniques which enable homogeneous assays to be carried out, as fluorescence is generated by the close proximity of the lanthanide to an acceptor molecule. The acceptor molecule used in this technique may either transmit or quench fluorescence, depending on the choice of assay [173].

A range of techniques including solution phase assays [174, 175], heterogeneous assays [176, 177] and SPA have also been used in enzyme inhibitor screening programmes as shown in Fig. 11.

Emerging assay technologies other than cell-based SPA and DEFRET include electrochemiluminescence (ECL), surface plasmon resonance (SPR), and microphysiometry (see Fig. 12).

Electrochemiluminescence utilizes labelling of one of the assay reagents. In this case, diffusion of a precursor molecule onto an electrode surface is followed by rapid electron transfer, which initiates the excitation of the reporter molecule and results in the emission of a photon of light at a specific wavelength. Use of an electrode allows reactions to take place in a variety of solvents, and gives the capability of increasing assay sensitivity to the $\mathrm{fmol\,l^{-1}}$ range by applying different electrode potentials [178].

Surface plasmon resonance is a label free, real time optical detection technique which facilitates quantification of molecular interactions. The basis of SPR is that a metal (usually gold) coated sensor chip is used to give rise to electronic oscillations, or surface plasmons, at a metal surface. These decay exponentially as a function of distance and their refractive index will change if a complex is formed on the sensor chip surface. This change can be detected and recorded using appropriate instrumentation [179].

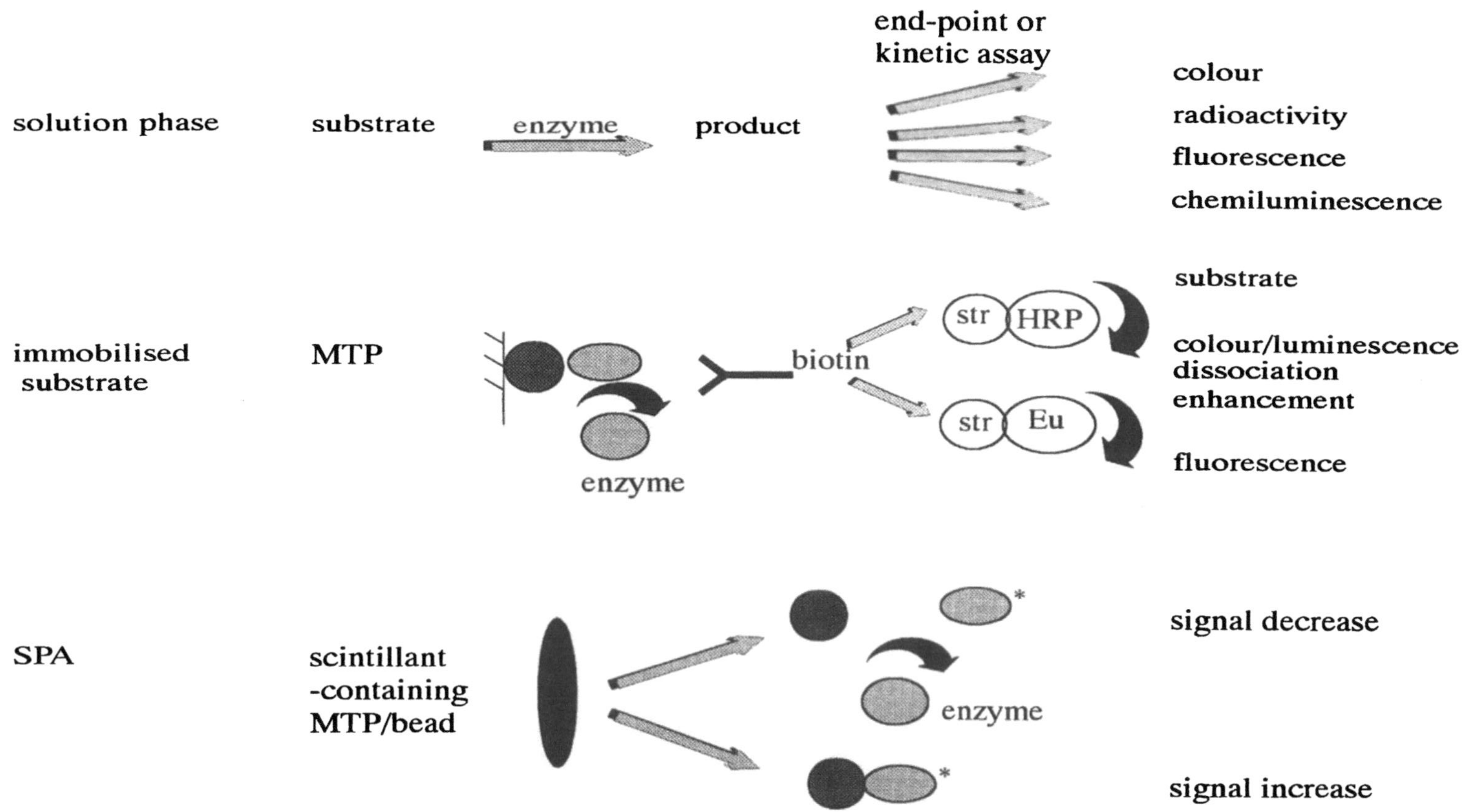

Fig. 11. Overview of a range of enzyme based assays used in screening programmes

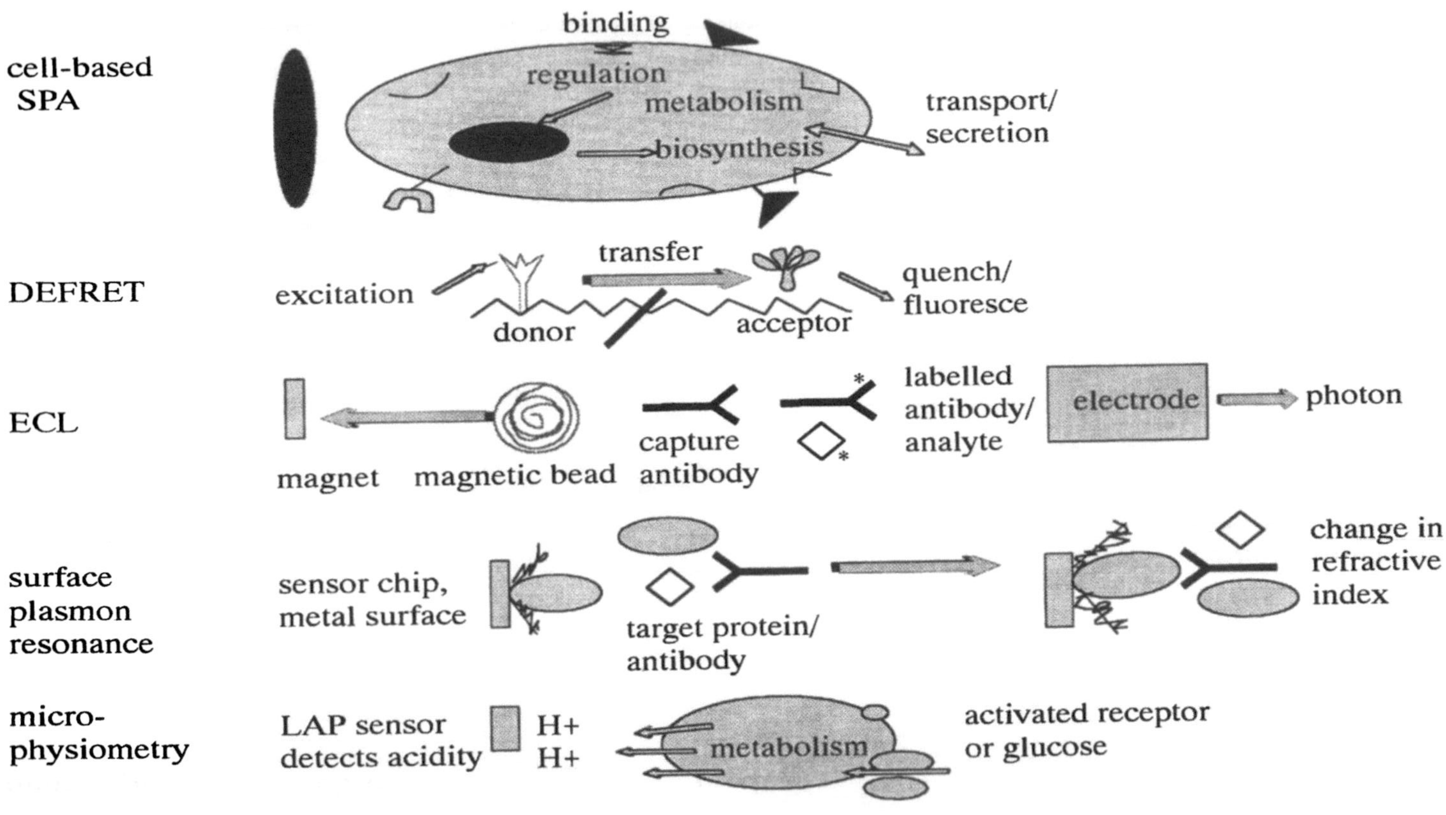

Fig. 12. Recently available and new screening technologies

Finally, the microphysiometer is a silicon-based device which can be used to detect and monitor the response of whole cells to a variety of chemical substances as well as facilitating the investigation of cell function and biochemistry. In this instance the machine uses a light addressable potentiometric sensor (LAPS) to measure the rate at which cells acidify their environment. The cells are retained in a flow chamber in aqueous diffusive contact with the pH-sensitive surface of a LAPS chip. Cellular acidification occurs mainly via glycolysis. However, when there is a ligand-receptor interaction at the cell surface there are substantial metabolic consequences in the cell which alter the acidification rate and can be measured using this instrument [180].

For assays based around 96-well microtitre plate technology, the major current trend in high throughput screening is to increase sample throughput in the race to find new lead compounds. To this end robotic operation of screens has become essential and several instrument companies (e.g. Robocon, Tomtek, Tecan, Beckman/Wallac Oy) now offer complete or custom-built robotic systems to the pharmaceutical industry. At Xenova, we have developed our own screening robot which has allowed us to retain flexibility of assay design and given us a versatile system which is under in-house control and has a maximum theoretical capacity of 1.2 million samples per year.

5.4 Rapid Delivery of New Drug Leads from Natural Products Sources

In the increasingly competitive world of drug discovery it is essentially that, for natural products screening, the available technology is amalgamated in a firmly engineered process to minimise the time to lead candidate selection. At Xenova, the average cycle time of our projects is one year with the first chemicals being identified from approximated six months after the start of each screen.

In the next era of drug discovery, companies who wish to remain at the forefront of screening technology will now need to develop radically different approaches. Integration of robotics into screening programmes, advances in bioinformatics, and re-engineering of drug discovery processes have all helped to ensure maximum use of the microtitre plate. The next generation of assays may be based on high density spotting techniques and paper technology, advanced plastics research, inkjet methods, image analysis systems, biosensors and nanotechnology among many others. For natural products discovery, maximisation of secondary metabolite production and diversity, the integration of chemical fingerprinting and other novel chemical technologies, and focusing effort on assay preformance and data analysis facilitates the selection of quality hits early in the process. Valuable natural products chemistry resources can then be focused on fewer, better characterised samples, enabling identification of new, biologically active, chemical entities which are currently beyond the capability of chemical synthesis.

6 References

1. Berdy J (1974) Adv Applied Microbiol 18: 309
2. Berdy J (1980) Process Biochem 15: 28
3. Nisbet LJ, Porter N (1989). In: Baumberg S, Hunter I, Rhodes M (eds) Soc Gen Microbiol Symp 44. Cambridge Univ Press, p 309
4. Reading C, Cole M (1977) Antimicrob Agents Chemother 11: 852
5. Albert SAW, Chen J, Kuron G, Hunt V, Huff J, Hoffman C, Rothtock J, Lopez M, Joshua H, Harris E, Patchett A, Monoghan R, Currie S, Stapley E, Albers-Schonberg G, Hensens O, Hirshfield J, Hoogsteen K, Liesch J, Springer J (1980) Proc Natl Acad Sci USA 77: 3957
6. Borel JF, Feurer C, Gubler HU, Stahelin H (1976) Agents Actions 6: 468
7. Kino T, Hatanaka H, Hashimoto M, Nishiyama M, Goto T, Okuhara M, Kohsaka M, Aoki H, Imanaka H (1987) J Antibiotics 40: 1249
8. Omura S (1992) J Indust Microbiol 10: 135
9. Rosteck PR (1994) Trends in Cardiovasc Med 3(5): 201–206
10. Umezawa H (1989) in Wagman GH, Cooper R (eds) Natural products isolation. Separation methods for antimicrobials, antivirals and enzyme inhibitors. Elsevier, Amsterdam, p 481
11. Iwami M, Nakayama O, Terano H, Kohsaka M, Aoki H, Imanaka H (1987) J Antibiotics 40: 612
12. Kayakiri H, Takase S, Shibata T, Okamoto M, Terano H, Hashimoto M, Tada T, Koda S (1989) J Org Chem 54: 4015
13. Yamashita T, Iijima M, Nakamura H, Isshiki K, Naganawa H, Hattori S, Hamada M, Ishizuka M, Takeuchi T (1991) J Antibiotics 44: 557
14. Kawatsu M, Yamashita T, Osono M, Ishizuka M, Takeuchi T (993) J Antibiotics 46: 1687
15. Kawatsu M, Yamashita T, Osono M, Masuda T, Ishizuka M, Takeuchi T (1993) J Antibiotics 46: 1692
16. Fujita T, Inoue K, Yamaoto S, Ikumoto T, Sasaki S, Toyama R, Chiba K, Hoshino Y, Okumoto T (1994) J Antibiotics 47: 208
17. Fujita T, Inoue K, Yamamoto S, Ikumoto T, Sasaki S, Toyama R, Yoneta M, Chiba K, Hoshino Y, Okumoto T (1994) J Antibiotics 47: 216
18. Sasaki S, Hashimoto R, Kiuchi M, Inoue K, Ikumoto T, Hirose R, Chiba K, Hoshino Y, Okumoto T, Fujita T (1994) J Antibiotics 47: 420
19. Sakamoto K, Tsujii E, Miyauchi M, Nakanishi T, Yamashita M, Shigematsu N, Tada T, Izumi S, Okuhara M (1993) J Antibiotics 46: 1788
20. McAlpine JB, Karwowski JP, Jackson M, Mullally MM, Hochlowski JE, Premachandran U, Burres NS (1994) J Antibiotics 47: 281
21. Ueno M, Amemiya M, Someno T, Masuda T, Iinuma H, Naganawa H, Hamada M, Ishizuka M, Takeuchi T (1993) J Antibiotics 46: 1658
22. Ueno M, Amemiya M, Iijima M, Osono M, Masuda T, Kinoshita N, Ikeda T, Iinuma H, Hamada M, Ishizuka M, Takeuchi T (1993) J Antibiotics 46: 719
23. Ueno M, Someno T, Sawa R, Iinuma H, Naganawa H, Ishizuka M, Takeuchi T (1993) J Antibiotics 46: 979
24. Ameyima M, Ueno M, Osono M, Masuda T, Kinoshita N, Nishida C, Hamada M, Ishizuka M, Takeuchi T (1994) J Antibiotics 47: 536
25. Ameyima M, Someno T, Sawa R, Naganawa H, Ishizuka M, Takeuchi T (1994) J Antibiotics 47: 541
26. Tanida S, Takizawa M, Takahashi T, Tsubotani S, Harada S (1989) J Antibiotics 42: 1619
27. Tsubotani S, Tanida S, Harada S (1991) Tetrahedron 47: 3565
28. Ishida-Okawara A, Kimoto Y, Watanabe K, Tokuda K, Shibata M, Masuda K, Takano Y, Kawaguchi K, Akagawa H, Nilubol N, Hotta K, Yazawa K, Mizuno S, Suzuki K (1991) J Antibiotics 44: 524
29. Kohama T, Miyaoka H, Torikata A, Inukai M, Kaneko I, Shiraishi A (1993) J Antibiotics 46: 1494
30. Kohama T, Enokita R, Okazaki T, Miyaoka H, Torikata A, Inukai M, Kaneko I, Kagasaki T, Sakaida Y, Satoh A, Shiraishi A (1993) J Antibiotics 46: 1503
31. Kohama T, Nakamura T, Kinoshita T, Kaneko I, Shiraishi A (1993) J Antibiotics 46: 1512

32. Karwowski JP, Jackson M, Rasmussen RR, Humphrey PE, Poddig JB, Kohl WL, Scherr MH, Kadam S, McAlpine JB (1993) J Antibiotics 46: 374
33. Hochlowski JE, Mullally MM, Spanton SG, Whittern DN, Hill P, McAlpine JB (1993) J Antibiotics 46: 380
34. Abe K, Yoshida M, Horinouchi S, Beppu T (1993) J Antibiotics 46: 728
35. Abe K, Yoshida M, Naoki H Horinouchi S, Beppu T (1993) J Antibiotics 46: 735
36. Hosokawa N, Iinuma H, Naganawa H, Hamada M, Takeuchi T (1993) J Antibiotics 46: 676
37. Karwowski JP, Jackson M, Sunga G, Sheldon P, Poddig JB, Kohl WL, Kadam S (1994) J Antibiotics 47: 862
38. Hochlowski JE, Whittern DN, Hill P, McAlpine JB (1994) J Antibiotics 47: 870
39. Kadam S, McAlpine JB (1994) J Antibiotics 47: 875
40. Uehara Y, Ki PM, Fukazawa H, Mizuno S, Nihei Y, Nishio M, Hanada M, Yamamoto C, Furumai T, Oki T (1993) J Antibiotics 46: 1306
41. Hino M, Iwami M, Okamoto M, Yoshida K, Haruta H, Okuhara M, Hosoda J, Kohsaka M, Aoki H, Imanaka H (1989) J Antibiotics 42: 1578
42. Hino M, Takase S, Itoh Y, Uchida I, Okamoto M, Hashimoto M, Kohsaka M (1989) Chem Pharm Bull 37: 2864
43. Kishimura Y, Kawashima A, Kagamizono T, Yamagishi M, Matsumoto K, Kawashima Y, Hanada K (1992) J Antibiotics 45: 892
44. Patel M, Hegde V, Horan A, Barret T, Bishop R, King A, Marquez J, Hare R, Gullo V (1989) J Antibiotics 42: 1063
45. Hegde VR, King AH, Patel MG, Puar MS, McPhail AT (1987) Tet Lett 28: 4485
46. Lauer U, Anke T, Hansske F (1991) J Antibiotics 44: 59
47. Otsuka T, Shibata T, Tsurumi Y, Takase S, Okuhara M, Terano H, Kohsaka M, Imanaka H (1992) J Antibiotics 45: 348
48. Ingber D, Fujita T, Kishimoto S, Sudo K, Kanamaru T, Brem H, Folkman J (1990) Nature 348: 555
49. Otsuka T, Ohkawa T, Shibata T, Oku T, Okuhara M, Terano H, Kohsaka M, Imanaka H (1991) J Microbiol Biotechnol 1: 163
50. Otsuka T, Takase S, Terano H, Okuhara M (1992) J Antibiotics 45: 1970
51. Nozaki Y, Hida T, Iinuma S, Ishii T, Sudo K, Muroi M, Kanemaru T (1993) J Antibiotics 46: 569
52. Omura S, Fujimoto T, Otoguro K, Matsuzaki K, Moriguchi R, Tanaka H, Sasaki Y (1991) J Antibiotics 44: 113
53. Omura S, Matsuzaki K, Fujimoto T, Kosuge K, Furuya T, Fufita S, Nakagawa A (1991) J Antibiotics 44: 117
54. Ito M, Maruhashi M, Sakai N, Mizoue K, Hanada K (1992) J Antibiotics 45: 1559
55. Ito M, Tsuchida Y, Mizoue K, Hanada K (1992) J Antibiotics 45: 1566
56. Toda S, Yamamoto S, Tenmyo O, Tsuno T, Hasegawa T, Rosser M, Oka M, Sawada Y, Konishi M, Oki T (1993) J Antibiotics 46: 875
57. Tsujii E, Tsurumi Y, Miyata S, Fujie K, Kawakami A, Okamoto M, Okuhara M (1992) J Antibiotics 45: 698
58. Shigematsu N, Tsujii E, Kayakiri N, Takase S, Tanaka H, Tada T (1992) J Antibiotics 45: 704
59. Chandler G, Elcock C, Depledge P, Wrigley S, Mous J, Malkovsky M, Moore M, Gammon G (1993) Int J Immunopharmacol 15: 361
60. Gammon G, Chandler G, Depledge P, Elcock C, Wrigley S, Moore J, Cammarota G, Sinigaglia F, Moore M (1994) Eur J Immunol 24: 991
61. Kondo H, Nakajima S, Yamamoto N, Okura A, Satoh F, Suda H, Okanishi M, Tanaka N (1990) J Antibiotics 43: 1533
62. Hori Y, Abe Y, Shigematsu N, Goto T, Okuhara M, Kohasaka M (1993) J Antibiotics 46: 1890
63. Shiomi K, Iinuma H, Hamada N, Naganawa H, Manabe M, Matsuki C, Takeuchi T, Umezawa H (1986) J Antibiotics 39: 487
64. Hori Y, Abe Y, Ezaki M, Goto T, Okuhara M, Kohsaka M (1993) J Antibiotics 46: 1055
65. Hori Y, Abe Y, Nakajima H, Takase S, Fujita T, Goto T, Okuhara M, Kohsaka M (1993) J Antibiotics 46: 1327
66. Roitman JN, Mahoney NE, Janisiewicz WJ, Benson M (1990) J Agric Food Chem 38: 538
67. Hori Y, Abe Y, Nakajima H, Shigematsu N, Takase S, Goto T, Okuhara M, Kohsaka M (1993) J Antibiotics 46: 1901
68. Weber W, Fischli W, Hochuli E, Kupfer E, Weibel EK (1991) J Antibiotics 44: 164

69. Wyss DF, Lahm HW, Manneberg M, Labhardt AM (1991) J Antibiotics 44: 172
70. Kojiri K, Ihara M, Nakajimi S, Kawamura K, Funaishi K, Yano M, Suda H (1991) J Antibiotics 44: 1342
71. Ihara M, Fukuroda T, Saeki T, Nishikibe M, Kojiri K, Suda H, Yano M (1991) Biochem Biophys Res Comm 178: 132
72. Miyata S, Hashimoto M, Masui Y, Ezaki M, Takase S, Nishikawa M, Kiyoto S, Okuhara M, Kohsaka M (1992) J Antibiotics 45: 74
73. Miyata S, Ohhata N, Murai H, Masui Y, Ezaki M, Takase S, Nishikawa M, Kiyoto S, Okuhara M, Kohsaka M (1992) J Antibiotics 45: 1029
74. Lam YKT, Williams DLJnr, Sigmund JM, Sanchez M, Genilloud O, Kongy L, Stevens-Miles S, Huang L, Garrity GM (1992) J Antibiotics 45: 1709
75. Morishita Y, Chiba S, Tsukuda E, Tanaka T, Ogawa T, Yamasaki M, Yoshida M, Kawamoto I, Matsuda Y (1994) J Antibiotics 47: 269
76. Yamasaki M, Yano K, Yoshida M, Matsuda Y, Yamaguchi K (1994) J Antibiotics 47: 276
77. Ohashi H, Akiyama H, Nishikori K, Mochizuki J (1992) J Antibiotics 45: 1684
78. Jackson M, Burres NS, Karwowski JP, Alder LA, Humphrey PE, Kohl WL, McAlpine JB (1994) J Antibiotics 47: 523
79. Hocklowski JE, Hill P, Whittern DN, Scherr MH, Ramussen RR, Dorwin SA, McAlpine JB (1994) J Antibiotics 47: 528
80. Shu YZ, Arcuri M, Kozlowski MR, Wang RR, Lam KS, Chang LP, Pirnik DM, Rose PM (1994) J Antibiotics 47: 1328
81. Pairet L, Wrigley SK, Chetland I, Reynolds E, Hayes MA, Holloway J, Ainsworth AM, Katzer W, Chen XM, Hupe DJ, Charlton P, Doherty AM (1995) J Antibiotics 48: 913
82. Hayashi K, Hashimoto M, Shigematsu N, Nishikawa M, Ezaki M, Yumashita M, Kiyoto S, Okuhara M, Kohsaka M, Imaraka H (1992) J Antibiotics 45: 1055
83. Shigematsu N, Hayashi K, Kayakiri N, Takase S, Hashimoto M, Taraka H (1993) J Org Chem 58: 170
84. Wong SM, Kullnig R, Dedinas J, Appell KC, Kydd GC, Gillum AM, Cooper R, Moore R (1993) J Antibiotics 46: 214
85. Wong S-M, Musza LL, Kydd GC, Kullnig R, Gillum AM, Cooper R (1993) J Antibiotics 46: 545
86. Sedlock DM, Barrow CJ, Borownell JE, Hong A, Gillum AM, Houck DR (1994) J Antibiotics 47: 391
87. Oleynek JJ, Sedlock DM, Barrow CJ, Appell KC, Casiano F, Haycock D, Ward SJ, Kaplita P, Gillum AM (1994) J Antibiotics 47: 399
88. Barrow CJ, Cai P, Snyder JL, Sedlock DM, Sun HH, Cooper R (1993) J Org Chem 58: 6016
89. Barrow CJ, Sedlock DM, Sun HH, Cooper R, Gillum AM (1994) J Antibiotics 47: 1182
90. Sun HH, Barrow CJ, Sedlock DM, Gillum AM, Cooper R (1994) J Antibiotics 47: 515
91. Barrow CJ, Sun HH (1994) J Nat Prod 57: 471
92. Ainsworth AM, Chicarelli-Robinson MI, Copp BR, Fauth U, Hylands PJ, Holloway JA, Latif M, O'Beirne G, Porter N, Renno DV, Richards M, Robinson N (1995) J Antibiotics 48: 568
93. Aoyagi T, Yamamoto T, Kojiri K, Morishima H, Nagai M, Hamada M, Takeuchi T, Umezawa H (1989) J Antibiotics 42: 883
94. Fredenhargen A, Fendrich G, Marki F, Marki W, Gruner J, Raschdorf F, Peter HH (1990) J Antibiotics 43: 1403
95. Yoshida T, Nakamoto S, Sakazaki R, Matsumoto K, Terui Y, Sato T, Arita H, Matsutani S, Inoue K, Kudo I (1991) J Antibiotics 44: 1467
96. Tanaka K, Matsutani S, Matsumoto K, Yoshida T (1992) J Antibiotics 45: 1071
97. Itazaki H, Nagashima K, Kawamuwa Y, Matsumoto K, Nakai H, Terui Y (1992) J Antibiotics 45: 38
98. Hamano K, Kinoshita-Okami M, Hemmi A, Sato A, Hisamoto M, Matsuda K, Yoda K, Harayama H, Hosoya T, Tanzawa K (1992) J Antibiotics 45: 1195
99. Nihei Y, Yamamoto H, Hasegawa M, Hanada M, Fukagawa Y, Oki T (1993) J Antibiotics 46: 193
100. Nishio M, Hasegawa M, Suzuki K, Sawada Y, Hook DJ, Oki T (1993) J Antibiotics 46: 193
101. Ohkuma H, Tomita K, Hoshino Y, Suzuki K, Hasegawa M, Sawada Y, Konishi M, Hook DJ, Oki T (1993) J Antibiotics 46: 705

102. Nikei Y, Hasegawa M, Suzuki K, Yamamoto S, Hanada M, Furumai T, Fukagawa Y, Oki T (1993) J Antibiotics 46: 900
103. Aoyagi T, Aoyawa T, Kojima F, Hattori S, Honma Y, Hamada M, Takeuchi T (1992) J Antibiotics 45: 1587
104. Aoyama T, Naganama H, Muraoka Y, Aoyagi T, Takeuchi T (1992) J Antibiotics 45: 1703
105. Fujie K, Shinguh Y, Hatanaka H, Shigematsu N, Murai H, Fujita T, Yamashita M, Okamoto M, Okuhara M (1993) J Antibiotics 46: 908
106. Fujita T, Hatanaka H, Hayashi K, Shigematsu N, Takase S, Okamoto M, Okuhara M (1994) J Antibiotics 47: 1359
107. Fujita T, Shinguh Y, Yamazaki A, Nakahara K, Okamoto M, Okuhara M (1994) J Antibiotics 47: 1365
108. Omura S, Van Der Pyl D, Inokoshi J, Takahashi Y, Takeshima H (1993) J Antibiotics 46: 222
109. Shiomi K, Yang H, Inokoshi J, Van der Pyl D, Nakagawa A, Takeshima H, Omura S (1993) J Antibiotics 46: 229
110. Imoto M, Kakeya H, Sawa T, Hayashi C, Hamada M, Takeuchi T, Umezawa K (1993) J Antibiotics 46: 1342
111. Kakeya H, Imoto M, Takahashi Y, Naganawa H, Takeuchi T, Umezawa K (1993) J Antibiotics 46: 1716
112. Okabe T, Yoshida E, Chieda S, Endo K, Kamiya S, Osada K, Tanaka S, Okura A, Suda H (1994) J Antibiotics 47: 289
113. Tanaka S, Okabe T, Nakajima S, Yoshida E, Suda H (1994) J Antibiotics 47: 294
114. Yanagisawa M, Sakai A, Adachi K, Sano T, Watanabe K, Tanaka Y, Okuda T (1994) J Antibiotics 47: 1
115. Ohtsuka T, Itezono Y, Nakayama N, Sakai A, Shimma N, Yokose K (1994) J Antibiotics 47: 6
116. Aoki M, Itezono Y, Shirai H, Nakayama N, Sakai A Tanaka Y, Yamaguchi A, Shimma N, Yokose K, Seto H (1991) Tet Lett 32: 4737
117. Fuji N, Yamashita Y, Chiba S, Uosaki Y, Saitoh Y, Tuji Y, Nakano H (1993) J Antibiotics 46: 1173
118. Horiguchi T, Hayashi K, Tsubotani S, Iinuma S, Harada S, Tanida S (1994) J Antibiotics 47: 545
119. Funabashi Y, Horiguchi T, Iinuma S, Tanida S, Harada S (1994) J Antibiotics 47: 1202
120. Kawada S, Yamashita Y, Uosaki Y, Gomi K, Iwasaki T, Takiguchi T, Nakano H (1992) J Antibiotics 45: 1182
121. Uosaki Y, Kawada S, Nakano H, Saitoh Y, Sano H (1993) J Antibiotics 4: 235
122. Okada H, Suzuki H, Yoshinari T, Arakawa H, Okura A, Suda H, Yamada A, Uemura D (1994) J Antibiotics 47: 129
123. Ishii T, Hida T, Ishimaru T, Iinuma S, Sudo K, Muroi M, Kanamaru T, Okazaki H (1991) J Antibiotics 44: 589
124. Hida T, Ishii T, Kanamaru T, Muroi M (1991) J Antibiotics 44: 600
125. Nakayama O, Yagi M, Tanaka M, Kiyoto S, Okuhara M, Kohsaka M (1989) J Antibiotics 42: 1221
126. Nakayama O, Arakawa H, Yagi M, Tanaka M, Kiyoto S, Okuhara M, Kohsaka M (1989) J Antibiotics 42: 1235
127. Aoyagi T, Aoyama T, Kojima F, Matsuda N, Maruyama M, Hamada M, Takeuchi T (1992) J Antibiotics 45: 1385
128. Aoyama T, Naganawa H, Muraoka Y, Nakamura H, Aoyagi T, Takeuchi T, Iitaka Y (1992) J Antibiotics 45: 1391
129. Aoyama T, Kojima F, Yamazaki T, Tatee T, Abe F, Muraoka Y, Naganawa H, Aoyagi T, Takeuchi T (1993) J Antibiotics 46: 712
130. Aoyama T, Kojima F, Abe F, Muraoka Y, Naganawa H, Takeuchi T, Aoyagi T (1993) J Antibiotics 46: 914
131. Yamazaki T, Tatee T, Aoyama T, Kojima F, Takeuchi T, Aoyagi T (1993) J Antibiotics 46: 1309
132. Aoyama T, Zhao W, Kojima F, Muraoka Y, Naganawa H, Takeuchi T, Aoyagi T (1993) J Antibiotics 46: 1471
133. Ogita T, Sato A, Enokita R, Suzuki K, Ishii M, Negishi T, Okazaki T, Tamaki K, Tanzawa K (1992) J Antibiotics 45: 1723
134. Tanzawa K, Ishii M, Ogita T, Shimada K (1992) J Antibiotics 45: 1733

135. Aoyagi T, Suda H, Uotani K, Kojima F, Aoyama T, Horiguchi K, Hamada M, Takeuchi T (1992) J Antibiotics 45: 1404
136. Aoyama T, Naganawa H, Suda H, Uotani K, Aoyagi T, Takeuchi T (1992) J Antibiotics 45: 1557
137. Tomada H, Nishida H, Masuma R, Cao J, Okuda S, Omura S (1991) J Antibiotics 44: 136
138. Nishida H, Tomoda H, Cao J, Okuda S, Omura S (1991) J Antibiotics 44: 144
139. Tomada H, Huang X, Nishida H, Masuma R, Kim YK, Omura S (1992) J Antibiotics 45: 1202
140. Nishida H, Huang X, Tomoda H, Omura S (1992) J Antibiotics 45: 1669
141. Tomoda H, Kim YK, Nishida H, Masuma R, Omura S (1994) J Antibiotics 47: 148
142. Kim YK, Tomoda H, Nishida H, Sunazuka T, Obata R, Omura S (1994) J Antibiotics 47: 154
143. Omura S, Tomoda H, Sunazuka T (1995) in Kuhn W, Fiedler HP (eds) Sekundarmetabolismus bei Mikroorganismen. Altempto Verlag, Tubingen, p 37
144. Huang X, Tomoda H, Nishida H, Masuma R, Omura S (1995) J Antibiotics 48: 1
145. Huang X, Nishida H, Tomoda H, Tabata N, Shiomi K, Yang D, Takayanagi H, Omura S (1995) J Antibiotics 48: 5
146. Kuroda K, Yoshida M, Uosaki Y, Ando K, Kawamoto I, Oishi E, Onuma H, Yamada K, Matsudo Y (1993) J Antibiotics 46: 1196
147. Hasumi K, Shinohara C, Iwanaga T, Endo A (1993) J Antibiotics 46: 1782
148. Shinohara C, Hasumi K, Takei Y, Endo A (1994) J Antibiotics 47: 163
149. Nuber B, Hansske F, Shinohara C, Miura S, Hasumi K, Endo A (1994) J Antibiotics 47: 168
150. Dawson MJ, Farthing JE, Marshall PS, Middleton RF, O'Neill MJ, Shuttleworth A, Stylli C, Tait RM, Taylor PM, Wildman HG, Buss AD, Langley D, Hayes MV (1992) J Antibiotics 45: 639
151. Sidebottom PJ, Highcock RM, Lane SJ, Procopiou PA, Watson NS (1992) J Antibiotics 45: 648
152. Bergstrom JD, Kurtz MM, Rew DJ, Amend AM, Karkas JD, Bostedor RG, Bansal VS, Dufresne C, Van Middlesworth FL, Hensens OD, Liesch JM, Zink DL, Wilson KE, Onishi J, Milligan JA, Bills G, Kaplan L, Nallin Omstead M, Jenkins RG, Huang L, Meinz MS, Quinn L, Burg RW, Kong YL, Mochales S, Mojena M, Martin I, Pelaez F, Diez MT, Alberts AW (1993) Proc Nat Acad Sci USA 90: 80
153. Nakanishi S, Ando K, Kawamoto I, Matsuda Y (1993) J Antibiotics 46: 1775
154. Nakanishi S, Kita K, Uosaki Y, Yoshida M, Saitoh Y, Mihara A, Kawamoto I, Matsuda Y (1994) J Antibiotics 47: 855
155. Hegde VR, Miller JR, Patel MG, King AH, Puar MS, Horan A, Hart R, Yarborough R, Gullo V (1993) J Antibiotics 46: 207
156. Bryans J, Charlton P, Chicarelli-Robinson MI, Collins M, Latham, Shaw I, Trew S in preparation
157. Aoyagi T, Yoshida S, Nakamura Y, Shigihara Y, Hamada M, Takeuchi T (1990) J Antibiotics 43: 143
158. Aoyagi T, Yoshida S, Matsuda N, Ikeda T, Hamada M, Takeuchi T (1992) J Antibiotics 44: 573
159. Nagai M, Ogawa K, Muraoka Y, Naganawa H, Aoyagi T, Takeuchi T (1992) J Antibiotics 44: 956
160. Aoyagi T, Nagai M, Ogawa K, Kojima F, Okada M, Ikeda T, Hamada M, Takeuchi T (1991) J Antibiotics 44: 949
161. Toda S, Obi Y, Numata K, Hamagishi Y, Tomita K, Komiyama N, Kotake C, Furumai T, Oki T (1992) J Antibiotics 45: 1573
162. Toda S, Kotake C, Tsuno T, Narita Y, Yamasaki T, Konishi M (1992) J Antibiotics 45: 1580
163. Abe M, Harpel JG, Metz CW, Nunes I, Loskutoff DJ, Rifkin DB (1994) Anal Biochem 216: 276–284
164. Lerner MJ (1993) In: 1st International Conference on Advanced Pharmaceutical Substance Screening, Conference Proceedings, Winfried Scheirer, Sandoz Research Institute, Vienna
165. Cook ND (1995) In: Rapid Screening Technologies for the Discovery of Novel Drug Candidates, Conference Proceedings, IBC Technical Services, London
166. Holland JD, Singh P, Brennand JC, Garman AJ (1994) Anal Biochem 222(2): 516–518
167. Pernelle C, Clerc FF, Dureuil C, Bracco L, Torque B (1993) Biochem 32(43): 11682–11687
168. Checovich WJ, Bolger RE, Burke T (1995) Nature 375: 254–256
169. Sportsman R, Bukar R, Lee S, Dilley H (1995) In: 2nd European Conference on High Throughput Screening and on Exploiting Molecular Diversity, Conference Proceedings, Society for Biomolecular Screening, Danbury, USA

170. Dickson EFG, Pollak A, Diamandis EP (1995) Pharmacol and Therapeutics 66(2): 207–235
171. Hill DC (1995) In: Rapid Screening Technologies for the Discovery of Novel Drug Candidates, Conference Proceedings, IBC Technical Services, London
172. MacAllan D (1995) In: *DELFIA* Research Meeting, Conference Proceedings, Wallac, UK
173. Mathis G (1995) In: 2nd European Conference on High Throughput Screening and on Exploiting Molecular Diversity, Conference Proceedings, Society for Biomolecular Screening, Danbury, USA
174. Walker JM, Winder JS, Kellam SJ (1993) Applied Biochim and Biotechnol 38(1–2): 141–146
175. Gopalakrishna R, Chen ZH, Gundimeda U, Wilson JC, Anderson WB (1992) Anal Biochem 206: 24–35
176. Farley K, Mett H, McGlynn E, Murray B, Lydon NB (1992) Anal Biochem 203: 151–157
177. Sadick M, Beresini M, Yen R, Golloway A, Yeh S, Wong WL(1995) In: 2nd European Conference on High Throughput Screening and on Exploiting Molecular Diversity, Conference Proceedings, Society for Biomolecuar Screening, Danbury, USA
178. Yang H, Leland JK, Yost D, Massey RJ (1994) Biotechnol 12: 193–194
179. Brigham-Burke M, Edwards JR, O'Shannessy DJ (1992) Anal Biochem 205: 125–131
180. McConnel HM, Owicki JC, Parce JW, Miller DL, Baxter GT, Wada HG, Pitchford S (1992) Science 257: 1906–1912

Received February 1996

Photobioreactors: Design and Performance with Respect to Light Energy Input

Otto Pulz and Karl Scheibenbogen
IGV Institut für Getreideverarbeitung, Arthur Scheunert-Allee 40–41, 14558 Bergholz/Rehbrücke, Germany

Photosynthesis is the most important natural process for the neogenesis of biological material. More than 40 000 species of oligocellular algae constitute a unique biochemical fund in the plant world. As the energy supply to the phototrophic microorganisms is transported via photons and as the physics of light distribution is a technical challenge, the technique of biomass production of single cell algae as against heterotrophic organisms is in a stage of intensive development. In contrast to area intensive open pond technology, space saving closed cultivation systems could be gradually established in the market of industrial photobioreactors. The present review surveys outdoor cultivation systems and indoor photobioreactors with respect to their light utilization and productivity. To evaluate reactor performance independently of their design and location, growth yields were recalculated from published data. It became obvious that closed systems may out-perform raceway ponds by about 300%, which represents productivities over 100 g dry algal biomass per m^2 per day. Thus, economic aspects for the cultivation of phototrophs, even in moderate climates, are better satisfied, but microalgal biotechnology is still of minor importance because of high biomass production costs. Considering indoor systems with internal illumination arrangements, the growth potential can be increased by a further 100%. Due to the high technical expenses, this technology will be confined to the production of high value products, e.g. pharmaceuticals. Consequently, very defined axenic cultivitation has to be guaranteed by all means available.

Advances in Biochemical Engineering/
Biotechnology, Vol. 59
Managing Editor: T. Scheper

1 Introduction

Together with some photoactive bacteria, mainly procaryotic cyanobacteria as well as mono- and oligocellular species of eucaryotic classes of Phycophyta (algae) belonging to the lower plants, are assigned to phototrophic microorganisms. The algae of predominantly aquatic habitat occur in a multitude of shapes and varieties and together with the phytoplankton of the oceans constitute approximately one third of the total phytomass production on the earth. Although the areal concentration of aquatic primary producers is only 0.8% of the terrestric phytomass (standing crop) of an average of 12.3 kg d.wt. m^{-2}, their growth constant and thus their photosynthetic activity exceeds the corresponding efficiencies of higher plants very considerably [1]. Consequently, recalculated protein production potential by phototrophic microorganism amounts to 96 g protein m^{-2} d^{-1}, which is considerably higher than protein supplied by agricultural plants like soya [2]. In this connection, microplankton and nanoplankton with cell diameters less than 60 μm represents more than 90% of the total photosynthetic activity in the seas [3].

Although marine species of higher Phaeophyceae (brown algae), some reaching lengths of 100 m and more, have been used for human nutrition for centuries, and unicellular organisms or cellular groups are of special biotechnological interest.

The importance of this microalgae and of microalgal products in general is reflected in two application fields:

a) *Biomass production* for the preparation of
- feed for aquacultures (breeding of larvae and bivalves),
- diets or additives for human and animal nutrition (health food, feed additive),
- vitamins, pigments, antioxidants and related biosubstances (carotenoids, colours, S.O.D.),
- fatty acids and carbohydrates (precursers for pharmaceuticals);

b) *Ecological impact* to
- environmental technology (e.g. heavy metal fixation, CO_2 elimination, life-support systems),
- agriculture and eco-systems (soil inoculum, minimization of eutrophy in waters).

2 Short Review of History

By the middle of this century, the high content of utilizable proteins of more than 50% gave rise to the idea of using microalgae as a protein source for human nutrition.

First attempts of mass cultivation were made in Germany during World War 2 when they tried to produce liquid fuel from Diatomeae. In the early 1950s, this led to efforts to convert waste carbon dioxide produced in the Rhine industrial zone to *Scenedesmus* and *Chlorella* biomass. The Soeder group in particular, and the summarizing works performed in the Carnegie Institute in Washington have laid the foundation for large-cale cultivation of phototrophic organisms as practiced today [4, 5].

In the following period, highly active centres were established, mainly in East Europe (Czechoslovakia, U.S.S.R., Bulgaria) and in Israel. In the middle of the 1970s, *Chlorella* was produced commercially in Japan for the first time. The philosophy on health food in Asia might have been decisive for accepting the relatively high price. The U.S. company Earthrise made one of the first attempts to establish algal production in the West. Today, the market is dominated by the filamentous blue-green alga of the genus *Spirulina* which was historically collected by the native population around the Chad lake and at Lake Texcoco. Due to the high protein and vitamin content, *Spirulina* is still used as meat supplement [6].

In the U.S.A. in the 1950s, algae were already involved in environmental technologies. The quality of waste waters could be improved significantly by the mass cultivation of phototrophic microorganisms. The resulting biomass was fermented to methane at a conversion rate of 50–70%. This process was subject to much interest during energy crises [7]. Beside environmental large scale cultivation, Oswald and coworkers also created the idea to utilize microalgae for the development of bioregenerative life-support systems [8]. Today, systems for spacecraft [9] and bioreactors are available, and may be used for air regeneration in submarines [10].

The still important field of aquaculture was opened in the early 1970s, when the large-scale cultivation of phytoplankton in a mixture of waste water and salt water was established. The yielded plankton mass was fed to marine bivalves and other freshwater organisms [11].

Despite first attempts of large-scale production of valuable substances from microalgae, e.g. production of β-carotene from *Dunaliella* in the Australian Hutt-Lagoon or in the Negev desert in Israel, it is only recently that much attention was paid to microalgae as a source for the production of fine chemicals [12, 13].

Table 1 gives a short market analysis of a multitude of valuable substances from microalgal biomass of economic and scientific interest [14].

The chemical and pharmaceutical industry is interested principally in products such as

- natural colours (carotenoids, phycobiliproteins),
- vitamins (tocopherols etc.),
- polysaccharides with immunomodulatory properties,
- fatty acids (polyunsaturated, e.g. eicosapentaenic acid).

However, there is a lack of cultivation systems which allow for high biomass concentrations for economic production of valuable substances with respect to sterile conditions.

Table 1. Microalgae as a source of valuable products

Product	Value US$/kg	Market US$ (Mio.)	Content % DM
COLORING Substances			
Beta Carotin	>1000	10 - 100	10
Xanthophylls	200 - 500	10 - 100	1
Astaxanthin	>1500	>50	1 - 6
Canthaxanthin	1000		>1
Phycocyanin	>100	>10	1 - 5
Phycoerhythrin	>10.000	2	1 - 5
BIOMASS			
Health food	10 - 20	2	100
Feed additive	1 - 130	100 - 1000	100
Soil inoculum	>100		100
PUFA			
Arachidonic acid	400 - 10.000	>1000	2 - 30
Eicosapentaenoic a.	400 - 10.000	>1000	2 - 30
Docosahexaenoic a.	400 - 10.000	>1000	2 - 30
ANTIOXIDANTS			
Tocopherol	25	60	>2
S.O.D.	>1.000		>1

3 Influence of Light on Microalgae

3.1 Photosynthetic Efficiency

Microalgae have an extremely high productivity compared with higher plants. Some strains of high-temperature adapted green algae, e.g. *Chlorella pyrenoidosa*, are able to double their cell number every $2\frac{1}{2}$ hours [15]. As opposed to higher plants, photosynthetically inactive cells (such as roots or trunks) are missing, and microalgae are a favoured subject of research for the determination of complex photosynthetical processes, especially with regard to quantum yield. Although the antenna complex of well-mixed microalgal cultures does not react to increased radiation intensity to the extent used by phanerogames to convert efficiently high quantum yields to utilizable energy, this is compensated largely by the mobility and the constant supply of carbon dioxide, minerals and water [16].

Photosynthetic production of biomass is generally represented by the following equation:

$$\alpha CO_2 + \beta H_2O + \gamma NH_3 + \text{minerals} + nh\upsilon \rightarrow \alpha[\text{CHON minerals}] + O_2$$

where $nh\upsilon$ is the number of energy quanta of a certain frequency which is required for the production of 1 mol O_2. Thereby, α must not have the value 1, but effects such as dark respiration or the energy required for metabolism maintenance will reduce the yield of fixed carbon with regard to the oxygen quantity produced. In the classical model of photosynthesis, the Z-scheme, 8 mol quanta absorbed are required for the evolution of 1 mol O_2, at which approx. 3 mol ATP are produced, cyclic photophosphorylation included. They also are required for reducing 1 mol CO_2 to glucose in the dark reaction. The real requirement of quanta for photosynthesis, however, is controversial and given by different authors as 6–16 quanta [17, 18].

The data of the calculated photosynthetic efficiency (PE), the quotient of the energy accumulated in the biomass and the light energy absorbed, are correspondingly different. A theoretical maximum PE of approx. 29% is calculated at 8 mol light quanta ($\lambda = 575$ nm) per 1 mol fixed CO_2. With regard to the solar spectrum the conversion of the irradiated sunlight is corrected theoretically to 18.1% [19]. This value, however, remains questionable under real cultivation conditions [20], all the more since the real conversion rates do not exceed 10% [21]. After extensive examination, Oswald [22] calculated a maximum utilization of sunlight of approx. 5–6% for outdoor plants, which was also reached in a continuously mixed culture in Richmond, California [7]. Algae, therefore, by far exceed agricultural crops, such as C_4 maize with a PE of 1.3%, fast growing tropical plants as sugar cane or bullrush millet with up to 4.3% [23] and are even superior to technical equipment such as classical photocells or solar collectors [19, 24].

Furthermore, a number of light-regulated mechanisms is available which make the metabolic cycles of photosynthesis more flexible, although, requiring additional energy. The information and energy content of light influence the thioredoxin system or the development of secondary messengers [25]. Thus, quantity and quality of the light are of decisive influence to a defined cultivation of phototrophic microorganisms.

3.2 Qualitative Light Requirements of Phototrophic Microorganisms

The qualitative light requirements of the organisms are mainly determined by their photosynthetically active pigments. The pigment spectrum of algae and higher plants is within the visual range from 380 to 700 nm, some phototrophic purple bacteria also absorb in the near IR range.

Investigations of the growth behaviour of algae at light irradiation of different quality have led to controversial results. Prokop et al. [26] have cultivated the green alga *Chlorella sorokiniana* at monochromatic conditions of high light intensity and confirmed with highest growth rates in the yellow-red spectral range the chromatic growth maxima of Kubin et al. [27]. Kohl and Nicklisch [28] have pointed out that the frequently postulated theory of a chromatic adaptation according to a specific distribution in depth of marine algal species begins to waver. Chromophyta, adapted to the blue light of the oceans, under irradiation by shortwave light, have a higher pigment content and a higher growth rate, but a green mutant of the red macroalga *Gracilaria tikvahiae*, under green and white light, did not show growth differing from that of the red wild type. Numerous other algae adapted to low light intensity have shown a growth independent of the PAR spectrum. Similar results were achieved by Richardson et al. [29] who emphasized the distinct influence of light quality on certain metabolic processes – the production of arachidonic acid [30], the UV-induction of mycosporine-ike amino acids [31], and the effect on morphology and pigment synthesis [32, 33] being typical examples – whereas the importance of light quantity to microalgal growth was rated even higher by them.

3.3 Quantitative Light Requirements of Microalgae

As the activity of photosynthesis within light-induced electron transport is limited by the redox reaction of the plastoquinone pool, the cell division rate approaches a limit with increasing light energy. A comparison of the various mathematical models, which define the light intensity dependent growth curves, revealed the exponential model according to Mitscherlich [166]

(Eq. 1) to be the best description of the real course for specific growth constant μ [28]:

$$\mu = \mu_{max}\left(1 - \exp(-\ln(2))\,\frac{I - i_0}{k_i - i_0}\right) \tag{1}$$

The light quantity i_0 represents the minimum metabolism maintenance energy. Grima et al. [34] have modified the hyperbolic model according to Baly [167] where the empirical exponent N (Eq. 2) takes into account the boundary conditions of cultivation (ΔI, I_0, pigment content, reactor diameter, etc.) and m, the specific maintenance rate:

$$\mu = \left(\frac{\mu_{max}\cdot I^N}{I^N + k_i^N}\right) - m \tag{2}$$

The calculation of the light energy value I at a random point within the cultivation system is rather difficult, because dynamic absorption and scattering processes on the cells require a complex mathematical approach, such as the highly numeric Monte Carlo method for three-dimensional space [17]. On the basis of the hypothesis of Schuster [35], Binouis et al. [36] have developed a simplified model with the aim of predicting the growth of phototrophic biomass in a tubular photobioreactor of the MELISSA-project (Eq. 3):

$$4\Pi J_r = \frac{I_0}{Z}\,\frac{2\cosh\delta z}{\cosh\delta + \alpha\sinh\delta}$$

$$\langle\mu\rangle = \frac{1}{\Pi(r_1 - r_2)^2}\int_{r_1}^{r_2} 2\Pi r\,\mu_{max}\,\frac{4\Pi J_r}{k_1 + 4\Pi J_r}\,dr \tag{3}$$

The total energy received, $4\Pi J$, is defined by the parameters α and δ which represent pigment absorption and cell dispersion, Z representing the normalized radius. This approach is suited to rapid on-line calculations, and the error between real and model values could be reduced to $<10\%$. The mean volumetric integral $\langle\mu\rangle$ could be further integrated over the wave length spectrum because of the significant influence of the emission of photosynthetically active radiation (PAR) of the light sources [37]. There are more general considerations existing, e.g. the model presented by Frohlich et al. [38], who pay attention to vessel geometry, first limiting inorganic nutrient and light inhibition by defining a modulus, but they all take complex numerical solutions. As may be seen from the various approaches, identical radiation conditions are given only if the geometries of the radiation field and of the radiation spectrum are identical [28].

There are significant variations of the optimum energy and PAR quantum flux density for the growth of microalgae, from species to species. Richardson et al. [29] have examined several species as to the genotypical light intensity. With Dinophyceae *Amphidinium carterae* living in the depths of the oceans they established cell division at extremely low light intensity (approximately 1–2 $\mu E\,m^{-2}\,s^{-1}$) and photoinhibition already at 80 $\mu E\,m^{-2}\,s^{-1}$. On the other hand, the dinoflagellate *Gonyaulax polyedra* does not grow at light intensities below 30 $\mu E\,m^{-2}\,s^{-1}$ and reaches maximum growth only at 130 $\mu E\,m^{-2}\,s^{-1}$. Richardson generally predicts good growth of diatoms, dinoflagellates and cyanobacteria at low quantum flux densities, while green algae clearly belong to the high light intensity group, for some *Chlorella* strains optimum radiation intensities of not below 300 $\mu E\,m^{-2}\,s^{-1}$ being recorded. In the data collection Richardson has respected the light/dark cycle (15/9 h), because at continuous illumination some varieties will change the original growth behaviour [39]. That means that the range of photosynthetically relevant radiation intensities at permanent light is generally even more restricted.

3.4 Effects of Changing Light Conditions on Algal Growth

Algae are able to adapt to significantly higher radiation intensities for a short period. The midday sun emits radiation densities of more than 2000 $\mu E\,m^{-2}\,s^{-1}$ PAR. In dynamic cultures, the radiation load is compensated for by reflection and absorption as well as by mixing effects [40]. Grobbelaar [41] observed an increase of productivity at high turbulences and suggested a possible synergistic relation between the light/dark cycle in the seconds-range and mechanisms of material transport. Laws [42] has cited seven publications in which an increased rate of photosynthesis at flash-light is described. He and his coworkers [43] have developed a simple device for the generation of highly turbulent fluxes where algal cells are exposed to the ambient light at a frequency of 60 Hz; the production rate of the alga *Phaeodactylum* was almost doubled. This corresponds to the flash-light effect postulated as growth promoter by Warburg as early as 1919 [44] and later by Kok [45]. Terry [46] has calculated an increase of the photosynthetic efficiency at a low frequency. Weller and Frank [47] and Phillips and Myers [48], however, did not establish any increase of the rate of photosynthesis, which was rather significantly lower at low frequencies and approached the continous light course only at higher flash-light rates. Recent work with LEDs as pulsed light sources in the range 0.1–100 Hz have confirmed this course. In this connection, Grobbelaar and Neddal [49] mention that it became apparent that algae would not acclimatize to certain frequencies, as theory of electron transport mechanism suggested. However, using the same intermittant light regime will result in growth rates higher and photoinhibition rates lower than in equivalent continuous light in some strains of Chlorophyceae [50]. As a result of the analysis of several references, Giger and Sager [51] support the hypothesis of Rabinowitch [52] that enzymes of photosynthesis are limited in concentration only because of their size to a maximum of 1 per 300 chlorophyll molecules; as a consequence the maximum productivity rate of light-collecting pigments inevitably will be higher than that of the subsequent enzyme reactions. The quantum utilization at intermittant light will thus not be improved. According to an extensive theory [53], the cyclic excitation of arbitrary reactions will lead to an increase of productivity rates by a maximum of 10–20%. Finally, the question for the flash-light effect now as before remains controversial. According to the literature published, a growth promoting effect can neither be supposed nor excluded.

Besides physics of illumination, the biochemical adaptation to the prevailing light conditions also plays an important role. Beale and Appleman [54] and Herron and Mauzerol [55] observed a significantly increased chlorophyll production of *Chlorella* and a mixotrophic culture of yellow *Chlorella* mutants, respectively, at low light intensity, so that the statistical probability of an absorption of the quanta striking the cell surface is confirmed. Sukenik et al. [56] have adapted the green alga *Dunaliella salina* to extremely intensive light. Contrary to "normal cells" an inhibition of photosynthesis could not be detected even in diluted culture at 2200 $\mu E\, m^{-2} s^{-1}$ PAR. Ben-Amotz [57] established an increase of the β-carotene content of the same alga at 8-fold at similar light intensity (control 50 $\mu E\, m^{-2} s^{-1}$), the carotene/chlorophyll ratio having increased by a factor of 35. Falkowski [58] has investigated the pigment production or pigment degradation during transition transformation and has calculated the first-order kinetics for chlorophyll a. The adaptation of pigment development and growth of *Dunaliella salina* from 45 $\mu E\, m^{-2} s^{-1}$ to 500 $\mu E\, m^{-2} s^{-1}$ (scalar measured values) and vice versa was finished after 80–100 h.

3.5 Utilization of Light-Source Energy

In general, photobioreactors can be supplied with light energy by natural or artificial illumination. The utilization of solar energy is by far the most important source for the production of algal biomass. Beside open pond technology, sunlight can also be applied in semi-closed systems [59] or, with appropriate technical expense, in completely closed plants [60]. Nevertheless, a number of disadvantages are inherent to this cost-effective energy, e.g. sunlight is not available at any time (day/night rhythm), its quality and quantity is subject to significant fluctuations determined by geography, weather and season, and thus exactly reproducible light conditions are not feasable. Furthermore, the solar emission spectrum comprises the range 120 nm to 100 μm, i.e. more than 50% of the radiation energy cannot be utilized photosynthetically. Therefore, artificial light sources are used almost exclusively for the defined production of microalgae on a laboratory scale.

Fluorescent lamps, a cost-effective means, are used most frequently for the cultivation of phototrophic organisms. The bands emitted from the mercury vapour can be converted to continuous radiation by modifying the composition of the fluorescent material, which allows for an optimum adaptation to the effective spectrum of photosynthesis. Efficiency and applicability of the fluorescent spectrum are easily determined by appropriate parameterization [37]. The proportion of disadvantageous IR and UV-radiation is very small. A surface quantum flux density of 300–500 $\mu E\, m^{-2} s^{-1}$ allows for a very close contact with the suspension of organisms so that the relatively high efficiency of 15% (compared with 5% for tungsten lamps) can be utilized quantitatively.

Compared with fluorescent lamps, incandescent and discharge lamps produce a significantly lighter emitting surface, frequently with increased IR-emittance; in most cases this requires the external installation of the fluorescent material. Due to the small sizes concerned, they are suited very well to a focussed light transfer via light-guiding systems with optical fibres. Light yield and colour spectra vary significantly from type to type. The sodium high pressure lamps known from street lighting produces a very high conversion of electrical power to visible light ($>25\%$), but are limited in the discontinuous spectrum to the yellow-orange range. Despite this, some researchers have achieved good results with this type of lamps [27, 61].

Of the commercial lamps, metal vapour and xenon high-pressure lamps have the highest light intensity and are used for sunlight simulation because of the well-balanced light distribution. During optimized pulsating operation xenon flash-light lamps can reach an efficiency of 50–80% [62]. However, for a corresponding scale-up the instrumental expenditure (operating equipment, cooling, etc.) will add significantly to the costs; therefore, this type of lamps is not described in the literature for plants exceeding the laboratory scale.

Light emitting or laser diodes could be an interesting prospect [63] because they can convert electrical energy into radiation energy at up to 80% efficiency. Lee and Palsson [64] have reached very high cell densities with *Chlorella* with the aid of a densely packed array of monochromatic red LEDs (680 nm). However, prices and complexity of the required control electronics do not allow for a commercial application.

Another alternative is currently developed by the U.S. company Fusion. In spherical glass beads sulfur atoms are excited by microwaves to light emission. The efficiency in VIS amounts to approximately 30%, interfering UV and IR-radiation being emitted only to a very small extent. Commercial application is to be expected in the next few years [65].

By direct use of fluorescent colours in the algal suspension and in the reactor jacket, light spectra were modified so that their photosynthetical utilization became more effective [26]. However, the high toxicity and low stability of the colours were a handicap; nevertheless, yield increases up to 20% were achieved for a short time.

4 Problems of Cultivating Phototrophic Microorganisms

The basic ability of photoautotrophic microorganisms to utilize light quanta in the visual range as an energy source for metabolism is handicapped by high population densities. As against heterotrophic microorganisms, such as yeasts or bacteria where the distribution of the organic molecules as energy carriers by mixing is solved technologically, the supply of photons is dependent on surface area. Light dissipation in the cell suspension is realized by numerous quantum interactions with exponential decrease along the boundary normal [66]. Only the conversion processes with the assimilation pigments of the microorganisms are used as energy receptors for the development of biomass, i.e. the radiation energy is lost largely as heat, fluorescence etc.

The large energy density at the boundary surface contrasts with the insufficient supply to the cells at only a few mm layer thickness. Despite a few commercial applications a significant break-through has not occurred. This is mainly due to the large space requirements for open plants and the expensive cultivation technique for closed plants, which allows only for a relatively low biomass concentration of 1–2 g dry matter per litre down to 0.4 g in open systems. Mainly extremophilic species of low contamination risk like osmotolerant *Dunaliella* are cultivated; the requirements of G.M.P.-complying production mostly being no issue [67, 68].

The biological and technical potential of algal biotechnology has been described in detail in several reviews [69–71]. Nearly all the commercial plants are based on open pond technologies, their construction and productivity is known from detailed descriptions [5, 72]. These systems, however, seem to have reached their limits. The gap between the theoretical biological potential of microalgal biomass and the productivities actually achieved may be reduced by developing closed cultivation systems [73]. It is rather difficult to compare the open pond technology with closed systems and indoor photobioreactors because of the strong variations of the prevailing boundary conditions. The general review suggests that open systems are predominating in mass cultivation because of cost considerations, and photobioreactors become more and more interesting for the preparation of valuable substances and for special applications [13, 16, 40, 74]. Emphasis of the present review is laid on approaches to minimize light limitations within the photobioreactors, with a short glimpse of the latest developments of small-scale laboratory applications and immobilization processes previously described by Lee [75] and Brouers et al. [76], respectively. The results of the 1st and 2nd European Workshops on Microalgal Biotechnology, Germany, the 6th European Congress on Biotechnology, Italy, and the 7th International Conference on Applied Algology, South Africa, will be respected, too.

5 Cultivation Techniques

5.1 Open Cultivation Systems

Open cultivation systems comprise natural or artificial ponds, raceway ponds, and so-called inclined surface systems. They represent the classical processes of the production of algal biomass. All of them require large areas. If appropriate areas are available in regions of balanced and sunny climate (e.g. waste land near seas) at reasonable cost, the investment efforts will be relatively low even at scaling up to 100 ha [16].

Natural and artificial ponds are generally used for the cultivation of fast growing, naturally occurring or extremophilic species, as otherwise certain contaminants will dominate. The cultivation of the extremely halophilic and highly light-tolerant green alga *Dunaliella salina* for β-carotene production in the Hutt-Lagoon in Western Australia is a good example. In the shallow unlined ponds of more than 50 ha the algae are harvested continuously through a widely branched tubing system and the nutrient-enriched sea water is added through pumps. Evaporation regulates the temperature of the suspension and produces hypercritical salt concentrations so that the growth of other species is inhibited. Due to the lack of any stirring mechanisms the material transport for CO_2 supply and biomass localization is minimized, which is the cause of relatively low productivities not exceeding 1 g d.wt. m^{-2} d^{-1} [69, 77].

At optimal temperatures obtained by utilizing the cooling water of power stations, *Spirulina* was also cultivated successfully in Europe. The yields amount to 24 tonnes dry biomass per ha shallow lagoons per 6 months, which corresponds to a growth rate of 1.3 g d.wt. m^{-2} d^{-1} [78].

In pools of the described dimension, the problem of stirring is of fundamental importance because large amounts of energy are required for the prevention of concentration gradients and algal sedimentation. Floating bodies with impellers which were pulled over the surface of the ponds [23] or raceway ponds were energy-saving solutions. The latter are characterized by parallel, loop-like channels of a length of several kilometers; several paddle wheels combined with aerating units produce an economic motion of the algal suspension and a uniform nutrient supply. The production of biomass in a plant of several hectares thus has a relatively low operating cost [16].

Weissman et al. [40] have investigated the influence of thorough mixing of the suspension on algal production in a demonstration plant, but did not observe any significant increase of productivity with the suspension velocities of 1–30 cm s^{-1} reached. As already mentioned, Grobbelaar [41] pointed to a synergistic relation between light/dark cycles and mechanisms of material transport which was verfied by his experiments with highly turbulent suspension flows. The energy input through stirring in order to maintain the necessary average lighting of the algal cells within the range of 10^{-1} to 10^{-2} seconds is rather expensive in larger plants. The simultaneously occurring increased shear forces have a counterproductive effect to numerous algal species [79].

The growth of biomass in raceway ponds is also dependent on the prevailing regional climate. Mean growth rates of 15–25 g d.wt. m^{-2} d^{-1} are recorded in tropical regions [23] and in California [16], in South Europe up to 28 g d.wt. m^{-2} d^{-1} [80], in Central Europe and Japan mean values of only 12–15 g d.wt. m^{-2} d^{-1} [59, 72, 81], related to Chlorophyceae and suspension depths of approximately 20 cm. At an average radiation of the surface of the ponds of 16.7 MJ m^{-2} d^{-1} (PAR) a theoretical productivity of 130 g d.wt. m^{-2} d^{-1} is calculated on the basis of a photosynthetic efficiency of the sunlight of approximately 18%. The actually calculated average conversion of 1–2% in large plants implies that raceway ponds are a limited cultivation system because of the self-shadowing and deficiencies in the material transport. In this connection, Benemann [72] refers to mixotrophic algal growth during the treatment of highly organic loads of sewage. The peak values up to 100 g d.wt. m^{-2} d^{-1} biomass or above 40 g d.wt. m^{-2} d^{-1} as the annual average support the thesis of an expressed light limitation. Furthermore, he refers to a number of contamination problems caused by zooplankton, fungi, or heavy metals in industrialized areas; it is true that productivity could be increased by covering the ponds, but not with respect to the contamination prevention [82].

In the inclined surface plant of an active surface of 900 m^2, developed by Setlik et al. [83] in the Czech town Trebon, the algal suspension flows over terraces of a defined inclination so that layer thicknesses below 1 cm and turbulent flow prevent any shadowed volumes. With the example of *Scenedesmus* the authors achieved peak productivities of 19 g d.wt. m^{-2} d^{-1}. Pulz [59] reports a *Chlorella* productivity of 24 g d.wt. m^{-2} d^{-1} which was achieved in a smaller plant of this type of a total volume of 5 m^3; this exceeds the productivity of raceway ponds by 100%. This superiority with regard to the suspension volume used is expressed even more clearly by the ratio of 0.95 g d.wt. l^{-1} d^{-1} and approx. 0.01 g d.wt. l^{-1} d^{-1} for raceway ponds, so that the constructive and energetic additional expenditures seem to have been worthwhile. The fundamental influence of the illuminated surface with regard to the total volume of the suspension (O/V-ratio) is obvious. Inclined surface systems with a small dark volume of collecting pond with thin suspension layers and a corresponding O/V-ratio of 20–100 m^{-1} (as against 3–10 m^{-1} in raceway ponds) fulfil the requirements, although the investment and energy costs are increased drastically. Nevertheless, a larger plant (5000 m^2) of this type was established in Rupite, Bulgaria, for the successful production of *Chlorella* with peak values of 37 g d.wt. l^{-1} d^{-1} during summer [72, 84]; it was established in a modified arrangement in sunny regions of the Third World (e.g. the Peruvian-German algal project [85]).

5.2 Closed and Semi-Closed Outdoor Photobioreactors

Attempts to make the cultivation conditions independent of environment were associated with an immense increase in investment and operating costs at corresponding scale-up, as against the open systems. This was justified only in the production of high-value products (e.g. radioactively labelled substances [16]). However, closed reactors have a number of advantages:

- low CO_2 losses,
- reduced risk of contamination
- temperature regulation,
- controllable hydrodynamics,

- reproducible cultivation conditions
- higher flexibility with regard to environmental influences,
- significantly smaller space requirements.

The first item can constitute a rather significant cost factor. Soeder [86] has calculated costs of US-$ 2000 for CO_2 per ton algal dry mass produced at optimum conditions. Weissman et al. [40] reported losses of more than 80% in open systems, while in the tubular reactors for comparison the CO_2 is better utilized, independently of the pH-value and salinity. However, the authors concluded that the tubular systems are not superior to raceway ponds because of the reduced suitability to up-scaling, the high energy consumption and the frequently observed hypercritical oxygen concentrations. Nevertheless, great attention has been paid to tubular systems, during the last years.

5.2.1 Tubular Photobioreactors

The groups of Materassi and coworkers in Italy and Gudin and Chaumont in the South of France have invested much effort in the development of outdoor reactors. Generally these tubular systems are arranged in a horizontal serpentine form and made of glass or plastic tubes. The culture suspension was recirculated either by a pump or – more preferably – by air-lift technology. Temperature was controlled by floating or submerging the tubes on or in a pool of water, oxygen degassing being guaranteed by flexible tube elements. With a 100 m^2 culture unit (tube diameter 6 cm) and *Porphyridium cruentum* a productivity of 20–25 g d.wt. $m^{-2} d^{-1}$ (corresponding to approx. 0.4 g d.wt. $l^{-1} d^{-1}$) was achieved during 2 months of steady-state continuous culture. The company Heliosynthèse SA in the south of France developed this type further and has launched commercial production [13, 87–89]. A manifold tubular reactor was established in Israel, using reduced diameters (32 mm) of polycarbonate tubes for the cultivation of *Spirulina* and *Anabaena* [82]. In Italy, too, several attempts were made to use closed tubular systems. Torzillo et al. [90] achieved a maximum productivity of 25 g d.wt. $m^{-2} d^{-1}$ (*Spirulina*) in a 10 m^3 serpentine bioreactor with intermitted culture circulation and obtained further improvements by constructing a two-plane tubular photobioreactor with mean daylight productivities of about 30 g d.wt. $m^{-2} d^{-1}$ corresponding to 1.5 g d.wt. $l^{-1} d^{-1}$ [91]. Using strongly curved outdoor tubular reactors with high flow rates (0.97 m s^{-1}) generating Reynolds numbers above 4000 within the tubes (2.6 cm inner diameter), the growth rate could be increased by about 17%, as against straight tubes with 1.2 g d.wt. $l^{-1} d^{-1}$, i.e. 23 g d.wt. $m^{-2} d^{-1}$ areal productivity [92]. For closed tubular reactors in Italy an overall mean volumetric productivity of 0.8 g d.wt. $l^{-1} d^{-1}$ *Spirulina* per annum has been summarized [93].

Lee et al. [94] developed with their α-shaped 300 l tubular loop reactor a similar approach to increase turbulences and yielded up to 72.5 g d.wt. $m^{-2} d^{-1}$ *Chlorella pyrenoidosa* biomass (about 2.9 g d.wt. $l^{-1} d^{-1}$).

Dvorin [95] reported about a plant of 130 m^3 size in Tadshikistan (Fa. Sagdiana PGT Javan) which operated with a hydrodynamic regime generating Reynolds numbers over 8000 for efficient mass transfer. In 1991, productivity of green algal biomass however was far below 0.1 g d.wt. l^{-1} d^{-1}. Another plant to produce *Chlorella* biomass commercially was established in Turkmenistan in 1978. With a 10 m^3 tubular device an annual capacity of 1 tonne, i.e. 1 g d.wt. l^{-1} d^{-1} and 0.7 g d.wt. l^{-1} d^{-1} *Spirulina* was achieved [96, 97]. Tredici and Zitelli [98] investigated the scale-up of photobioreactors to commercial size. They stated that incorrect evaluation of the efficiencies can lead to economic disaster. In case of PhotoBioreactors Ltd. (PBL, Spain) an inclined and near horizontal tubular system, according to full-scale design patented by Pirt et al. [99], was built in 1990, using 200 km of polyethylene tubing (12 mm in diameter). However, incorrect management of the system led to poor growth and PBL was shut before starting operations.

The helical tubular system commonly called Biocoil seems to be a promising method. The idea was set up in the U.K. [100] and is based on an arrangement of coiled polyethylene tubes (30–60 mm in diameter) around an open circular framework. Algal suspension is recirculated either by pumps or according to the air-lift principle. In Australia, 40–100 l laboratory reactors have been intensively studied and successfully scaled up to 1 m^3 outdoor pilot plants [101, 102]. Laboratory scale reactors still are subject to intensive microalgal research [103]. Productivities are comparable to other tubular systems, but the completely closed Biocoil could be continuously operated for several months with high service ability and simple maintenance occupying less space than traditional tubular configurations. Several 5 m^3 reactors are planned to be established in the U.S.A. for aquaculture purposes [104]. Thus, the Biocoil has to be considered as one of the few commercially available outdoor photobioreactors.

There are several further approaches in small-scale tubular technology. Anderson and Eakin [105] experimented with a highly sophisticated angular adjustment of solar collector module to produce polysaccharides (20–26 g d.wt. m^{-2} d^{-1}) from *Porphyridium*. At the end of the 1980s, James and Al-Kahrs [106] have designed for aquaculture applications a 200 l turbidostatic system with transparent vertically arranged tubes (diameter 30 cm) where the algal suspension is aerated by an air stream and illuminated by fluorescent lamps at a maximum of 330 $\mu E\, m^{-2}\, s^{-1}$. The productivity achieved with *Chlorella* amounted to 0.26 g d.wt. l^{-1} d^{-1} (20–26 g d.wt. m^{-2} d^{-1}). Similar results were achieved by Pulz [59] in a tubular fermenter (diameter of the glass tubes 2.5 cm) where the algal suspension (50 l) is pumped around the light source by a radial centrifugal pump (Fig. 3). Taking the glass tube surface as the basis of calculation the surface-related growth rates of 14 g d.wt. m^{-2} d^{-1} do not exceed the productivities of open systems.

The fundamental role of the ratio of illuminatable surface and suspension volume was described by Myers and Graham in 1961 [107]. They cultivated in a simple cylinder which was illuminated only from above by focused light of 1750 $\mu E\, m^{-2}\, s^{-1}$. The surface/volume ratio (d/r^2 ratio of tubes) was increased by an opalescent cone from 1.92 m^{-1} to 19.0 m^{-1}, at a loss of effective volume of only 15%, but a doubling of the productivity of *Chlorella* from 0.15 g d.wt. $l^{-1} d^{-1}$ to 0.31 g d.wt. $l^{-1} d^{-1}$.

Miyamato et al. [108] used a very cost-effective variant to cultivate microalgae in vertically arranged glass tubes ($\phi_A = 5$ cm, height $= 2$ m) with mass produced fluorescent lamps, and achieved an O/V-ratio of 80 m^{-1} which corresponds to the ratio of the Biocoil reactor. At natural illumination, the maximum productivities amounted to 0.3 g d.wt. $l^{-1} d^{-1}$ for cyanobacteria *Nostoc* and to 0.6 g d.wt. $l^{-1} d^{-1}$ for the green alga *Monoraphidium*, the heating of the suspension having caused significant complications.

Jüttner [109] has solved the problem of overheating by the installation of another internal tube which contained a cooling liquid and by reducing the dark volume. This increased the growth rate by a factor of 1.5. This technically expensive idea, however, has only limited capability for up-scaling.

5.2.2 Plate Type Reactors

If light energy has to be available continuously to the cells, a lamination of the photobioreactor directed to the light source seems to be the best solution. This basic principle of laminar concept has been practiced by plants for millions of years. Compared with tubular systems, plate type geometries seem to have an identical configuration potential and high surface/volume ratio, but some advantages with respect to compactness (narrow U-turns, wall thickness). The simplest approach is found in the application of widely used polyethylene bags. With 50 l turbidostatic cultures of *Tetraselmis*, Trotta [110] achieved yields of 20–30 g d.wt. $m^{-2} d^{-1}$. Samson and Leduy [111] described transparent vertically arranged plates for cultivation. Illumination on both sides of the tanks (10 cm wide) and simple aeration of the blue-green alga *Spirulina maxima* with CO_2-enriched air have resulted in a maximum increase of 1.17 g d.wt. $l^{-1} d^{-1}$.

Ramos de Ortega et al. [112] were one of the first to test the horizontal meandering channels which, then, were further developed by Fallowfield [113]. Tredici and Materassi [80] and Pulz [59] also favoured this idea by arranging vertically the rectangular channels in the plate geometry. In contrast to the air-lift driven vertical alveolar panels (VAPs) of Tredici and Materassi, Pulz uses a closed system with horizontal flow and pumps (PBR). The flow rate must be high enough to maintain the turbulent flow in the channels (32×27 mm) and the distance to the temperature-regulated degassing container must be short enough to avoid critical gradients (e.g. O_2, CO_2-concentrations, T). Parallel plates can be packed closely so that 42 plates (6 m^3 plant) will require only approximately 100 m^2 at a surface/volume ratio of 60 m^{-1}.

Such a photobioreactor design, which was developed at the JGV (Inst. for Cereal Processing) thus meets the requirements for improved areal outputs: compact principle distributing over-saturating solar energy to as large a number of cells as possible, narrow light path and strong turbulence streaming to move the algal cells in and out of the photic volume at as high a frequency as possible [114]. Despite some light limitation caused by self-shading of the plates, even under the rather unfavourable climatic conditions of Germany, maximum biomass productivity (*Chlorella*) of 1.3 g d.wt. l^{-1} d^{-1} was obtained. This corresponds to an areal growth of 130 g d.wt. m^{-2} d^{-1}, which is considerably higher as reported for raceway ponds and conventional tubular reactors [115]. Beside the Biocoil reactor, this plate type photobioreactor, available as PBR-series (B. Braun Biotech Int.) from 10 l up to a capacity of 6000 l, is the only system which is sold commercially [116]. With vertical alveolar panels the O/V-ratio may be increased to 80 m^{-1}, the productivities related to the thin artificially illuminated plates of a surface of 0.3 m^2 amounting to 20–30 g d.wt. m^{-2} d^{-1} (*Spirulina*), 3 g d.wt. l^{-1} d^{-1}(*Tetraselmis*) respectively [117].

The latest development in the field of sloped photobioreactor systems was presented by Doucha et al. [118] who proposed optimized large-scale modules of 1000 m^2. The culture area is made of several 50 m long and 5–10 m wide meandering sloped glass lanes (inclination 1.6%) which produce a layer thickness of just 6 mm. Though productivity dropped to 14.5 g d.wt. m^{-2} d^{-1}, the increase of the volume productivity to 3 g d.wt. l^{-1} d^{-1} seems to be very high, and the harvesting density of 30 g d.wt. l^{-1} achieved – one of the highest cell concentration ever reported – allows for effective downstream processing.

5.3 Experimental Photobioreactors

5.3.1 Laboratory-Scale Reactors

On the basis of numerous research works in the field of the preparation of valuable substances from phototrophic microorganisms, Lee [75] has summarized the development of closed photobioreactors. Therefore, mainly the trends of the last ten years will be considered.

The design of the 0.5 m^2 serpentine tubular system of Pirt et al. [99], though known for more than ten years, now as before belongs to the most frequently cited developments in the field of photobioreactor construction. It consists of 52 glass tubes, 1 m length with a 1 cm bore, connected in series by silicone rubber U-bends, and an air-lift system generating an effective surface/volume ratio of 127 m^{-1}. Biomass concentrations in excess of 20 g d.wt. l^{-1} could be maintained in continuous flow cultures and very high maximum productivities over 50 g d.wt. m^{-2} d^{-1} (about 5 g d.wt. l^{-1} d^{-1}) could be achieved representing a PE of 40% with respect to *Chlorella* growth. Both theory of design and performance was described and discussed in full detail, and thus the patent, still representing one of the best analyses of fundamental engineering of tubular reactors, also covers the design of a full-scale system.

A number of further tubular laboratory systems was described. The superiority of helical air-lift systems as against stirred reactors was demonstrated with a maximum biomass yield (*Porphyridium cruentum*) of 1.7 g d.wt. l^{-1} d^{-1}

[119]. Watanabe and Hall [120] designed a cone-shaped helical tubular structure and thus improved the PE of the top-illuminated photobioreactor in contrast to standard geometry. Within the MELISSA-project, Binois et al. [36] applied a tubular reactor with *Spirulina* and *Rhodospirillaceae* as a life-support system for space equipment. The experimental work is continued with tubular systems at outdoor conditions, high growth rates up to 2.7 g d.wt. l^{-1} d^{-1} (*Phaeodactylum tricornutum*) having been achieved [121].

Thanks to the high technical level of cylindrical reactors of classical biotechnology, as offered by a number of known fermenter manufacturers, this type of reactor is commonly applied for inoculation purposes and experimental cultivations in laboratory scale. In stirrer reactors, fully sterilizable 1–10 l glass containers, surrounded by fluorescent lamps, well defined cultivation conditions may be established. All the cylindrical reactors are characterized by a high luminance at the surface, related to a large dark volume. Strong turbulences are required in order to achieve sufficient productivities and to avoid photoinhibition [34, 122], which may lead to inhibitory growth effects by the corresponding shear forces.

A reactor type, representing the various designs with internal illumination, will be described which includes most of the typical features [123]. The cylindrical reactors are made of stainless steel with diameters between 20 and 60 cm and heights between 35 and 100 cm, correponding to volumes of 20–250 l. At least four transparent pipes are inserted into the cylinder from the top through holes in the lid, holding fluorescent tubes which can easily be removed during sterilization without opening the reactor. A T-shaped multifunctional stirrer agitates the suspension and supplies sterile air. Light supply was sufficient to produce more than 10 mg exopolysaccharides from cyanobacteria on a daily basis [124]. For the production of *Spirulina* biomass light input was found to be insufficient though the concept of fluorescent tubes in combination with thin layers was considered to be advantageous with regard to growth yield [122]. The idea to use internally fixed fluorescent lights was already set up in 1978, when a 3 m^3 vessel was presented [125], but no data on algal growth are available.

For fully sterile operation, Schulze and Stahl [126] presented a photobioreactor with a tumbler head equipped with plunger lamps. Due to the special tumbling motion which ensures high gas exchange, troublesome glide ring seals can be avoided.

Quite another concept is described by Semenenko et al. [61], the sterilizable laboratory photobioreactor ФБР-10. The algal suspension is brought to rotation by a cylindrical insert and pressed by centrifugal force to the tempered wall in defined layer thickness. Due to the special design of the rotor a high turbulence within the suspension is generated which will result in a good exchange of material without disturbing the limits of shear force sensibility. The luminescent material is in the centre of the reactor, it is protected by a water-chilled quartz retort; a great part of the heat generated is absorbed and discharged immediately. The growth rates of several microalgal species confirm the high efficiency of this concept. With *Chlorella*, a growth of

4.9 g d.wt. l^{-1} d^{-1}, and with slow growing *Dunaliella* a rate of 2 g d.wt. l^{-1} d^{-1} was attained. The cell concentrations achieved in batch cultivation are also rather high; at approx. 12 g *Chlorella* or *Spirulina* dry mass per litre effective cell density growth inhibition was already reached. The results were determined in a 10-l reactor. Due to the fact that the algal suspension as a whole has to be brought into rotation, the energy input is increasing significantly with increasing volume, so that with the planned 25-l fermenter the limit of technical usefulness will be practically reached.

Even higher cell concentrations were achieved by Lee and Palsson [64] in a 100-ml reactor and an ultrafiltration unit. Intensive illumination was realized by an array of LEDs with monochromatic light at 680 nm. This thin layer system is best suited to studies on mass transfer processes in microalgal cultures [127] as well as on photosynthetic efficiency and the flash-light effect [128].

The transfer of light from external sources to the reactor chamber was also subject to research over the last few years. The patent of Meyer et al. [129] covers a plate-type acryl-glass unit which, as immersion block, distributes the light in the suspension at mathematically defined conditions. At the Faculté Polytechnique in Mons, Belgium, small quantities of isotopically labelled organic compounds are produced from *Spirulina* in this type of reactor. The trend to use light guiding systems for photobioreactors has led to the development of fibre optic arrangements.

5.3.2 Fibre Optic Reactors

Javanmardian and Palsson [63] have solved the problem of uniform light dissipation by using the inner walls of the medium vessel itself as dissipation unit. The 0.6-l acryl-glass fermenter consists of concentrically fitted cylinders which are supplied with light from the exterior via fibre optics. The fibres have their ends in channels with wedge-shaped notches which distribute the incoming light over the whole surface of the cylinder. In this reactor, the light dissipation system takes approx. 40% of the reactor volume but a rather high O/V-ratio of 320 m^{-1} is reached. At a PAR luminance of 30–60 $\mu E\, m^{-2}\, s^{-1}$ and with a strain of *Chlorella vulgaris* the maximum productivity amounted to 1.5 g d.wt. l^{-1} d^{-1} which approaches the value of 1.9 g d.wt. l^{-1} d^{-1} at approx. 30 $\mu E\, m^{-2}\, s^{-1}$ theoretically calculated by Javanmardian. Regarding the growth rates of other reactors of up to 5 g d.wt. l^{-1} d^{-1} and the relatively low PAR-intensities, this system has the disadvantage of rather low light input. Nevertheless, dry masses of more than 20 g d.wt. l^{-1} result from batch cultivation with an ultrafiltration unit.

Concepts such as the reactor designed by Mori [130] make use of the idea to use fibres directly as radiation surface because, due to their small diameter, they produce a very high O/V-ratio. This system, which may also be used for solar radiation, consists of a light collecting unit, a Fresnel lens, which filters out IR and UV-radiation by aberration and supplies the residual light into a fibre system, as well as the fibre system transferring the light into the reactor chamber. The latter consists of vertically fixed side-emitting acryl fibres (diameter 3 mm) covered by transparent protective rods, thus forming tubular radiation bodies with a diameter of 1 cm (calculated average radiation density approximately 75 $\mu E\, m^{-2}\, s^{-1}$ at the surface of the radiation bodies during afternoon sun). With 91 densely packed tubes Mori established an O/V-ratio of 755.4 m^{-1} against 2.4 l algal suspension, although nearly 60% of the reaction space was occupied by the lillumination body. With this arrangement, at an initial concentration of 5 g d.wt. l^{-1} *Chlorella pyrenoidosa*, Mori has achieved a productivity of 9.6 g d.wt. $l^{-1}\, d^{-1}$, recalculated by calibrating pH decrease with growth rate, which represents not quite an undisputed peak value. With a xenon high pressure lamp, used for sunlight simulation, the growth rate was reduced to 5.3 g d.wt. $l^{-1}\, d^{-1}$ despite the increased (by 23%) light input.

The idea to utilize sunlight by collective devices for biological carbon dioxide fixation was investigated by Hirata et al. [131], although growth rates obtained with *Chlorella* sp. have proven to be distinctly lower. Nevertheless, Fresnel lens systems are an interesting subject of controlled and defined intake of natural light in closed vessels. They are used by Eriksen et al. [132] for constant daylight illumination of algal cells in their tubular "lumostats".

The biophotoreactor developed by Matsunaga et al. [133] for the preparation of glutamate by immobilized cells of the cyanobacteria *Synechococcus* sp. is operated only with artificial light. The reactor consists of a column (length 90 cm, diameter 7 cm) which contains 661 parallel fixed light diffusing optical fibres (LDOF). A LDOF (diameter 1 mm) consists of a PMMA core and a fluorescein sheath. Internal tensions are produced by additional hot/cold treatment and cause the diffusion effect. The light of a 400-W metal-vapour lamp produces at the surface of the fibres an emitting density of 20 $\mu E\, m^{-2}\, s^{-1}$; for 1.8 l algal suspension an O/V-ratio of 692 m^{-1} is established, the fibres taking 13.5% of the reaction space. Takano et al. [134] have used this photoreactor for the determination of the growth of cyanobacteria adapted to low light intensity as liquid culture. They determined a maximum growth of 2 g d.wt. $l^{-1}\, d^{-1}$ and a final concentration of 11.2 g d.wt. l^{-1} in batch cultivation with continuous filtration. Main interest was laid on the conversion of CO_2 to biomass. The maximum conversion rate amounted to 4.44 g $CO_2\, l^{-1}\, d^{-1}$ and by far exceeded the value 0.1–2.6 g d.wt. $l^{-1}\, d^{-1}$ of traditional reactors [135], which despite the low radiation density confirms the excellent efficiency of this concept. This type of photobioreactor was further used successfully for the cultivation of *Isochrysis*, the production of docosahexanoic acid respectively [136].

Pulz et al. [115] reported on the development of heat sterilizable LDOFs which emit light evenly along their length. The simultaneous demand for sterilization and scale-up suggested basing the whole lighting system on conventional fermentation vessels. A unit of 1600 fibres of 1 mm diameter was constructed for 10-l fermenters representing a surface/volume ratio of 150 m^{-1}. External light supply was established by a special reflector with a small, pulsed light source which can produce light energy quantities up to 400 J s^{-1} [137], but no data of microalgal growth are published yet.

5.3.3 Immobilized Cell Systems and Hydrogen Production

In their review, Robinson et al. [138] stated the reasons for immobilizing algae:

- use as biocatalysts and biotransformations,
- production of energy (hydrogen, electricity) and polysaccharides,
- for co-immobilized systems,
- bioaccumulation of waste material,
- towards prolonging longevity of cultures.

Immobilization techniques range from different matrices – agar, alginate or carrageenan – to carrier material like polyurethane foam blocks or glass beads. The design of immobilized cell photobioreactors depends on the purpose for which organisms are employed. Packed-bed and fluidized-bed reactors as well as air-lift systems are mentioned in the literature, but parallel plate types are preferred for effective utilization of sunlight. However, growth rates of immobilized phototrophs are generally lower than those of free cells. The authors emphasize that contents of pigments and storage substances, as well as production of extracellular products, especially hydrogen, could be improved remarkably in most cases.

During the last few years, hollow fibre bioreactors have aroused considerable interest in the fixation of phototrophic organisms because of the very high surface/volume ratios obtainable. The product potentials from immobilized cyanobacteria were summarized by Hall et al. [139], ranging from fine chemicals and fuels to water purification. The photosynthetic production of ammonia and hydrogen with *Anabaena* fixed in hollow fibre systems and good carbon dioxide uptake rates are described in detail [140, 141], though effective light distribution is still a problem. Optical guiding systems could present solutions. Applications on the basis of side light optical fibres for glutamate secretion with *Synechococcus* [133] or of composite agar layer/microporous membrane structures for hydrogen photoproduction with *Rhodospirillum* [142] were examined. For the illumination of the included organisms the gel plates are shielded to the side of the nutrient solution by a microporous membrane and to the other side by a steel plate. Small boreholes hold the ends of the fibers by which the immobilisate is evenly illuminated.

Planar systems, such as the plate loop reactor proposed by Tramm-Werner et al. [143] are suited very well to self-immobilization. With purple bacteria *Rhodobacter capsulatus* a biofilm dry weight of over 20 g d.wt. m^{-2} could be achieved in a 20-l system. Depending on light intensity and quality, a H_2-production rate of 2 l m^{-2} h^{-1} seems to be possible. Light penetration into cultures of *Rhodobacter sphaeroides* was examined by Miyake [144]. He reports a maximum conversion of light energy into photobacterially produced hydrogen of 7%. This is in agreement with the extended analyses of Rechenberg [145] who calculated similar efficiencies in the sunny conditions of the Sahara desert within his so-called Heliomites, i.e. tubular cone-shaped photobioreactors. Thus, photosynthetic systems for H_2-production could compete with photoelectric alternatives. However, due to the complexity of the system and with view to the operational costs, industrial H_2-production remains a long-term challenge.

Buchholz and Gerbsch [137] suggested for high value products immobilization of algae in transparent hollow micro-spheres by a polyelectric system comprising sodium cellulose sulfate. By the defined porosity of membrane, extracellular and intracellular products are trapped and enriched during downstream processing.

As already mentioned, self-immobilization of cells on planar surfaces led to the idea to use inexpensive foil material as carrier. Pulz and Broneske [116] investigated an ultrathin layer system in which adhesive forces between two vertically arranged foils cause suspension layer thicknesses of 0.2–1 mm. While growth in the suspension was insignificant, *Chlorella* dry matter concentrations at the foil surface amounted to 40 g d.wt. m^{-2}. By simple fluid-mechanical harvesting methods, together with high up-scalability, this principle can be used to yield very concentrated biomass suspensions, extracellular products like exopolysaccharides [146], degradation of waste substances, and for intense CO_2/O_2 gas exchange processes.

It is noted additionally that not all the cultivation systems are suited to all species. In particular, the dinoflagellates, which are supposed to bear a high potential for pharmaceutical substances, appear to be too fragile to be cultivated in common systems. As was shown by Silva et al. [147], hydrodynamic stress derived from simple gas bubbling negatively affects the growth of *Dunaliella* flagellate. Dialysis culture systems have been proposed for shear-force sensitive organisms and are still being explored [12]. Numerous examples are suggested to use permeable membranes for a sufficient diffusion of nutrients. Membrane technology is an interesting means for the design of separate chamber reactors for sensitive organisms or other special purposes [148–150].

5.4 General Aspects of Phototrophic Mass Cultivation

Open pond technology is subject to the natural light/dark cycle, i.e. respiratory metabolic processes will take place at night as reverse effect of photosynthesis. The reduction of productivity depends on the cell concentration of the suspension [151]. For *Spirulina* an optimal areal density (OAD) of 50–75 g d.wt. m^{-2} is given [152]. Tredici et al. [153] have examined the OAD in their VAP-reactors with steady-state *Spirulina* cultures and determined respiratory losses of 18–30% with increasing biomass concentration. Due to the high surface/volume ratio of the VAPs, the thesis of Richmond and Grobbelaar [154] seems to be confirmed, saying that OAD is shifted towards lower values with increasing surface/volume ratio. Therefore, high cell densities are not desirable in cases of long dark phases.

Lee [155] proposed heterotrophic growth for *Chlorella* in order to avoid non-productive night periods. Oxygen-consuming mixotrophic growth during the day would be a further effect of sugar addition releasing the cultures from symptoms of toxic O_2-concentrations. Ogawa and Aiwa [156] demonstrated that, compared with *Chlorella vulgaris* which was cultivated at autotrophic conditions, the specific growth rate is two times higher at mixotrophic conditions and is in range with heterotrophic growth, up to concentrations of 10 g glucose l^{-1}. Yokoi et al. [157] reported that high density cultivation up to 30 g d.wt. l^{-1} is possible by the use of carbon hydrate feed, even with higher plant cells (*Oryza sativa*). Beside the high microalgal cell concentrations, the content of polyunsaturated fatty acids could be increased three-fold in heterotrophic cultures [158]. A variety of algal strains is able to utilize organic carbon sources [159]. The main problem is to find less expensive carbon sources for enhancing the economic production of algae. Organic wastes and animal manure were tested for their potential to yield phototrophic biomass [160, 162, 163].

The fact that both continuous and intermittant lighting result in stable culture productivities prevents a definite comparison of closed reactors to outdoor systems. For a more appropriate evaluation of the performance of photobioreactors, independently of their design or arrangement, it is suggested to consider the photosynthetic efficiency (PE) or the growth yield Y, as g dry biomass per mole absorbed photons [98]. That only makes sense if cultivation units for maximum biomass production are considered, because both PE and Y are dependent on the quantum yield which is much higher at low radiant energy corresponding to poor growth. Nevertheless the economic importance of energy conversion at artificially illuminated reactors is magnified with increasing scale up. In Table 2 growth yield values Y_{PAR} on the basis of absorbed photosynthetically active radiation are summarized for several open and closed systems, which were all operated at high irradiance levels. Though depending on cultivated species, this bioenergetic consideration allows for a rough estimation of the energy conversion performance.

It seems that outdoor systems are exhausted by values of 0.5 g E^{-1} corresponding to an areal productivity of 25–30 g m^{-2} d^{-1}. Considerations of light intensive summer periods and sophisticated cultivation techniques, e.g. the tubular reactor of Lee et al. [94] promise a possible doubling of productivity. Most of the tubular indoor reactors merely compensate the natural dark cycle. An outstanding exception is represented by the results of Pirt et al. [99] with 3.18 g E^{-1}, the highest conversion rate of all the studies examined. Common laboratory cultivation units and plate type reactors are in range or slightly succeed tubular performance. However, through the use of light distribution units, like side light optics within the reactors, Y_{PAR} increased approximately fourfold with regard to outdoor results. Photons emitted by internal light

Table 2. Recalculations of Y_{PAR} (g d.wt. per Einstein 400–700 nm) from published data

Growth Yield of different photobioreactor types

system	type	locality	species	Y_{PAR}	ref.
outdoor systems (based on areal productivity; PAR constitutes 43% of total solar radiation)	**ponds**	Italy	*Spirulina*	0.27*	[80]
		Italy	*Tetraselmis*	0.50	[72]
		California	*not specified*	0.90**	[165]
	tubular	South France	*Scenedesmus*	0.30	[79]
		Italy	*Spirulina*	0.47*	[80]
		Singapore	*Chlorella*	1.35*	[94]
		indoors	*Spirulina*	0.51***	[120]
	inclined	Italy	*Scenedesmus*	0.53	[80]
		arrangement			
indoor systems (based on continously illuminated surface)	**tubular reactors**	vertical / pump	*Scenedesmus*	0.38	[72]
		not specified	*Isochrysis*	0.61	[121]
		vertical / stirred	*Chlorella*	0.93	[107]
		helical / air lift	*Porphyridium*	1.36	[119]
		vertical / air lift	*Chlorella*	3.18	[99]
	non tubular with external illumination	lab. container	*Spirulina*	0.98	[17]
		lab. container	*Chlorella*	1.02	[17]
		lab. container	*Chlorella*	1.45	[15]
		alveolar panel	*Eustigmatophy.*	0.62	[117]
		vertical plate	*Chlorella*	1.32	[+]
	non tubular with internal illumination	immersion cyliders	*Chlorella*	1.9	[63]
		immersion tubes	*Chlorella*	2.5	[130]
		fluorescent light	*Chlorella*	2.4	[122]
		side light fibers	*Synechococcus*	2.0	[133]

energy conversion factors according to. [29]

* assuming 24.5 MJ per m² and day total solar radiation according [92]

** mixotrophic growth in domestic sewage

*** light / dark cycle 12 h / 12 h

+ Pulz O, unpublished data

sources are effectively absorbed and transformed into biomass energy. This is in agreement with latest data published. A set of cultivation experiments demonstrated good energy utilization with uniform light distribution at moderate quanta fluxes [122]; beside technical characteristics of gas exchange rate and quality of mixing, productivity is simultaneously reciprocal to layer thickness and proportional to the surface/volume ratio [163].

Lee and Low [164] added some interesting aspects by short-time bioenergetic investigation of outdoor *Chlorella* cultures. During the morning hours, Y_{PAR} reached values up to 5 g E^{-1} with a flat, horizontally arranged 10 l reactor. At midday, growth yield decreased to approximately 1 g E^{-1} by photoinhibition, and failed to increase to values of 2 g E^{-1} during the rest of the day. These effects strongly suggest that cells are primarily limited by their inability to respond properly to the daily cycle of varying illumination.

With respect to an overall evaluation of reactor performance, the absolute upper limit for volumetric productivities of microalgae seems to be confined to the range of 5–10 g l^{-1} d^{-1}. According to the growth yield potentials, it would require an input of PAR of at least 20 μE l^{-1} s^{-1} (ca. 4 W l^{-1}). Thus it became apparent that light limitation remains the main technological challenge for designing industrial photobioreactors.

6 Conclusions

Today, commercial algal mass production is mainly performed in open ponds. However, as was discussed in this article, area intensive cultivation reactors are limited by a variety of natural restrictions and are technologically exhausted by the efforts of the last decades. Using minimal investment efforts, each improvement in mixing and illumination of the phototrophic cells extensively increased the operational cost. This is the crux of microalgal biotechnology: the price for microalgal biomass has to be reduced by every means. This could be done either by increasing areal productivity or by the isolation of novel products which could justify the high technical expense. New developments of closed systems are directed to the high-value product market. Industrial photobioreactors like the Biocoil or the plate-type PBR move slowly into this market because they combine reliable production of high biomass quality and simple maintenance with inexpensive technology. Nevertheless their photosynthetic efficiency is far from being optimized and their installation is still capital intensive. Further efforts should be made to apply inexpensive materials to arrangements running automatically at low operational costs. Ultrathin layers and immobilization of algae on inexpensive material could be a very interesting alternative for suspension culture, since this configuration promotes the main factor which induces quick algal growth – nutrition and light supply, respectively.

Though artificial illumination, in contrast to freely available natural light, allows a better control of growth parameters, energy costs will restrict the application only to highly valuable bioproducts. With respect to the agreements of the pharmaceutical industry, there is a need for closed, fully sterilizable reactors which at present are limited to a working volume of a few litres. The development of such a technology with scale-up suitability, e.g. internal illumination by light distribution systems like optical fibres with highest growth yields, would open up completely new perspectives with respect to heterotrophic and mixotrophic growth. The use of carbohydrate-rich byproducts as substrate for some algal strains could help to close the gap with classical biotechnological processes of mass cultivation. Beside technological obstacles, there is a demand for expanded screening of pharmaceutically interesting biosubstances with thousands of species still waiting for their potentials to be discovered. Today, the budget for high quality phototrophic biomass sometimes exceeds the price of the product itself, as is the case of arachidonic acid from *Porphyridium.*

In general it is not possible to favour a certain cultivation technique, but the diversity of photobioreactor concepts will increase because of the variety of species demanded and optimized growth conditions. The very sensitive cultivation needs of dinoflagellates would be in contrast to stress induced product accumulation, as one would proceed in antioxidant production. Considering the individual process steps, the reactor design should result in an integration of down-stream procedures. Depending on the cultivation purpose, highly concentrated algal suspension, flocculated biomass or cell disruption by enhanced shear forces may reduce the production costs drastically.

7 References

1. Mueller HJ (1984) Ökologie. VEB Fischer Verlag, Jena, Germany (ger)
2. Böhm H, Pulz O (1991) Phototrophe Mikroorganismen. In: Rutloff H (ed) Lebensmitteltechnologie. Akademie Verlag Berlin, p 165–180, 1976, Neuauflage 1991
3. Graham JM (1991) J Protozool 38: 66
4. Burlew JS (ed) (1953) In: Algal culture from laboratory to pilot plant. Carnegie Institution of Washington, Washington DC, p 357
5. Soeder CJ (1986) An historical outline of allied algology. In: Richmond A (ed) Handbook of microalgal mass culture. CRC Press, Boca Raton, Florida, p 25
6. Richmond A (1983) Phototrophic microalgae. In: Dellweg H (ed) Biotechnology. Verlag Chemie, Deerfield Beach, p 109
7. Oswald WJ (1988a) Microalgae and waste-water treatment. In: Borowitzka MA Borowitzka LJ (eds) Microalgal biotechnology. Cambridge University Press, Cambridge, p 305
8. Shelef G, Oswald WJ, McGauhey PH (1970) J of the Sanitary Engineering Division of American Society of Civil Engineers 96: 91
9. Sogokenkyusho E (1991) Japanese patent 3.022.990
10. Roe SP (1994) An evaluation of marine biotechnology for atmosphere control in diesel- electric powered submarines. ECB6: Proceedings of the 6th European congress on biotechnology, Florence, p 1125

11. Bardach JE, Ryther JH, McLarney WO (1972) Aquaculture. The farming and husbundry of freshwater and marine organisms. New York: Wiley Interscience
12. Borowitzka MA (1992) J Appl Phycol 4:267
13. Chaumont D (1993) J Appl Phycol 5: 593
14. Pulz O (1994) presented at the 2nd Asian – Pacific Conference on Biotechnology, Singapore
15. Oh-Hama T, Miyachi S (1988) Chlorella. In: Borowitzka MA, Borowitzka LJ (eds) Microalgal biotechnology. Cambridge University Press, Cambridge, p 3
16. Benemann JR, Tillet DM, Weisman JC (1987) Trends in Biotechnology 5: 47
17. Aiba S (1982) Advances in Biochemical Engineering 23: 85
18. Raven J (1988) Limits to growth. In: Borowitzka MA, Borowitzka LJ (eds) Microalgal biotechnology. Cambridge University Press, Cambridge, p 331
19. Pirt SJ, Lee YK, Richmond A, Pirt MW (1980) J Chem Tech Biotech 30: 25
20. Pirt SJ (1983) Biotech Bioeng 25: 1915
21. Laws EA, Taguchi S, Hirata J, Pang L (1988) Biomass 16: 19
22. Oswald WJ (1978) the engineering aspects of microalgae. In: Laskins AI (ed) CRC Handbook of microbiology. CRC Press, Boca Raton, Florida, p 519
23. Venkataraman LV, Becker EW (1985) Biotechnology and utilization of algae. Sharada Press, Mangalore, India
24. Hall DO, Rao KK (1996) The potential application of cyanobacterial photosynthesis. Abstracts of the 7th International Conf of Applied Algology, Knysna, South Africa, p 40
25. Scheibe R (1996) Biologie in unserer Zeit 26: 27 (ger)
26. Prokop A, Quinn MF, Fekri M, Murad M, Ahmed SA (1984) Biotech Bioeng 26: 1313
27. Kubin S, Borns E, Doucha J, Seiss U (1983) Biochem Physiol Pflanzen 178: 193
28. Kohl JG, Nicklisch A (1988) Ökophysiologie der Algen. Akademie Verlag Berlin (ger)
29. Richardson K, Beardall J, Raven JA (1983) New Phytologist 93: 157
30. Radwan SS (1988) J Biosciences 43: 15
31. Karentz D, McEuen FS, Land MC, Dunlap WC (1991) Marine Biology (Berl) 108: 157
32. Kopecky J, Doucha J, Loest K, Pulz O (1996) Algological Studies 81: 53
33. Lopez-Figueroa F (1991) Z Naturforsch 46c: 542
34. Grima EM, Camacho FG, Perez JAS, Sevilla JMF, Fernandez FGA, Gomez AC (1994) J Chem Tech Biotech 61: 167
35. Schuster A (1905) Astrophys J 21: 1
36. Binois C, Lasseur C, Cornet JF, Doyle D (1994) The Engineering Society for Advancing Mobility Land Sea Air and Space SAE Technical paper series 941411
37. Simmer J, Tichy V, Doucha J (1994) J Appl Phycol 6: 309
38. Frohlich BT, Webster IA, Ataai MM, Shuler ML (1983) Biotech Bioeng Symp 13: 331
39. Brand LE, Guillard RRL (1981) J Experimental Marine Biology and Ecology 50: 119
40. Weissman J, Raymond PG, Benemann JR (1988) Biotech Bioeng 31: 336
41. Grobbelaar JU (1994) J Appl Phycol 6: 331
42. Laws EA (1983) Nova Hedwigia H83 supplement
43. Laws EA, Terry KL, Wickman J, Chalup MS (1983) Biotech Bioeng 25: 2319
44. Warburg O (1919) Biochem Z 100: 230 (ger)
45. Kok B (1956) Biochem Biophys Acta 21: 254
46. Terry KL (1986) Biotech Bioeng 28: 988
47. Weller S, Franck J (1941) J Phys Chem 45: 1359
48. Phillips JN, Myers J (1954) Plant Physiol 29: 152
49. Grobbelaar JU, Neddal L (1996) Growth of algae under turbulent conditions and intermitted light. Abstracts of the 7th Internat Conf of Applied Algology, Knysna, South Africa, p 40
50. Tichy V, Xiong F, Grobbelaar JU, Neori A, Nebdal L (1996) Algae in intermitted light. Abstracts of the 7th Internat Conference of Applied Algology, Knysna, South Africa, p 69
51. Sager JC, Giger W (1980) Agricult Meteorol 22: 289
52. Rabinowitch EI (1956) Photosynthesis and related Processes. vol 2, part 2, Intersience, NY
53. Schädlich HK (1993) Chem Ing Tech 65: 1482 (ger)
54. Beale SI, Appleman D (1971) Plant Physiol 47: 230
55. Herron HA, Mauzerall D (1972) Plant Physiology 50: 141
56. Sukenik A., Falkowski PG, Bennet J (1987) Biotech Bioeng 30: 970
57. Ben-Amotz A (1983) Nova Hedwigia 83 supplement: 132
58. Falkowski PG (1984) Photosynthetica 18: 62

59. Pulz O (1992) Cultivation techniques for microalgae in open and closed ponds. Proceedings of the 1st European workshop on microalgal biotech, Potsdam-Rehbrücke, Germany, p 61
60. Karube I, Takeuchi T, Barnes DJ (1992) Adv Biochem Eng Biotech 46: 64
61. Semenenko VE, Gabel BV, Tsoglin LN (1992) Closed type photobioreactor. Proceedings of the 1st European workshop on microalgal biotech, Potsdam-Rehbrücke, Germany, p 67
62. International Light Corp. (1995) An overview of Flashlamps and CW Arc Lamps Technical Bulletin No.3, ILC Technology, Sunnyvale, CA, USA
63. Javanmardian M, Palsson BO (1991) Biotech Bioeng 38: 1182
64. Lee CG, Palsson BO (1994) Biotech Bioeng 44: 1161
65. Fusion Lighting Inc. (1996) Product preview, Rockwell, Maryland, USA
66. Kirk JTO (1994) Light & Photosynthesis in Aquatic Ecosystems. 2nd edn. Cambridge University Press, Cambridge
67. Fogg GE (1991) New Phytol 118: 191
68. Grobbelaar JU, Soeder CJ, Stengel E (1990) Biomass 21: 297
69. Borowitzka MA, Borowitka LJ (1988) Microalgal Biotechnology. Cambridge University Press
70. Canell RJ (1990) Appl Biochem Biotechnol 26: 85
71. Becker EW (1994) Microalgae. Cambridge Universitiy Press
72. Benemann JR (1989) Microalgal Biotechnology: Products, Processes and Opportunities. OMEC, Intern Inc
73. Richmond A, Vonshak A (1991) Preface of Bioresource Technol 38: 83
74. Richmond A (1990) Large scale microalgal culture and applications. In: Round FE, Chapman DJ (eds) Progress in physiological research. vol 7. Biopress, Bristol
75. Lee YK (1986) Trend Biotechnol 4: 186
76. Brouers M, Dejong H, Shi DJ, Hall DO (1989) Immobilized cells. In. Cresswell RC et al. (eds) Algal and cyanobacterial biotechnology. Longman Scientific & Technical, NY, p 272
77. Borowitzka LJ (1991) Bioresource Technol 38: 251
78. Dujardin E, Sironval C, Bombart P, Brouers M (1992) Controlled algae cultures at Liege. Proceedings of the 1st European workshop on microalgal biotechology, Potsdam-Rehbrücke, Germany, p 87
79. Gudin C, Chaumont D (1991) Bioresource Technol 38: 145
80. Tredici MR, Materassi R (1992) J Appl Phycol 4: 221–231
81. Becker EW (1981) Process Biochem 8/9: 10
82. Richmond A, Boussiba S, Vonshak A, Kopel R (1993) J Appl Phycol 5: 327
83. Setlik I, Sust V, Malek I (1970) Algol Stud (Trebon) 1: 111
84. Fournadzhieva S, Pillarsky P (1993) Mass Culture and Application of Algae in Bulgaria. Abstract books of the 6th international conference on applied algology, Czech Republic, p 20
85. Heussler P, Castillo J, Morino S, Vasquez V (1978) Arch Hydrobiol 11: 254
86. Soeder CJ (1978) Archiv für Mikrobiologie 11: 259
87. Gudin C, Chaumont D (1983) Solar biotechnology study. In: Palz W, Pirrwitz D (eds) Proceedings of the workshop and E C contractor's meeting in Capri, D Reidel, Dordrecht, p 184
88. Chaumont D, Thepenier C, Gudin C (1988) Scaling up tubular photoreactors. In: Stadler T et al. (eds) Algal biotechnology. Elsevier, London, p 199
89. Chaumont D, Ferreira dos Santos P, Gudin C, Chaintron G, Assice D (1991) French patent 9.115.735
90. Torzillo G, Pushparaj B, Bocci F, Balloni W, Materassi R, Florenzano G (1986) Biomass 11: 61
91. Torzillo G, Carlozzi P, Pushparaj B, Montaini E, Materassi R (1993) Biotech Bioeng 42: 891
92. Carlozzi P, Torzillo G (1996) Appl Microbiol Biotechnol 45: 18
93. Zitelli G, Tomasello V, Tredici MR (1996) One-year cultivation of *Athrospira.* Abstracts of the 7th International Conference of Applied Algology, Knysna, South Africa, p 84
94. Lee YK, Ding SY, Low CS, Chang YC, Forday WL, Chew PC (1995) J Appl Phycol 7: 47
95. Dvorin SA (1992) A factory of commercial cultivation of microalgae. Proceedings of the 1st European workshop on microalgal biotechnology, Potsdam-Rehbrücke, Germany, p 91
96. Seitgeldiyev N (1992) Industrial photoreactors for producing *Chlorella.* Proceedings of the 1st European workshop on microalgal biotech, Potsdam-Rehbrücke, Germany, p 105
97. Seitgeldiyev N, AkyevAY (1995) A strategy for gas supply at large scale production.Proceedings of the 2nd European workshop on microalgal biotechnology, Bergholz-Rehbrücke, Germany, p 25
98. Tredici MR, Chini Zitelli G (1995) Scale-up of photoreactors to commercial size. Proceedings of the 2nd European workshop on microalgal biotech, Bergholz-Rehbrücke, Germany, p 21

99. Pirt SJ, Lee YK, Walach MR, Pirt MW, Balyuzi HHM, Bazin MJ (1983) J Chem Tech Biotech 33B: 35
100. Robinson LF (1987) European patent 0.239.272
101. Chrismadha T, Borowitzka MA (1994) Growth and lipid production of *Phaeodactylum*. In: Phang SM et al. (eds) Algal biotechnology in the asia-pacific region – conference proceedings 1992, Institute of Advanced Studies, University of Malaya, Kuala Lumpur, p 122
102. Borowitzka MA (1996) Tubular photobioreactors. Abstracts of the 7th International Conference of Applied Algology, Knysna, South Africa, p 25
103. Watanabe Y, Delanoue J, Hall DO (1995) Biotech Bioeng 47: 261
104. Borowitzka MA (1996) personal communication
105. Anderson DB, Eakin DE (1985) Biotech Bioeng 15: 533
106. James CM, Al-Kahrs AM (1990) Aquaculture 87: 381
107. Myers J, Graham J (1961) Plant Physiol 36: 342
108. Miyamato K, Wable O, Benemann JR (1988) Biotechnology letters 10: 703
109. Jüttner F (1977) Biotech Bioeng 19: 1679
110. Trotta P (1981) Aquaculture 22: 383
111. Samson R, Leduy A (1985) Canadian J Chem Eng 63: 105
112. Ramos de Orterga A, Roux JC, Fourcy A (1984) Culures intensives de chlorelles sous rayonnement solaire. 2eme colloque de l'Assoc Franc pour l'algol appl, chateau de Fontager
113. Fallowfield HJ (1991) UK patent 2.235.210 A
114. Richmond A (1996) Photobioreactor design. Abstracts of the 7th International Conference of Applied Algology, Knysna, South Africa, p 60
115. Pulz O, Gerbsch N, Buchholz R (1995) J Appl Phycol 7: 145
116. Pulz O, Broneske J (1995) Layer thickness in photoreactor design. Proceedings of the 2nd European workshop on microalgal biotechnology, Bergholz-Rehbrücke, Germany, p 38
117. Tredici M, Biagiolini S, Chini Zitelli G, Montaini E, Favilli F, Mannelli D, Materassi R (1994) Fully-controllable photobioreactors. ECB6: Proceedings of the 6th European congress on biotechnology, Florence, p 1011
118. Doucha J, Livansky K, Kostelnik K (1996) Thin-layer microalgal culture technology. Abstracts of the 7th Internat Conference of Applied Algology, Knysna, South Africa, p 32
119. Lee ETY, Bazin MJ (1990) New Phytol 116: 331
120. Watanabe Y, Hall DO (1996) Appl Microbiol Biotechnol 44: 693
121. Grima EM, Perez JAS, Camacho FG, Sevilla JMF, Fernandez FGA (1996) Outdoor chemostat productivity analysis. Abstracts of the 7th International Conference of Applied Algology, Knysna, South Africa, p 39
122. Scheibenbogen K, Peine G, Pulz O (1995) Influence of different light input parameter. Proceedings of the 2nd European workshop on microalgal biotech, Bergholz-Rehbrücke, Germany, p 27
123. Pohl P, Kohlhase M, Martin M (1986) Planta Medica 52: 416
124. Pohl P, Gehrmann C, Hoffmann M, Poluljach O (1992) Axenic mass cultivation. Proceedings of the1st European workshop on microalgal biotechnology, Potsdam-Rehbrücke, Germany, p 73
125. Böhm H (1978) GDR patent DD-PS 133.248
126. Schulze S, Stahl W (1994) BioEngineering 10: 28 (ger)
127. Lee CG, Palsson PO (1995) J Ferment Bioeng 79: 275
128. Matthijs HCP, Balke H, Vanhes UM, Kroon BMA, Mur LR, Binot RA (1996) Biotech Bioeng 50: 98
129. Meyer M, Levert JM, Vanthournh M (1990) World patent 9.015.953
130. Mori K (1986) Biotech Bioeng Symp 15: 331
131. Hirata S, Hayashitani M, Taya M, Tone S (1996) J Ferment Bioeng 81: 470
132. Eriksen NT, Poulsen BR, Iversen IJL, Geest T (1996) On-line optimisation of light intensity. Abstracts of the 7th Internat Conference of Applied Algology, Knysna, South Africa, p 32
133. Matsunaga T, Takeyama H, Sudo H, Oyama N, Ariura S, Takano H, Hirano M, Burgess JG, Sode K, Nakamura N (1991) Appl Biochem Biotechnol 28/29: 157
134. Takano H, Takeyama H, Nakamura N, Sode K, Burgess JG, Manabe E, Hirano M, Matsunaga T (1992) Appl Biochem Biotechnol 34/35: 449
135. Yamamura K (1993) Development of photosynthetic bioreactor. BIOTECHNICA, Hannover, conference abstract no 5
136. Burgess JG, Iwamoto K, Miura Y, Takano H, Matsunaga T (1993) Appl Microbiol Biotechnol 39: 456

137. Buchholz R, Gerbsch N (1995) New developments for microalgae cultivation. Proceedings of the 2nd European workshop on microalgal biotech, Bergholz-Rehbrücke, Germany, p 17
138. Robinson PK, Mak AL, Trevan MD (1986) Process Biochem 21: 122
139. Hall DO, Garbisu C, Kannaiyan S, Lichtl R, Markov S, Rao KK, Serra JL, Sopko B (1993) Photobioreactors with immobilized cyanobacteria for production of fuels and chemicals. Abstract books of the 6th internat conference on applied algology, Czech Republic, p S5/1
140. Wang SC, Jin MR, Hall DO (1991) Bioresource Technol 38: 85
141. Markov SA, Bazin MJ, Hall DO (1995) Enzyme Microbial Technol 17: 306
142. Mignot L, Junter GA, Labbe M (1989) Biotechnol Techniques 3: 299
143. Tramm-Werner S, Weng M, Hartmeier W, Modigell M (1996) Photobiological hydrogen production. pres at the 9th Europ bioenergy conf, Copenhagen (submitted for publication)
144. Miyake J (1994) Photosynthetic bacteria for solar energy conversion. ECB6: Proceedings of the 6th European congress on biotechnology, Florence, p 1019
145. Rechenberg I (1994) Photobiologische Wasserstoffproduktion in der Sahara.Werkstatt Bionik und Evolutionstechnik, vol 2. Frommann-Holzboog, Stuttgart (ger)
146. Pohl P, Schlösser UG, Fischer D (1996) Exopolysaccaride production by cyanobacteria. Abstracts of the 7th Internat Conference of Applied Algology, Knysna, South Africa, p 57
147. Silva HJ, Cortinas T, Ertola RJ (1987) J Chem Tech Biotech 40: 41
148. Märkl H, Zenneck C, Wilderer PA (1991) Wasser Abwasser 132: 414 (ger)
149. Jehle W, Staneff T, Steinwandel J, Wagner B (1994) The Engineering Society for Advancing Mobility Land Sea Air and Space SAE Technical paper series 941339
150. Reitzig W, Helmdach M, Broneske J, Pulz O (1995) Bioregenerative life-support systems. Proceedings of the 2nd European workshop on microalgal biotechnology, Bergholz-Rehbrücke, Germany, p 97
151. Hartig P, Grobbelaar JU, Soeder CJ, Groeneweg J (1988) Biomass 15: 211
152. Vonshak A, Abeliovich A, Boussiba S, Richmond A (1982) Biomass 2: 175
153. Tredici MR, Zitelli GC, Biagioliniu S (1992) Influence of areal density. Proceedings of the 1st European workshop on microalgal biotechnology, Potsdam-Rehbrücke, Germany, p 58
154. Richmond A, Grobbelaar JU (1986) Biomass 10: 253
155. Lee YK (1996) Could moxotrophic growth be the best mode for outdoor micoalgal cultures? Abstracts of the 7th Internat Conference of Applied Algology, Knysna, South Africa, p 46
156. Ogawa T, Aiwa S (1981) Biotech Bioeng 23: 1121
157. Yokoi H, Koga J, Yamamura K, Seike Y, Tanaka H (1993) J Ferment Bioeng 75: 48
158. Johns MR (1994) Heterotrophic culture of microalgae. In: Phang SM et al. (eds) Algal biotechnology in the asia-pacific region – conference proceedings 1992, Institute of Advanced Studies, University of Malaya, Kuala Lumpur, p 150
159. Droop MR (1974) Heterotrophy of carbon. In: Steward WPD (ed) Algal physiology and biochemistry. Blackwell, London, p 503
160. Chung P, Pond WC, Kingsburg JM, Walker EF, Krook L (1978) J Animal Science 47: 319
161. Adler PR (1996) Fermentation byproducts from organic wastes as carbon source for heterotrophic algae. Abstracts of the 7th Internat Conference of Applied Algology, Knysna, South Africa, p 19
162. Moi Phang S, Jeyaratnam S, Hashim MA (1996) Effect of carbon dioxide supplementation on *Chlorella* production in a high rate algal pond treating rubber effluent. Abstracts of the 7th International Conference of Applied Algology, Knysna, South Africa, p 56
163. Tsygankov AA, Laurinavichene TV, Bugatin VE, Gogotov IN, Hall DO (1995) Comparison of different photobioreactors in biomass production. Proceedings of the 2nd European workshop on microalgal biotechnology, Bergholz-Rehbrücke, Germany, p 48
164. Lee YK, Low CS (1994) The productivities of outdoor algal cultures. In: Phang SM et al. (eds) Algal biotechnology in the Asia-Pacific region – conference proceedings 1992, Institute of Advanced Studies, University of Malaya, Kuala Lumpur, p 150
165. Oswald WJ (1988b) Large-scale algal culture systems. In: Borowitzka MA, Borowitzka LJ (eds) Microalgal biotechnology. Cambridge University Press, Cambridge, p 357
166. Mitscherlich EA (1909) Landwirtschaftliches Jahrbuch 38: 537 (ger)
167. Baly ECC (1935) Proc Roy Soc London Ser B117: 218

Received June 1997

Kinetics and Energetics of Photosynthetic Micro-Organisms in Photobioreactors

Application to Spirulina Growth

J.-F. Cornet, C.G. Dussap, J.-B. Gros
Laboratoire de Génie Chimique Biologique Université Blaise-Pascal 24, Avenue des Landais 63177 Aubière Cedex, France

Advances in Biochemical Engineering/
Biotechnology, Vol. 59
Managing Editor: T. Scheper

This paper provides the basis for the description and quantitative analysis of cultures of photosynthetic micro-organisms in batch and in continuous photobioreactors. The methodology generally accepted for modeling submerged aerobic or anaerobic cultures is used and applied to the growth of the blue-green algae *Spirulina platensis*.

From analysis of the metabolic pathways inside the cell, stoichiometric equations are derived for the main metabolic events which must be considered for *Spirulina* growth, including exopolysaccharide formation. Together with the description of the reaction kinetics, it forms the basis for modeling.

As the rate of growth is closely related to the light energy available inside the culture medium, special attention is paid to the description of light energy transfer inside a dense liquid medium which absorbs and scatters the light, and the useful concept of working illuminated volume is introduced. The reaction kinetics accounts too for the mineral limitations (nitrogen, sulphur and phosphorus) as well as physical and physiological limitations by the carbon source.

Finally, a so-called biochemically structured model gives a unified vision of yields and rates of growth observed for various light energy inputs and different nutrients limitations. The proposed methodology enables one to define and calculate the thermodynamic efficiency for the light energy conversion process in photobioreactors.

List of Abbreviation and Symbols

$\mathcal{A}$	Volumetric rate of radiant energy absorbed ($W.m^{-3}$)
$\mathcal{A}_\lambda$	Volumetric rate of radiant energy absorbed for radiation of wavelength λ ($W.m^{-4}$)
A_i	Chemical affinity for the component i ($J.kmole^{-1}$)
a	Extinction coefficient of Lambert (m^{-1})
C	State vector ($kg.m^{-3}$)
C_i	Mass concentration of component i ($kg.m^{-3}$)
C'_i	Molar concentration of component i ($kmole.m^{-3}$)
D	Dilution rate (h^{-1})
E	Acceleration factor by chemical reaction (–)
Ea	Absorption mass coefficient ($m^2.kg^{-1}$)
Ea(λ)	Absorption mass coefficient for radiation of wavelength λ ($m^2.kg^{-1}$)
Es	Scattering mass coefficient ($m^2.kg^{-1}$)
Es(λ)	Scattering mass coefficient for radiation of wavelength λ ($m^2.kg^{-1}$)
F	Volumetric flow rate ($m^3.h^{-1}$)
F	Radiant energy flux vector ($W.m^{-2}$)
$\mathbf{F}_\lambda$	Radiant energy flux vector for radiation of wavelength λ ($W.m^{-3}$)
G	Molar gas flow rate ($kmole.h^{-1}$)
g_i	Partial molar free enthalpy for the component i ($J.kmole^{-1}$)
H′	Henry's constant ($Pa.m^3.kmole^{-1}$)
ΔH	Enthalpy ($J.kmole^{-1}$)
Δh^*	Activation enthalpy ($J.kmole^{-1}$)
Δh_d	Deactivation enthalpy ($J.kmole^{-1}$)
I	Specific radiant light intensity ($W.m^{-2}$)
I_λ	Specific radiant light intensity for radiation of wavelength λ ($W.m^{-3}$)
J	Mean specific intensity ($W.m^{-2}$)
J_λ	Mean specific intensity for radiation of wavelength λ ($W.m^{-3}$)
J_j	Specific molar conversion rate of reaction j (kmole.kg $DM^{-1}.h^{-1}$)
K	Chemical equilibrium constant for a given reaction
K_i	Partition coefficient between liquid and gas phases ($kg.m^{-3}.Pa^{-1}$)
K_i	Monod saturation constant for the component i ($kg.m^{-3}$)
K_J	Monod saturation constant for available radiant light energy ($W.m^{-2}$)
k	Extinction coefficient for medium only (m^{-1})

k_i, k_{-i}	Direct and reverse kinetic constants for the reaction i
k_J	Andrew inhibition constant for available radiant light energy ($W.m^{-2}$)
k_La	Volumetric coefficient of gas–liquid transfer (h^{-1})
L	Total length or optical thickness of a rectangular reactor (m)
L_2	Working illuminated length (m)
L_{ij}	Phenomenological coefficient ($kmole^2.kg\ DM^{-1}.J^{-1}.h^{-1}$)
M_i	Molar mass of component i ($kg.kmole^{-1}$)
m	Extinction coefficient of Lambert-Beer ($m^2.kg^{-1}$)
n_i	Number of moles of component i inside the bioreactor (kmole)
P	Pressure (Pa)
P_i	Partial pressure for the component i (Pa)
$p(\theta, \theta', \phi, \phi')$ or $p(\mathbf{u}, \mathbf{u}')$ or $p(\cos\Theta)$	Phase function (−)
q	Coupling coefficient (−)
q_i	Specific mass conversion rate of component i ($kg.kg\ DM^{-1}.h^{-1}$)
R	Reflexion (−)
R	Total radius or optical thickness of a cylindrical reactor (m)
R	Ideal gas constant ($R = 8.3143\ J.mole^{-1}.K^{-1}$)
R_2	Working illuminated radius (m)
Ri	Mass volumetric conversion rate of component i in reference to light limitation only ($kg.m^{-3}.h^{-1}$)
r	Radius (m)
r_i	Mass volumetric conversion rate of component i ($kg.m^{-3}.h^{-1}$)
r_i'	Molar volumetric conversion rate of component i ($kmole.m^{-3}.h^{-1}$)
S	Cross section (m^2)
T	Transmission (−)
T	Temperature (K)
t	Time (h)
$\mathbf{u}, \mathbf{u}'$	Unit vector (−)
V	Volume (m^3)
V_2	Working illuminated volume (m^3)
v	Specific photosynthesis or respiration rate ($kmole.kg\ DM^{-1}.h^{-1}$)
$\mathbf{X}$	Biotic state vector ($kg.m^{-3}$)
x	Generalized force ratio (−)
$\mathbf{Y}$	Abiotic state vector ($kg.m^{-3}$)
$Y_{i/j}$	Global mass conversion yield of substrate i into product j (−)
y_i	molar gas fraction of component i (−)

Z	Dimensionless length or radius (–)
Z_2	Dimensionless working illuminated length or radius (–)
z	Length (m)
z_i	Mass fraction of the component i (–)
$\langle\rangle = 1/V \iiint_V dV$	Mean volumetric integral (–)

Greek letters

γ	Illuminated fraction volume (–)
η, η_{th}	Thermodynamic efficiency (–)
θ, θ', Θ	Angle (rd)
ζ	Proportionality coefficient (–)
ε	Gas hold-up (–)
λ	Wavelength (m)
μ	Specific growth rate (h^{-1})
μ_M	Maximum specific growth rate (h^{-1})
ν_{ij}	Stoichiometric coefficient for the component i in reaction j (–)
σ	Dissipation function of entropy ($J.kg\ DM^{-1}.h^{-1}$)
ϕ, ϕ'	Angle (rd)
χ	Phenomenological stoichiometric coefficient (–)
$\omega, \omega', \Omega, \Omega'$	Solid angle (–)
$\bar{\omega}_0$	Albedo of single scattering = Es/(Ea + Es) (–)

Subscripts

ATP	Adenosine tri-phosphate
C	Carbohydrates or bicarbonate
CH	Chlorophyll *a*
COF	Cofactor
EPS	Exopolysaccharide
G	Glycogen or gas
L	Lipids or liquid
N	Nitrate
NA	Nucleic acids
Pr	Protein
PC	Phycocyanin
P	Phosphate
R	Respiration
r	r-direction
S	Sulfate
T	Total volume
X	Biomass
XA	Active biomass
XT	Total biomass
z	z-direction
ΦS	Photosynthesis

Superscripts

E	For the incoming flow in the reactor
′	For molar concentration
*	For gas–liquid equilibrium
+	Positive direction
−	Negative direction

Abbreviations

PFTR	Plug flow tubular reactor
WTR	Well-mixed tank reactor

1. Introduction

The use of activities of photosynthetic micro-organisms, either in a closed vessel under light irradiation or in an open outdoor pond, calls for a quantitative understanding of their behaviour and of the rate controlling processes. As for other bioprocesses, where different classes of models provide adequate framework for such endeavours, the cultivation of photosynthetic micro-organisms needs specific problems to be treated, leading to particular models. These models can be applied to process development, process optimization or process control and can also be useful guidelines for the design of new culture vessels and for the formulation of the composition of nutrient feed.

The problems to be overcome in modeling the behaviour of a photosynthetic microorganism growing in an illuminated reactor derive mainly from the complexity of the physiological regulation of metabolism and the physical transfer of light energy inside the medium. Considerable reduction of these two levels of complexity is required to obtain mathematical models that are tractable for the comprehensive study of photosynthetic cultures. The danger of such a simplification is that it can result in loss of realism and so lead to erroneous conclusions. Hence the real difficulty lies in determining the appropriate degree of detail for the description at a physiological level or at the physical level for the radiative transfer characterization.

This paper presents a basis for simulating batch and continuous cultures of photosynthetic micro-organisms under varying conditions of incident light energy and nutrient limitations. The cyanobacterium *Spirulina platensis* is taken as an example of such micro-organism because it presents classical photosynthesis with two photosystems, and it is one of the more studied and documented in literature. The state of the art in modeling this kind of photosynthetic culture includes the widely made assumption that growth rate is closely related to light energy availability inside the medium, thereby paying special attention to the physical problem of light energy transfer inside a dense culture that absorbs and scatters light. It is thus important to assess the effects of light energy consumption and light scattering caused by the presence of the micro-organisms. This produces an uneven light intensity distribution within the reactor and consequently a biochemical conversion rate inhomogeneity inside the culture medium. This creates a high degree of complexity for the modeling of photosynthetic micro-organism cultures.

Also, a variety of situations occur in which the compositional variables of the organisms are subject to changes that significantly affect behaviour. In such cases, representation of the culture variables requires appropriate knowledge of the biochemistry of the conversion processes and their regulation in the cell environment.

The characterization of *Spirulina platensis* growth kinetics thus includes both physiological and physical aspects. To handle these problems, the general approach for modeling submerged aerobic or anaerobic cultures is used. It separates stoichiometric analysis of the metabolic conversions and the establishment of reaction kinetic laws; the specific problem of radiative transfer is examined independently.

In the second section, the various classes of models are presented for the stoichiometric characterization of the main metabolic events that have to be considered for growth. The third section presents light energy diffusion models of varying mathematical complexities from the simple Lambert-Beer model to sophisticated gridding techniques. The effect of mineral limitations (nitrogen, sulphur and phosphorus) on *Spirulina* growth, including their influence on synthesis rates and resulting biomass composition, is examined in the fourth section. Various classes of physical or physiological limitations by the carbon source are presented in the fifth section. The sixth section describes a more refined technique of modeling growth that uses a biochemically structured model that affords a comprehensive representation of yields and rates of growth vs various light energy inputs and nutrient limitations. Finally, the thermodynamic efficiency for the light energy conversion process in photobioreactors is defined and calculated.

2 Modeling Photobioreactors: Application to Growth of *Spirulina platensis*

2.1 General Approach of Modeling Photobioreactors

Mathematical modeling of the growth kinetics of a photosynthetic micro-organism can be performed using the general approach for characterizing submerged cultures of micro-organisms. Formally, the macroscopic description of the culture, i.e. the mathematical representation of the state variables involves simultaneous handling of two types of mechanisms.

– physiological behavior including characterization of metabolic activity, determination of bioreaction stoichiometries, and the establishment of kinetic laws for the conversion of substrates into biomass or products;

– physical controlling steps including gas–liquid transfer rates, mixing characteristics of the bioreactors, thermodynamic equilibria and the light energy transfer characterization.

The physical characterization of the light energy transfer throughout the liquid volume of a photobioreactor is clearly vital for proper understanding of the rate controlling processes and must be carefully described. However, the final mathematical model of the photobioreactor is formed by the association of the transport phenomena models with the micro-kinetics of biological conversions.

Thus kinetic modeling of the metabolic activity must be performed in such a way that it can be coupled to the physical rate-controlling process of light energy transfer already described in dense media by radiative transfer equations. This places conditions on how the kinetic laws are expressed. A second constraint concerns the mass conservation laws. Generally speaking, it is always possible to build as many kinetic relations as there are substrates and products that are consumed or evolved. However, such modeling does not ensure mass conservation balances, which are the basic tool of macroscopic description; this includes the determination of mass conversion yields, particularly the biomass/CO_2 and O_2/CO_2 production yields in the case of photosynthetic growth. The representation of the physiological processes that is to be adopted will therefore be translated from the classical approach used for chemical reaction engineering. It separates the establishment of stoichiometric equations (which satisfy per se elemental conservation balances) and the determination of the kinetic laws related to each stoichiometric equation. This approach has been more extensively used in the last decade for modeling aerobic or anaerobic cultures [1]; it can be effectively applied to cultures of photosynthetic micro-organisms such as *Spirulina platensis* without significant modifications, affording stoichiometric models for growth and product formation.

2.1.1 Characterization of Growth by a Single Stoichiometric Equation

It should be emphasized that the metabolism of photosynthetic micro-organisms is too complex to be modeled in all its intricacies. A reduction in the complexity is therefore necessary. At ground state, the growth model will contain a single stoichiometric equation and a single kinetic law that includes the physical limitation by light energy transfer inside the medium.

The first difficulty is the determination of the elemental composition of dry cell material. This can be done in two ways: (i) experimental determination of the global content of carbon, oxygen, hydrogen, nitrogen, phosphorus and sulphur in the dry matter, or (ii) determination of the mass fractions of different classes of macromolecules such as proteins, lipids, carbohydrates, nucleic acids. The second difficulty is the calculation of the different stoichiometric coefficients. In general, the number of stoichiometric coefficients to be calculated is equal to the number N of components involved minus one. As there are as many conservation laws as elements, e.g. P elements, the number of independent experimental yields that have to be used to obtain the stoichiometric equation is equal to (N-P-1).

Consequently, a stoichiometric model will contain two types of coefficients that have to be identified from experimental results:

– the yields, in the knowledge that the determination of a conversion yield will affect all other conversion yields via the stoichiometric relation;

– the kinetic coefficients, which again must be split into two groups, (i) those related to enzyme kinetics and (ii) those specific to transport phenomena, which are generally dependent on reactor design and operating conditions.

2.1.2 Balance Equations

The general balance equation for any given component i inside a photobioreactor takes the straightforward form

$$\begin{bmatrix} \textit{Rate of accumulation} \\ \textit{of component } i \end{bmatrix} = \begin{bmatrix} \textit{Transport of} \\ \textit{component } i \end{bmatrix} + \begin{bmatrix} \textit{Conversion of} \\ \textit{component } i \end{bmatrix} \tag{1}$$

Because of spatial inhomogeneity of light energy transfer in a suspension of micro-organisms, the rates of conversion are not invariant, so that it is necessary to calculate average values throughout the liquid volume, indicated by the operator $\langle\ \rangle$, for macroscopic balances.

Two different kinds of reactors must be considered leading to different formulations of Eq. (1): Well-mixed Tank Reactor (WTR) or Plug Flow Tubular Reactor (PFTR). For a dissolved component with a constant liquid volume in a WTR, the general expression at Eq. (1) reduces to

$$V_L \frac{dC_i}{dt} = F(C_i^E - C_i) + \langle r_i \rangle V_L \tag{2}$$

where the concentration of species i, C_i, is constant throughout the entire reaction volume and equal to the concentration in the outgoing stream. For PFTR, Eq. (1) becomes

$$-F \frac{\partial C_i}{\partial V_L} + \langle r_i(z) \rangle = \frac{\partial C_i}{\partial t} \tag{3}$$

where $\langle r_i(z) \rangle$ is a mean reaction rate on a cross section S of the reactor (the term ∂V being obviously equal to $S\partial z$), enabling one to integrate Eq. (3) over the entire liquid volume of the reactor. In this case, C_i varies along the volume of the reactor but steady-state conditions often occur for the process, so the accumulation term vanishes.

After introducing the dilution rate $D = F/V_L$, the specific rate of conversion of component i, q_i, the molar specific rate of reaction J, and the stoichiometric coefficients $\nu_i (\nu_i > 0$ for a product; $\nu_i < 0$ for a substrate), we obtain

$$\langle r_i \rangle = \langle q_i \rangle C_x \tag{4}$$

$$\langle q_i \rangle = \nu_i M_i \langle J \rangle \tag{5}$$

which gives respectively for Eq. (2) and (3)

$$\frac{dC_i}{dt} = D(C_i^E - C_i) + \nu_i M_i \langle J \rangle C_x \tag{6}$$

$$-F \frac{\partial C_i}{\partial V_L} + \nu_i M_i \langle J(z) \rangle C_x(z) = \frac{\partial C_i}{\partial t} \tag{7}$$

The general expression of the kinetic rate takes the form $J = f(C_1, \ldots C_i, \ldots C_N,$ transport dynamics).

The mass yield $Y_{i/j}$ for the conversion of substrate i into product j is expressed as follows:

$$Y_{i/j} = \left|\frac{\langle q_j \rangle}{\langle q_i \rangle}\right| = \left|\frac{\nu_j}{\nu_i}\right| \frac{M_j}{M_i} \tag{8}$$

$Y_{i/j}$ is obviously independent of the reaction rate J, which confirms that $Y_{i/j}$ must be interpreted stoichiometrically rather than kinetically.

The mass balance relation for a gaseous component in a PFTR is written as follows:

$$M_i \frac{\partial n_i}{\partial t} = -F \frac{\partial C_i}{\partial V_T} + (1 - \varepsilon)\langle r_i(z) \rangle - M_i \frac{\partial G y_i}{\partial V_T} \tag{9}$$

where ε is the gas hold up, $\varepsilon = V_G/V_T$.

The number of moles of component i inside the culture device is expressed as follows:

$$n_i = \left(\frac{C_i(1 - \varepsilon)}{M_i} + \frac{p_i \varepsilon}{RT}\right) \tag{10}$$

At gas–liquid equilibrium, the liquid phase concentration is related to partial pressure using the partition coefficient

$$C_i = K_i p_i = K_i y_i P \tag{11}$$

so that

$$n_i = \left(\frac{K_i(1 - \varepsilon)}{M_i} + \frac{\varepsilon}{RT}\right) y_i P \tag{12}$$

Generally, the rate of accumulation of component i inside the apparatus $\partial n_i/\partial t$ can be neglected. Furthermore, the transport term by the liquid phase $F\partial C_i/\partial V_T$ is much lower than by the gas phase so that the balance equation reduces to the steady-state molar gas balance equation

$$\frac{dGy_i}{dV_T} = (1 - \varepsilon) \frac{\langle r_i(z) \rangle}{M_i} = (1 - \varepsilon)\nu_i \langle J(z) \rangle C_x(z) \tag{13}$$

For WTR, Eq. (13) reduces to

$$\frac{dGy_i}{dV_T} = (1 - \varepsilon)\nu_i \langle J \rangle C_x \tag{14}$$

with mean homogeneous concentrations and a mean spatial reaction rate $\langle r_i \rangle$.

In most cases, the molar fraction in the output flow of the reactor is close to the input one y_i^E, so the gas phase may be considered as perfectly mixed. Integration of Eq. (14) then leads to

$$Gy_i - G^E y_i^E = \nu_i \langle J \rangle C_x V_L \tag{15}$$

However, such assumptions (no accumulation in liquid phase, no transport by liquid flow) may be questionable in the case of total CO_2 at high pH values, the partition coefficient K_i taking much greater values than for other dissolved gases such as O_2, N_2, etc. This case of importance for the culture of photosynthetic micro-organisms will be discussed in more detail in Sect. 5.

2.1.3 Characterization of Growth and Product Formation by Several Stoichiometric Equations

The drawback of a single stoichiometric equation model is that it implies constant yield representation of the growth process, i.e. independent of external conditions such as light energy supply or nutrient limitations. Also, the biomass elemental formula is held constant.

A number of situations occur in which this kind of model fails for cultures of photosynthetic micro-organisms where yields and biomass composition are known to change as a function of the external conditions applied.

Models are therefore needed in which changes in the organism's composition are considered. A class of potentially useful models results from a single extension of the previous global approach, in which amounts and properties of biomass are specified by several variables. These models are generally termed compartment models and combine a better description of the behaviour of the cultures under different operating conditions with moderate complexity.

For such compartment models, the amount of biomass is not only specified by the total biomass concentration C_X, but also by a biotic state vector $\mathbf{X}$ [1], which contains the relative proportions of the components defining biomass. Accordingly, the various chemical compounds can be treated independently, each being synthesized according to a particular stoichiometric equation. The time course of cell composition can be properly represented provided the rate of synthesis for each part of the biotic vector is a function of the external conditions.

When using this approach for the growth of *Spirulina platensis*, the biotic state vector has to be characterized by proteins, carbohydrates, lipids, nucleic acids, sulfated glycogen contents and by a complex sulfated exopolysaccharide. A further degree of refinement allows for distinctions within the protein pool and the pigments (phycocyanins, chlorophylls), which is important when studying the effects of mineral limitations. The stoichiometries of synthesis for each of these macromolecules are established as for a single stoichiometric equation model: determination of the elemental formula from elemental analysis and/or from their chemical formula and calculation of the stoichiometric coefficients.

The general formulation of the mathematical model associated with this description can be easily deduced from the single stoichiometry model. Let subscript XA stand for the active biomass, i.e. the fraction of biomass that actually ensures growth and metabolite production, so excluding reserves, let $\langle J_j \rangle$ be the average molar specific rate of the jth reaction and ν_{ij} the stoichiometric coefficient of compound i in the jth reaction. Relations (Eqs. 1–8) established for a single stoichiometry model now become

$$\langle r_i \rangle = \langle q_i \rangle C_{XA} \tag{16}$$

$$\langle q_i \rangle = M_i \sum_j \nu_{ij} \langle J_j \rangle \tag{17}$$

For WTR:

$$\frac{dC_i}{dt} = D(C_i^E - C_i) + M_i C_{XA} \sum_j \nu_{ij} \langle J_j \rangle \tag{18}$$

For PFTR:

$$-F \frac{\partial C_i}{\partial V_L} + M_i C_{XA}(z) \sum_j \nu_{ij} \langle J_j(z) \rangle = \frac{\partial C_i}{\partial t} \tag{19}$$

$$Y_{i/k} = \frac{M_k}{M_i} \left| \frac{\sum_j \nu_{kj} \langle J_j \rangle}{\sum_j \nu_{ij} \langle J_j \rangle} \right| \tag{20}$$

$Y_{i/k}$ becomes a function of the kinetic rates of the different reactions where the compounds k and i are involved.

2.1.4 Biochemically Structured Models

The main shortcoming of the above approach is that it often leads to a high number of coefficients that have to be predicted and/or identified from experimental results. In the general case, they are, as above, unknown stoichiometric coefficients, calculated using conversion yields values, and the coefficients of the kinetic laws.

The internal structure of the metabolism can be used to relate some synthesis rates to others in order to reduce the degree of uncertainty of the model without altering its flexibility. This biochemically structured description entails analysis of the metabolic pathways responsible for the synthesis of each class of macromolecule, yielding stoichiometric equations that no longer contain undetermined coefficients, and involve energy carriers such as ATP and GTP and hydrogen carriers such as NADH, H^+, NADPH, H^+ and $FADH_2$.

These metabolic intermediates cannot be over produced nor consumed from the external medium, so that non-accumulation constraints, making a pseudo steady-state assumption for a functioning metabolism, must be considered, leading to as many independent relations between the specific rates as there are non-accumulated intermediates. For such a model to be operative, it is necessary to add to the anabolic reactions (synthesis of the constitutive macromolecules for the biomass growth) the energy-transducing reactions, namely, for photosynthetic bacteria, the water photolysis and the photophosphorylation mechanisms.

The number of independent kinetic laws that remain to be established is lower than for an unstructured compartment model. Also, it becomes much easier to assign the influence of a rate limiting process to a specific metabolic reaction, which increases the robustness and predictivity of such a model. This approach, which has become popular for representing aerobic culture processes [1], will be extensively used in Sect. 6 for modeling the growth of *Spirulina platensis* under different light-limiting conditions.

2.2 Main Components of Cell Material

Detailed composition of the main components of cell material, such as proteins, carbohydrates, lipids and nucleic acids is required to establish stoichiometries for growth. This provides a global formula for biomass either by averaging each compound from its molar fraction experimentally observed in specific operating conditions, or by taking into account the production rates of each compound given by a biochemically structured model for growth [1]. The first approach leads to a fixed stoichiometric equation for biomass growth that is only valid for limited applications, while the second approach enables allowance for marked changes in biomass composition when the micro-organisms are cultivated over a wide range of limiting conditions.

Based on ultrastructure [2, 3] and biochemical studies [4, 5], the following composition may be assumed for the active biomass of the cyanobacterium *Spirulina platensis*:

– the mass protein content is 60–74% with a phycocyanin content of about 25% as the main pigment in photosystem II;

– the mass carbohydrate content is 12–16%. Carbohydrates are mainly located in the cell wall, except for an intracellular sulfated glycogen occurring as carbon reserve [6];

– the mass lipid content is 8–12% located mainly in membranes of cells except for polyhydroxybutyrate as carbon reserve [7, 8];

– the mass nucleic acid content is about 4% (78% RNA and 22% DNA) [7];

– the carbon reserves for dark respiration are mainly sulfated glycogen [6] and polyhydroxybutyrate [8] in equimolar proportions.

S. platensis also synthesizes a complex sulfated exopolysaccharide [7, 9] which together with the active biomass gives the total biomass.

The relative mass fraction of each main component varies with the cultivation conditions, i.e. it depends on the limiting factor governing the biomass synthesis (light transfer, CO_2 transfer, mineral concentration, bicarbonate concentration etc.). The elemental formula for the constitutive macromolecules and the average composition of active biomass (excluding exopolysaccharide) are given in Table 1.

Table 1. Global elemental formula of different fractions of *S. platensis* dry matter [7]

	Elemental formula	Molar mass (kg/kmole)	Average range mass of fractions in cell dry mass (% w)	Mass fraction in this study (% w)
Proteins	$CH_{1.526}O_{0.327}N_{0.250}S_{0.007}$	22.482	60–74	68
Carbohydrates	$CH_{1.670}O_{0.711}S_{0.001}$	25.078	12–16	16
Lipids	$CH_{1.714}O_{0.204}P_{0.003}S_{0.003}$	17.167	8–12	12
Nucleic acids	$CH_{1.273}O_{0.710}N_{0.393}P_{0.104}$	33.359	4	4
Active biomass (mean composition)	$CH_{1.566}O_{0.405}N_{0.192}S_{0.005}P_{0.006}$	23.080	(−)	
Carbon reserves (glycogen-PHB)	$CH_{1.570}O_{0.690}S_{0.009}$	24.898	Dark metabolism	(−)
Exopolysaccharide	$CH_{1.650}O_{0.950}S_{0.015}$	29.330	Depends on light energy available	

2.3 Analysis of Metabolic Pathways

Metabolic analysis is also required for modeling stoichiometry and kinetics of growth, in addition to the above information on the composition of biomass in the main classes of macromolecules.

2.3.1 Photosynthesis

For cyanobacteria like *S. platensis*, photosynthesis occurs mainly in thylakoid membranes where the photosynthetic pigments are located: chlorophyll *a*, carotenoids and phycocyanins. Phycocyanins are aggregated in phycobilisomes, which are photon trapping antennas of photosystem II (PSII). This photosystem also contains chlorophylls *a*, a reaction center P680 and various electron transporters. The electrons evolved from the water photolysis by the P680 centre are transferred via the electron transport chain toward the reaction centre of the photosystem I (PSI), the P700. These electrons reduce ferredoxin and then $NADP^+$ into NADPH, H^+. The electron transport is coupled to ATP synthesis with a possibility of additional cyclic phosphorylation by deviation of electrons from ferredoxin to cytochrome f/b6 complex. These steps are summarized in the well-known Z-scheme for photosynthesis [7, 10] (Fig. 1). This scheme is a basic tool for the determination of the ratio of the production rates of ATP and reduced cofactors by photosynthesis.

2.3.2 Carbon Uptake

The primary carbon source used by *S. platensis* is bicarbonate ion HCO_3^- [11–14] which enters the cell by the active transport mechanism described by

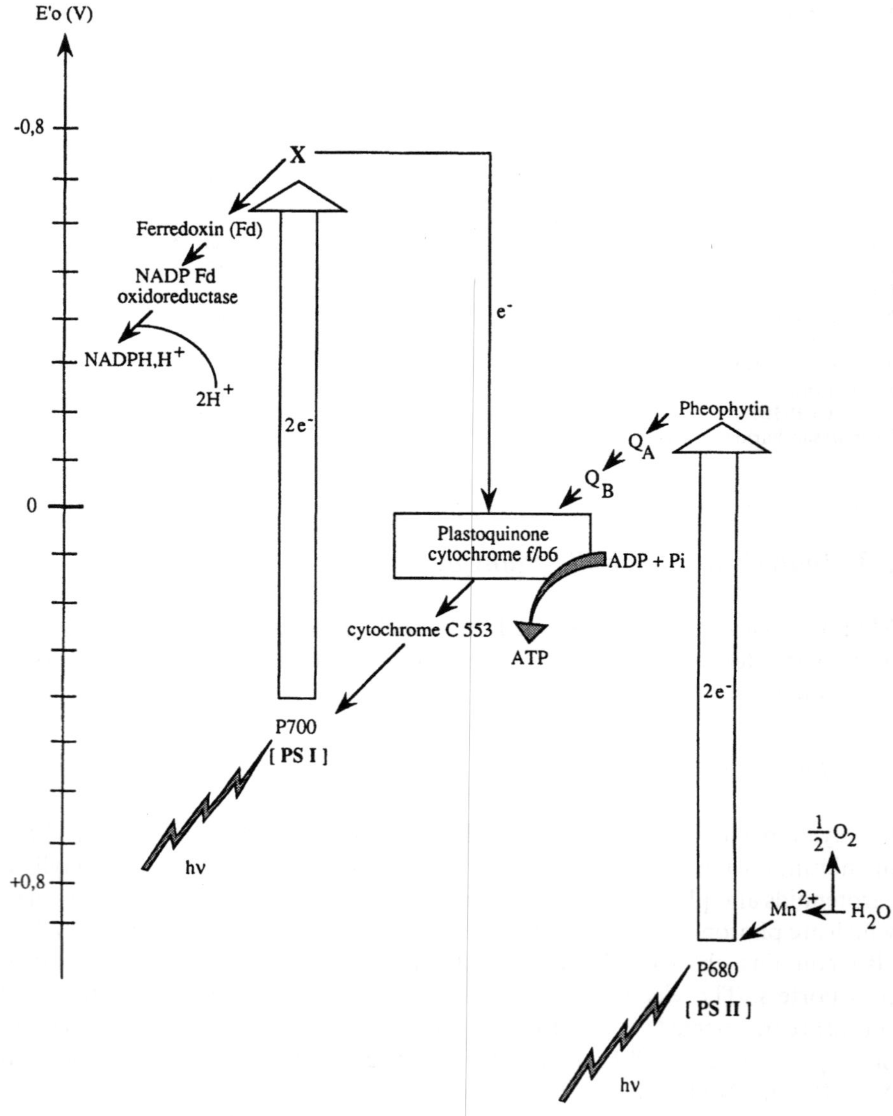

Fig. 1. Z-scheme for *Spirulina* photosynthesis

Volokita et al. [15] and Badger et al. [16]. The intracellular bicarbonate is then dehydrated via the carbonic anhydrase to CO_2, which is incorporated into the Calvin cycle via the Rubisco. This carbon is dispatched simultaneously for the synthesis of all the macromolecules in cells (proteins, carbohydrates, lipids and nucleic acids) from the 3-phosphoglycerate as a key intermediate metabolite, since in cyanobacteria the electron transport chains on thylakoids are the same for photosynthesis and respiration [17] and consequently the respiration is inhibited by light [7, 18]. This is a crucial difference compared to eucaryotic micro-organisms with chloroplasts for which all macromolecules are obtained from the Calvin cycle.

Interestingly, the photorespiration is cyanobacteria seems quasi inexistant [19–22] and has never been demonstrated in *S. platensis*. Also, special attention must be paid to the incomplete Krebs cycle in the metabolism of cyanobacteria, resulting in the amino acid synthesis via the glyoxylic shunt pathway.

2.3.3 Nitrogen Uptake

The preferred nitrogen source for *S. platensis* is the nitrate ion NO_3^-, which enters the cells by active transport. The nitrate is then reduced to ammonium NH_4^+ consuming 4 (NADPH, H^+) and is incorporated on glutamic acid in the amino acid metabolism. Depending on the strains, NH_4^+ may also be used as nitrogen source.

2.3.4 Respiration

In darkness, the intracellular carbon reserves are consumed for maintenance of cells [23]. For *S. platensis*, this is essentially sulfated glycogen [6] and polyhydroxybutyrate [8]. Respiration uses the same electron transport chains as photosynthesis in thylakoids and plasmic membrane [17]. Thus, the respiration is totally inhibited by light in *S. platensis* when the dissipated light level is greater than 5 $W.m^{-2}$ [7].

2.4 Stoichiometry of Growth

The above metabolic pathways analysis [24–27] enables determination of the molar quantities of ATP and reduced cofactors (NADPH, H^+) required for the synthesis of a given metabolite. When the detailed composition of each class of macromolecules is known (e.g. the aminogram for the *Spirulina proteins* [4, 5]), biochemically structured stoichiomeric equations can be established. This approach [1] is an interesting alternative to the classical unstructured single stoichiometric equation model for growth with fixed conversion yields for substrates and products, only valid for limited applications. It enables allowances to be made for modifications in biomass composition and conversion yields when different limiting conditions for growth are applied, by calculating the production rates for each biomass compound defined above [28].

2.4.1 Synthesis of Cell Material

From the general analysis of *S. platensis* metabolism pathways, the stoichiometric equations including ATP and cofactor balances for synthesis of the main

Table 2. Stoichiometric equations for the synthesis of the constitutive fractions of *S. platensis*

Carbohydrates: $P/2e^- = 1.57$

$$CO_2 + 2.037H_2O + 0.001H_2SO_4 + 3.330ATP + 2.127(NADPH, H^+) \xrightarrow{\langle J_C \rangle} CH_{1.670}O_{0.711}S_{0.001} + 3.330Pi + 3.330ADP + 2.127NADP^+ \quad (21)$$

Proteins: $P/2e^- = 1.20$

$$CO_2 + 1.252H_2O + 0.250HNO_3 + 0.007H_2SO_4 + 3.702ATP + 3.081(NADPH, H^+) \xrightarrow{\langle J_P \rangle} CH_{1.526}O_{0.327}N_{0.250}S_{0.007} + 3.702Pi + 3.702ADP + 3.081NADP^+ \quad (22)$$

Lipids: $P/2e^- = 1.13$

$$CO_2 + 1.203H_2O + 0.003H_2SO_4 + 3.013ATP + 2.659(NADPH, H^+) \xrightarrow{\langle J_L \rangle} CH_{1.714}O_{0.204}P_{0.003}S_{0.003} + 3.010Pi + 3.013ADP + 2.659NADP^+ \quad (23)$$

Nucleic acids: $P/2e^- = 1.30$

$$CO_2 + 1.203H_2O + 0.393HNO_3 + 3.776ATP + 2.909(NADPH, H^+) \xrightarrow{\langle J_{NA} \rangle} CH_{1.273}O_{0.710}N_{0.393}P_{0.104} + 3.672Pi + 3.776ADP + 2.909NADP^+ \quad (24)$$

Active biomass: $P/2e^- = 1.24$

$$CO_2 + 1.379H_2O + 0.192HNO_3 + 0.005H_2SO_4 + 3.568ATP + 2.874(NADPH, H^+) \xrightarrow{\langle J_{XA} \rangle} CH_{1.566}O_{0.405}N_{0.192}S_{0.005}P_{0.006} + 3.562Pi + 3.568ADP + 2.874NADP^+ \quad (25)$$

Exopolysaccharide: $P/2e^- = 1.73$

$$CO_2 + 2.220H_2O + 0.015H_2SO_4 + 3.330ATP + 1.920(NADPH, H^+) \xrightarrow{\langle J_{EPS} \rangle} CH_{1.650}O_{0.950}S_{0.015} + 3.330Pi + 3.330ADP + 1.920NADP^+ \quad (26)$$

Cofactor reduction:

$$2.779NADP^+ + 2.779H_2O \xrightarrow{\langle J_{COF} \rangle} 2.779(NADPH, H^+) + 1.389O_2 \quad (28)$$

ATP regeneration:

$$3.544(ADP + Pi) \xrightarrow{\langle J_{ATP} \rangle} 3.544ATP + 3.544H_2O \quad (29)$$

compounds in the biomass have been established by Cornet [7] and are given in Table 2.

The ATP demands have been calculated from the data of Oura [29] for the carbohydrates and nucleic acid synthesis, from the data of Stouthamer [30] for proteins and from information given by Tremolières [31] for lipids (Eqs. 21–24).

These equations of the four main components of biomass may be combined to obtain an overall equation for active biomass under different limiting conditions of growth. According to the average compositions reported in Table 1, the following stoichiometric equation represents the active biomass synthesis:

$$CO_2 + 1.379H_2O + 0.192HNO_3 + 0.005H_2SO_4 + 3.568ATP + 2.874(NADPH, H^+) \xrightarrow{\langle J_{XA} \rangle} CH_{1.566}O_{0.405}N_{0.192}S_{0.005}P_{0.006} + 3.562Pi + 3.568ADP + 2.874NADP^+ \quad (25)$$

The total *Spirulina* biomass produced requires a fifth equation for the exopolysaccharide synthesis, which is established from the exopolysaccharide composition [9] and from the data of Oura [29]:

$$CO_2 + 2.220H_2O + 0.015H_2SO_4 + 3.330ATP + 1.920(NADPH, H^+) \xrightarrow{\langle J_{EPS} \rangle} CH_{1.650}O_{0.950}S_{0.015} + 3.330Pi + 3.330ADP + 1.920NADP^+ \quad (26)$$

In the general case, the respective rates of exopolysaccharide (Eq. 26) and active biomass (Eq. 25) synthesis remain undetermined. The classical unstructured approach may be used by weighting the two equations with fractions only valid for a restricted range of operating conditions. Taking as an example a limitation by light energy transfer with an incident radiant energy flux lower than 20 $W.m^{-2}$, one can assume a molar fraction of 0.9 for active biomass and 0.1 for exopolysaccharide [7]. The combination of the synthesis equations (Eqs. 25 and 26) then gives, for the total biomass produced:

$$CO_2 + 1.463H_2O + 0.173HNO_3 + 0.006H_2SO_4 + 3.544ATP + 2.779(NADPH, H^+) \xrightarrow{\langle J_{XT} \rangle} CH_{1.575}O_{0.459}N_{0.173}S_{0.006}P_{0.006} + 3.538Pi + 3.544ADP + 2.779NADP^+ \quad (27)$$

The consumptions of ATP and reduced cofactors NADPH, H^+ can be further eliminated; the two stoichiometric equations for photosynthesis are

$$2.779NADP^+ + 2.779H_2O \xrightarrow{\langle J_{COF} \rangle} 2.779(NADPH, H^+) + 1.389O_2 \quad (28)$$

$$3.544(ADP + Pi) \xrightarrow{\langle J_{ATP} \rangle} 3.544ATP + 3.544H_2O \quad (29)$$

which gives, by addition of Eqs. (27–29):

$$CO_2 + 0.698H_2O + 0.173HNO_3 + 0.006H_2SO_4 + 0.006Pi \xrightarrow{\langle J_{XT} \rangle} CH_{1.575}O_{0.459}N_{0.173}S_{0.006}P_{0.006} + 1.389O_2 \quad (30)$$

This equation corresponds to the single stoichiometric equation for *Spirulina* growth, only valid for low radiant light energy inputs (< 20 $W.m^{-2}$). It always holds for photosynthesis, since as discussed in Sect. 2.3.2, respiration of *S. platensis* is inhibited by light. Unlike the case of respiration, the elemental mass balance of this equation does not require experimentally determined conversion yields and it is fully predictive, since the six stoichiometric coefficients are calculated from the six conservation balances of the C, H, O, N, S and P elements.

Considering the ionic forms of the salts and taking into account the H^+ balance, Eq. (30) is rewritten as

$$HCO_3^- + 0.698H_2O + 0.173NO_3^- + 0.006SO_4^{2-} + 0.006HPO_4^{2-} + 0.197H^+ \xrightarrow{\langle J_{XT} \rangle} CH_{1.575}O_{0.459}N_{0.173}S_{0.006}P_{0.006} + 1.389O_2 + OH^- \quad (31)$$

The C-molar mass of such a biomass is 23.730 kg/C-kmol^{-1}. The theoretical values of mass conversion yields of bicarbonates $Y_{C/XT}$, nitrates $Y_{N/XT}$, sulfates $Y_{S/XT}$ and phosphates $Y_{P/XT}$ in total biomass can then be calculated:

$$Y_{C/XT} = \frac{M_{HCO_3^-}}{23.730} = 2.571 \text{ kg } HCO_3^-.\text{kg biomass}^{-1} \tag{32}$$

$$Y_{N/XT} = \frac{0.173 M_{NO_3^-}}{23.730} = 0.452 \text{ kg } NO_3^-.\text{kg biomass}^{-1} \tag{33}$$

$$Y_{S/XT} = \frac{0.006 M_{SO_4^{2-}}}{23.730} = 0.024 \text{ kg } SO_4^-.\text{kg biomass}^{-1} \tag{34}$$

$$Y_{P/XT} = \frac{0.006 M_{HPO_4^{2-}}}{23.730} = 0.024 \text{ kg } HPO_4^{2-}.\text{kg biomass}^{-1} \tag{35}$$

All the main stoichiometric equations are summarized in Table 2.

2.4.2 Respiration

Likewise, a stoichiometric equation can be obtained for biomass respiration in darkness. It is still possible to calculate the four stoichiometric coefficients from the four mass balances on C, H, O and S, since respiration energy allows only the maintenance of the micro-organism from oxidation of the carbon reserves, i.e. sulfated glycogen and PHB, and is not coupled with biomass synthesis. This leads to the following equation [7]:

$$CH_{1.570}O_{0.690}S_{0.009} + 1.060 O_2 \xrightarrow{\langle J_R \rangle} CO_2 + 0.780 H_2O + 0.009 H_2SO_4 \tag{36}$$

3 Growth Kinetics Under Physical Limitation by Radiant Light Energy Transfer

3.1 Kinetic Data for Photosynthesis and Respiration

3.1.1 Photosynthesis

The culture pH for *S. platensis* ranges between 8.3 and 11 [32] with a slight optimum at 9.5 for photosynthesis.

The optimum temperature is 35–36 °C. The following normalized formula was established for the effect of temperature on *Spirulina* photosynthesis rate [7] in the range 20–45 °C, giving a maximum activity for 36 °C:

$$\frac{v_{\Phi S}}{v_{\Phi S_{max}}} = \frac{A \exp(-\Delta h^*/RT)}{1 + B \exp(-\Delta h_d/RT)} \tag{37}$$

$$\Delta h^* = 8.30 \cdot 10^7 \text{ J.kmol}^{-1}$$

$$\Delta h_d = 2.16 \cdot 10^8 \text{ J.kmol}^{-1}$$

where $v_{\Phi S}$ is the photosynthesis rate at temperature T and at a given radiant light energy flux, in kmol O_2.kg phycocyanins^{-1}.h^{-1} and $v_{\Phi S_{max}}$ the maximum photosynthesis rate at optimal temperature and in the same radiant light conditions. Δh^* and Δh_d are respectively the activation and de-activation enthalpies. The values calculated with Eq. (37) are in close agreement with those obtained by Zarrouk [32] on growth rates and by Lee [33] on energy conversion yields vs temperature.

Photosynthetic saturation curves are required to relate growth rates or photosynthetic activities to the available radiant light energy in the medium. Figure 2 shows the curve obtained by Cornet [7] for *Spirulina* at 36 °C and pH 9.5. Saturation for photosynthesis is reached at 100–130 W.m^{-2} available radiant light energy and photoinhibition at 300 W.m^{-2}. For a radiant light energy of 1250 W.m^{-2}, the photosynthetic activity is reduced by 30%. These values agree closely with the data reported by Vonshak et al. [34]. This photoinhibition corresponds to exposure times of about 1 min, so it can be considered as a negligible phenomenon in most stirred tank reactors where light exposure times for cells at high fluxes are lower than 1 s. The saturation phenomenon is closely related to the fall in quantum yield for high photon flow rates on

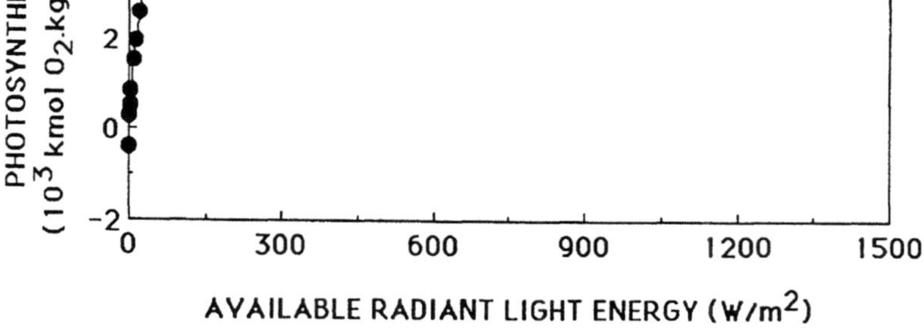

Fig. 2. Photosynthetic activity saturation curve for *S. platensis* vs the available radiant light energy in the culture medium

photosystems, impairing thermodynamic photosynthetic efficiency. The photoinhibition is related to damage to reaction centre P680 or to a polypeptide D1 in photosystem II.

The maximum photosynthetic rate obtained for *S. platensis* is $7.1 \cdot 10^{-3}$ kmol O_2.kg biomass^{-1}.h^{-1}, which corresponds to 0.05 kmol O_2.kg phycocyanins^{-1}.h^{-1}.

3.1.2 Respiration

The respiration rates in cyanobacteria have been found to be 10 to 50 times lower than photosynthesis rates [35], respiration corresponding to a loss in total cell mass for maintenance of micro-organisms. Also, it must be noted that the time necessary to start respiration when the cells change over from light to dark is about 1 min [7]. This is most important to know as respiration in photobioreactors can be neglected if the residence time of cells in darkness is short enough.

The effect of temperature on the respiration rate for *S. platensis* has been investigated and the following formula established in the range 20–45 °C [7]:

$$v_R = v_0 \exp(-\Delta H/RT) \tag{38}$$

where v_R is the respiration rate at temperature T, in kmol O_2.kg phycocyanins^{-1}.h^{-1}; v_0 is a constant equal to $1.71 \cdot 10^{11}$ kmol O_2.kg phycocyanins^{-1}.h^{-1} and ΔH is equal to $8.72 \cdot 10^7$ J.kmol^{-1}.

At 36 °C, the respiration rate is $2.5 \cdot 10^{-4}$ kmol O_2.kg phycocyanins^{-1}.h^{-1} or $3.8 \cdot 10^{-5}$ kmol O_2.kg biomass^{-1}.h^{-1} [7].

3.1.3 Compensation Point

The compensation point is defined as the available radiant light energy at which photosynthesis compensates for respiration, which means that the net exchange rates for O_2 and CO_2 are equal to zero. For *S. platensis*, this point is 1 W.m^{-2} [7].

3.2 Coupling Growth Kinetics and Radiative Transfer

3.2.1 What Variables Should be Used to Describe the Available Radiant Light Energy in a Culture Medium? Radiative Transfer Definitions

The attenuation of light in culture media containing cells creates a heterogeneous radiation field, responsible for local kinetics. It is therefore necessary to consider the effects on growth rate of the light intensity available at each point of the reactor.

Light energy dissipation inside a liquid medium considered as non-emitting and non-fluorescing depends on two independent phenomena: absorption by pigments and scattering by whole cells. The scattering of radiant light energy makes the mathematical description of light transfer extremely complex, since the available energy at any point of the reactor derives both from the main light source and from all directions as light scattered by the suspension.

The radiative transfer theory [36, 37] provides analytical tools to calculate local intensity or local energy flux in such complex media. As the specific radiant light intensity for wavelength λ, I_λ, depends on the observation direction (Fig. 3), variables integrated over all space directions may be used in mathematical modeling. The radiant energy flux $\mathbf{F}_\lambda$ and mean specific intensity J_λ for the wavelength λ are defined by the following integrals over the solid angle ω (Fig. 3):

$$\mathbf{F}_\lambda = \int_{4\pi} I_\lambda \cos\theta \, d\omega \tag{39}$$

$$J_\lambda = \frac{1}{4\pi} \int_{4\pi} I_\lambda \, d\omega \tag{40}$$

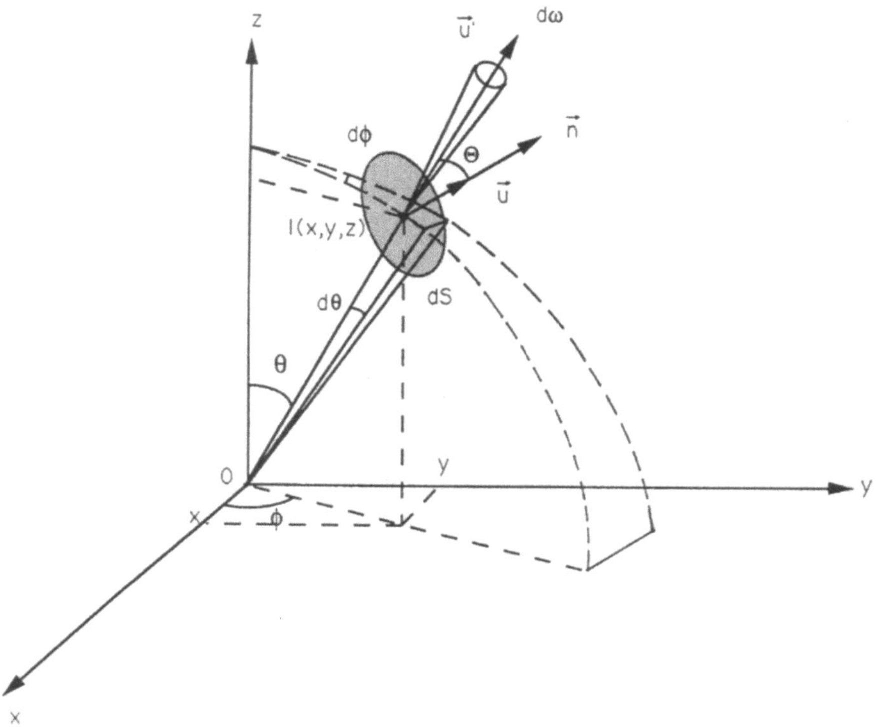

Fig. 3. Definitions of specific intensity I, radiant energy flux vector **F** and available radiant energy 4πJ for radiative transfer theory

In Eq. (39), θ is the angle from the normal $\mathbf{n}$ of a surface reference dS to the direction of I_λ. Since θ is in the integral, the flux $\mathbf{F}_\lambda$ depends on the direction of $\mathbf{n}$; hence $\mathbf{F}_\lambda$ is a vector.

For the entire spectrum, the following expressions for specific intensity I, mean specific intensity J, and radiant energy flux $\mathbf{F}$ are

$$I, J, \mathbf{F} = \int_0^\infty I_\lambda, J_\lambda, \mathbf{F}_\lambda \, d\lambda \tag{41}$$

The integrated variables J_λ(J) or $\mathbf{F}_\lambda$($\mathbf{F}$) may be used to describe the availability of light in the medium; but given the vectorial nature of the radiative flux $\mathbf{F}$, which does not represent the microorganism's light energy environment, it is preferable to work with the mean specific intensity J_λ. More exactly, the total available radiant energy at each point of the medium is described by the scalar quantity $4\pi J_\lambda$ (or $4\pi J$ over the entire spectrum). This is justified a posteriori by the calculation of the local volumetric rate of radiant energy absorbed by the cells $\mathscr{A}_\lambda$ or $\mathscr{A}$ at each point in the medium in which this quantity appears (this calculation will be presented in Sects. 3.3 and 3.4).

Moreover, if a bounded volume V is defined, such as a photoreactor or an open pond, the mean volumetric rate of radiant energy absorbed in this volume for the wavelength λ or the entire spectrum must be defined as follows:

$$\langle \mathscr{A}_{(\lambda)} \rangle = \frac{1}{V} \int_V \mathscr{A}_{(\lambda)} \, dV \tag{42}$$

This mean volumetric rate in radiant energy absorbed is crucial for the characterization of absorbing and scattering media, as it represents the radiative energy exchanged from the photonic phase to the material phase in the medium [38]. Also, this rate appears as a macroscopic quantity that must be experimentally determined on a given volume to validate theoretical calculations for local available radiant energy profiles [39].

However, $\langle \mathscr{A} \rangle$ is a macroscopic term and therefore unsuitable for modeling coupling between kinetics and radiative transfer, which is typically a local problem because the quantum yield of the reaction on the local available light energy and parameters of the model have to be identified from local kinetic laws and volumetric integration over the working illuminated volume. This is an important special feature of photobioreactors compared to chemical photoreactors. Hence the use of $\langle \mathscr{A} \rangle$ must be restricted to energy aspects [38].

3.2.2 Kinetic Laws for Photosynthesis and Photoinhibition:

The kinetic laws for photosynthesis saturation curves vs available radiant light energy are well-known and have been thoroughly discussed [40, 41]. The most often encountered law in the literature is the Monod-type law [42–46] in which the growth rate is related to the available radiant energy for the entire spectrum $4\pi J$ by

$$\mu = \mu_M \frac{4\pi J}{K_J + 4\pi J} \tag{43}$$

If the effects of photoinhibition described above for high photon density fluxes are considered, Andrew's kinetic law may be used, in the form

$$\mu = \mu_M \frac{4\pi J}{K_J + 4\pi J + \dfrac{4\pi J^2}{k_J}} \tag{44}$$

These laws clearly show that the growth rate is highly dependent on the available radiant energy $4\pi J$ and is therefore linked to the radiant light transfer problem. The main problems arise in assessing available local radiant light energy ($4\pi J$) profiles and determining mean biomass volumetric growth rate over the entire volume under consideration. These aspects are closely related to the geometry of the system and its light supply conditions.

3.2.3 Coupling Growth Kinetics and Light Transfer Models: Working Illuminated Volume

Light energy limitation is the most commonly encountered condition during cultures of photosynthetic bacteria or algae, even when light energy is available throughout the culture volume. The existence of a radiant energy profile in the illuminated zone means that a mean growth rate must be calculated with a mean volumetric integral, assuming a kinetic law for growth. When the Monod law is used, the mean mass volumetric growth rate is given, from Eq. (43), by

$$\langle r_x \rangle = \frac{1}{V} \int_V \mu_M C_x \frac{4\pi J}{K_J + 4\pi J} dV \tag{45}$$

This mean growth rate must be numerically calculated using the appropriate equation for $4\pi J$ as discussed below. Parameters μ_M and K_J need to be identified, which can be done with a specified light energy absorption and scattering model. Equation (45) illustrates two extreme situations already discussed by Cornet et al. [46]:

– if the biomass concentration is low, metabolism is limited by light energy but growth is exponential with an apparent maximum specific growth rate;

– if the biomass concentration is high, the growth rate is strongly dependent on the absorption and scattering characteristics of the medium and proportional to the incident energy flux. It becomes independent of biomass concentration resulting in linear growth.

These developments can be further simplified by introducing the concept of a working illuminated volume [46], dividing the entire volume into an illuminated zone in which growth takes place and a dead zone with no metabolic activity. The working illuminated volume is defined as the volume in which growth is exponential, the other part of the reactor being inactive. The average mass volumetric growth rate is expressed by

$$\langle r_x \rangle = (1-\gamma)\frac{1}{V_1}\int_{V_1} r_{x1}\,dV + \gamma\frac{1}{V_2}\int_{V_2} r_{x2}\,dV \tag{46}$$

where subscripts 1 and 2 refer to the dark and illuminated zones respectively and γ stands for the ratio of illuminated volume to total volume. Assuming that the cell residence time in darkness is short enough to exclude respiration, i.e. $r_{x1} = 0$, Eq. (46) becomes

$$\langle r_x \rangle = \gamma\frac{1}{V_2}\int_{V_2} r_{x2}\,dV \tag{47}$$

where

$$r_{x2} = \mu_M C_x \frac{4\pi J}{K_J + 4\pi J} \tag{48}$$

As the lowest value for the available local energy at which photosynthesis is efficient is the compensation point, the working illuminated volume can be defined as the volume in which the available local energy is higher than this compensation point (1 $W.m^{-2}$ for *S. platensis*).

The determination of the working illuminated volume is a prerequisite for identifying kinetic parameters such as μ_M and K_J. This approach provides unicity for identified parameter values over a wide range of incident radiant energy fluxes, insuring the robustness of the model. This robustness is also set by the choice of the physical light transfer model, giving a high accuracy for the calculation of three-dimensional $4\pi J$-distribution and for the working illuminated volume determination.

3.3 The Simplified Lambert-Beer Law for Absorption of Light

Lambert's law has usually been used to express the exponential attenuation of radiant light energy resulting from the sole absorption by the culture medium [43–45, 47]. Considering a steady-state non-flow system in which emission is unimportant and the radiation travels only in the z-direction, a differential radiant energy balance over a cross-sectional area yields [46, 48].

$$-\frac{dF_{z\lambda}}{dz} = \mathscr{A}_{z\lambda} \tag{49}$$

To use this equation, we need information about the volumetric absorption rate $\mathscr{A}$. For a collimated light beam passing through a medium in which refraction and scattering by suspended particules are neglected and light field remains parallel, a conventional expression is [48]

$$\mathscr{A}_{z\lambda} = a(\lambda)F_{z\lambda} \tag{50}$$

in which $a(\lambda)$ is known as the extinction coefficient. Basically, this simply states that the probability for photon absorption is proportional to the concentration of photons; these conditions are commonly encountered in spectrophotometry.

In this case Eq. (49) may be integrated with the boundary condition $z = 0$, $F_{z\lambda} = F_{0\lambda}$, giving the well-known Lambert's law

$$F_{z\lambda} = F_{0\lambda} \exp(-a(\lambda)z) \tag{51}$$

In the Lambert-Beer law, the extinction coefficient $a(\lambda)$ is expressed as a sum of two terms [49]:

$$a(\lambda) = k(\lambda) + m(\lambda)C_x \tag{52}$$

in which $k(\lambda)$ is a coefficient for the absorption of the medium, often negligible, and $m(\lambda)$ is the extinction coefficient for biomass.

Importantly, for a collimated unidirectional light beam, the radiative flux F_λ is equal to the available radiant light energy $4\pi J_\lambda$ inside the medium. Neglecting the attenuation of light by the medium, Eqs. (51) and (52) then give for the whole spectrum

$$4\pi J_z = F_0 \exp(-mC_x z) \tag{53}$$

Equation (53) is an oversimplified equation for obtaining available radiant energy profiles in an absorbing and non-scattering medium, but can be used in Eqs. (47) and (48) to obtain a first approximation of volumetric biomass growth rates. Integration of these equations for one-dimensional applications along the z-axis yields the following analytical solution for the mass volumetric growth rate:

$$\langle r_x \rangle = \mu_M C_x \frac{1}{mC_x L} \ln\left(\frac{F_0 + K_J}{F_0 \exp(-mC_x L_2) + K_J}\right) \tag{54}$$

in which L is the total considered optical thickness and L_2 the working illuminated length. Remembering that the compensation point for *Spirulina platensis* is obtained for $4\pi J = 1\ \mathrm{W.m^{-2}}$, the working illuminated length L_2 is given from the Lambert-Beer law by

$$L_2 = \frac{\ln F_0}{mC_x} \tag{55}$$

Equation (54) was already obtained by Ogawa et al. [43] but the integration was performed over the total volume V, giving different values for the parameter K_J vs the incident radiant energy flux F_0. The calculation of the working illuminated length L_2 enables one to determine and use a fixed value of K_J for any value of F_0.

In addition to this analysis, considering a macroscopic energy balance of light over the entire volume V, it is easily shown that the volumetric rate of radiant energy absorbed $\mathscr{A}$ is given by the following relation:

$$\frac{\langle \mathscr{A} \rangle}{F_0/L} = 1 - \exp(-mC_xL) \tag{56}$$

which tends to 1 as the biomass C_x or the length L tend to infinity.

3.4 General Radiative Transfer Theory

Although the well-known Lambert-Beer law is widely used in the literature and affords simple solutions [49], it is inefficient for determining the available radiant light energy profiles with high accuracy because of the neglect of light scattering by the micro-organisms [38, 39, 46, 50–54]. It is improved by experimental determinations of the mean volumetric rate of radiant energy absorbed in the medium, which does not tend to 1 because of the scattering of light by the biomass [39]. More sophisticated models are then required to obtain radiant energy profiles with high accuracy.

To allow independently for absorption by pigments and light scattering by cells, a two-parameter model is required instead of the extinction coefficient of Lambert-Beer, which aggregates the two phenomena. The general theory of radiative transfer originally developed by Chandrasekhar [36] in astrophysics has been successfully applied to modeling radiant light energy transfer in photobioreactors [38, 39, 46, 50, 53, 54].

However, although used by Spadoni et al. [50] and Aiba [39] for kinetic aspects and Cornet et al. [38] for energy aspects, solving the general form of the radiative transfer equation for a finite medium is an integral-differential problem in a six-dimensional Euclidean space needing sophisticated numerical tools. The computation method used has to be linked to time-dependent kinetic equations, entailing high complexity and long calculation time [38, 39, 50].

To obtain more tractable models for photosynthetic micro-organisms kinetics, a compromise was sought between the simple-but-inefficient and accurate-but-complex-models described above. This can be achieved whenever operating conditions (e.g. the lighting system of a photoreactor or plant geometry) allow a one-dimensional approximation for light energy transfer.

Solutions of one-dimensional radiative transfer equations with different boundary conditions are well-known in rectangular coordinates, e.g. using P–N or discrete ordinate methods [36, 37, 53, 54]. The respective accuracies of these techniques in relation to boundary conditions and medium characteristics have been discussed [36, 37, 53–56]. The authors have already proposed a simple two-flux model based on the assumptions of Schuster [57] for one-dimensional geometries in different systems of coordinates, that has the advantage of offering analytical solutions. In this paper, the simple two-flux method will be recalled and other more sophisticated methods will be briefly presented and discussed in cases where information on incident radiation field and optical characteristics of the medium is abundantly available.

The radiant energy flux in a non-emitting medium containing cells is given by the integral-differential equation of radiative transfer in its integrated form over the entire spectrum and all the directions from the specific intensity, as follows [36, 37]:

$$-\int_{4\pi} (\mathbf{u}.\nabla I)\, d\omega = -\nabla.\mathbf{F} = (Ea + Es)C_x \int_{4\pi} I d\omega - EsC_x \frac{1}{4\pi} \times \int_{4\pi}\int_{4\pi} p(\theta, \theta', \phi, \phi') I d\omega' d\omega \quad (57)$$

where Ea and Es are the mean absorption and scattering mass coefficients on the visible spectrum respectively, ω is a solid angle and $p(\theta, \theta', \phi, \phi')$ the phase function characteristic of the medium. This phase function is also called scattering diagram, because it describes how the radiation is scattered from the incident direction.

By calculating the integral terms in Eq. (57) the light energy balance on a differential volume is established [37, 48] giving the local volumetric rate of radiant energy absorbed by the cells at each point in the medium $\mathscr{A}$:

$$-\nabla.\mathbf{F} = EaC_x 4\pi J = \mathscr{A} \quad (58)$$

Equation (58) is the general form of Eq. (49) used for the establishment of Lambert's law in unidirectional applications. It confirms the above considerations concerning the radiative transfer definitions in demonstrating that the available radiant energy $4\pi J$ multiplied by a constant factor EaC_x is equal to the volumetric rate of radiant energy absorbed $\mathscr{A}$. Introducing the mean volumetric integral, the volumetric rate of radiant energy absorbed in the total volume is defined as

$$\langle \mathscr{A} \rangle = EaC_x \frac{1}{V} \int_V 4\pi J\, dV \quad (59)$$

The problem lies in the necessity to calculate local $4\pi J$-profiles from integration of the local specific intensity I over the solid angle ω (Eq. 57), because Eq. (58) cannot be directly solved in most cases.

One-dimensional geometries simplify the problem by reducing the knowledge required to the projection of the radiant energy flux vector along one axis only. If the operating conditions validate this assumption, Eq. (57) may be simplified and easily solved in different systems of coordinates. This requires accurate knowledge of the light transfer parameters (absorption and scattering coefficients, phase function) and the boundary conditions (incident light flux) for the considered medium.

3.4.1 Light Transfer Parameters and Incident Fluxes Determinations

3.4.1.1 Absorption and Scattering Coefficients Ea, Es

The absorption and scattering coefficients for a given photosynthetic micro-organism can be determined spectrophotometrically for each wavelength or by direct radiation measurements for mean coefficients on the visible spectrum.

The general procedure requires the measurement of both transmission T(λ) and reflexion R(λ) of the sample [58, 59]. The two unknowns Ea(λ) and Es(λ) can then be calculated from the two non-linear equations giving T(λ) and R(λ) with the chosen level of complexity. This means that the values of coefficients Ea and Es depend on the degree of sophistication chosen for describing the radiative transfer problem (one dimension, three dimensions, phase function and so on). Thus the consistency of the model relies on using together, correctly, coefficients and assumptions. With regard to the great difficulty in measuring spectrophotometrically the reflexion of a sample, different simplications of the preceding approach may be made.

If the characteristic size of the micro-organism (e.g. its diameter) is very large compared to the wavelength, one can apply the hypothesis of Shibata [58], i.e. the ratio of the transmission to the reflexion is independent of the wavelength. This method was used by different authors [39, 46] for a great number of micro-organisms. Practically, this hypothesis would be reasonable only for characteristic sizes greater than 100 μm, i.e. it does not apply for most of the photosynthetic micro-organisms.

In a more general approach, a new relation between transmission and reflexion is needed. For example, we have proposed a relation based on the Lorentz-Mie theory for light scattering by particles [60, 61].

The radiative transfer problem can then be solved using the resulting wavelength-dependent coefficients or using mean coefficients, averaged on the visible spectrum:

$$\mathrm{Ea, Es} = \frac{1}{\Delta\lambda} \int_{\Delta\lambda} \mathrm{Ea}(\lambda), \mathrm{Es}(\lambda)\, d\lambda \qquad (60)$$

3.4.1.2 Phase Function for Scattering

The phase function for scattering or scattering diagrams is a very important parameter for modeling a radiative transfer problem because it describes how an incident ray is scattered by the micro-organism in the θ or ϕ direction over the 4π steradians of a solid angle (for one-dimensional approximation with hemispherical symmetry, only the θ direction is considered).

The more simple approach is to postulate that the scattering is independent of the solid angle, i.e., that the field of radiation is isotropic. In this case, the phase function p(θ, ϕ) is equal to unity. This assumption is currently made [37–39, 51, 57], especially in the simple two-flux model discussed hereafter.

In the general case, the phase function may be obtained experimentally and empirically expanded as a series in a Legendre polynomial of the form [36]

$$p(\theta, \theta') = p(\cos \Theta) = \sum_{l=0}^{\infty} \omega_l P_l(\cos \Theta) \tag{61}$$

For particular applications to micro-organisms, the phase function has been shown to be sharply peaked in the forward direction [62–64]. So, it is possible to use an approximation given by geometrical optics for large specularly reflecting surfaces [37], or use empirical approximate phase functions such as the forward-scattering phase function or the Henyey-Greenstein phase function [37].

In all case, the more rigorous approach is to compute the "exact" series given by the Lorentz-Mie theory if all the electromagnetic constants of the micro-organism are known [63].

The performances and the respective accuracies of these different approaches have been investigated by different authors [54, 59, 65–69] with the conclusion that, for micro-organisms, the best engineering approach is to use any approximate phase function giving a marked forward scattering diagram. Nevertheless, the use of isotropic phase function is possible as a first approximation with a sufficient accuracy [39, 59].

3.4.1.3 Incident Radiant Energy Fluxes

One of the most important problem in describing the radiant energy transfer in a photobioreactor is to express correctly the boundary conditions, that is the incident radiant energy flux onto the reactor. Different cases exist, still depending on model complexity.

If the incident flux is really homogeneous, it may be easily measured, but if the homogeneity is only a first approximation, experimental difficulties arise in the correct assessment of the mean incident flux. Different methods may then be used as for example chemical actinometry [39, 59, 70], or integral measures with a spherical sensor. In this last case, we have proposed the following relation to determine the mean radiant incident flux on a cylindrical photoreactor radially illuminated [51]:

$$\frac{E}{F_R} = \frac{R\left[\pi I_0\left(\delta \frac{r_b}{R}\right)\right]}{r_b[\mathrm{ch}(\delta R) + \alpha\,\mathrm{sh}(\delta R)]} \tag{62}$$

in which I_0 is a zero-order modified Bessel function of first species, E the total radiant energy caught by the spherical sensor (radius r_b) at the centre of the reactor (radius R), $\alpha = \sqrt{Ea/(Ea + Es)}$ and $\delta = \alpha(Ea + Es)C_x$.

The homogeneous incident radiant flux as a boundary condition is only feasible in the case of simple two-flux model for radiative transfer. For higher order solutions or discrete ordinate methods it is necessary to describe more accurately the incident radiation. It is then possible from tedious experimental measurements or from modeling the emission of light sources [68] (as an example, natural photobioreactors illuminated by sun have well-known boundary conditions for incident fluxes [54, 65]).

3.4.2 Analytical Approximized One-Dimensional Solutions for the Radiative Transfer Equation: Schuster's Hypotheses

Analytical solutions for the equation of radiative transfer are possible from only two main assumptions:

- the incident flux is considered as homogeneous;
- the phase function is isotropic.

With an isotropic radiation field, the intensities do not depend on the azimuthal angle, so the solid angle $d\omega = 2\pi \sin\theta \, d\theta$ and Eq. (57) along a single axis reduces to [51]

$$-\nabla . \mathbf{F} = (\mathrm{Ea} + \mathrm{Es}) C_x 2\pi \int_0^{\pi} I \sin\theta \, d\theta - \tfrac{1}{2} \mathrm{Es} C_x 2\pi \int_0^{\pi}\!\!\int_0^{\pi} I \sin\theta' \sin\theta \, d\theta' \, d\theta \quad (63)$$

with the definition of the radiative flux as

$$F = 2\pi \int_0^{\pi} I \cos\theta \sin\theta \, d\theta \quad (64)$$

A solution of Eq. (63) that neglects the angular distribution of the incident intensity at the external walls and the angular distribution of the scattered intensity inside the medium has been proposed by Schuster for rectangular coordinates [57] and used by Cornet et al. [46]. As the main approximation is the one-dimensional approximation, neglecting these distributions does not significantly impair the resulting radiant energy profile. Also, it is often difficult experimentally to determine the n boundary conditions to be set if this assumption is not made and n-flux or discrete ordinate methods are used. Scattered intensity is assumed to be emitted by suspended particles in the main direction of the radiation either positively or negatively. The field of radiation is then split into two opposite and different fluxes F^+ and F^- on each hemisphere in θ: $0 - \pi/2$ and $\pi/2 - \pi$, assumed to be parallel to the single axis; this is the two-flux approximation, which gives the following system in rectangular geometries:

$$\begin{cases} \dfrac{dF^+}{dz} = -\mathrm{Ea} C_x F^+ + \tfrac{1}{2} \mathrm{Es} C_x (F^- - F^+) \\ \dfrac{dF^-}{dz} = \mathrm{Ea} C_x F^- + \tfrac{1}{2} \mathrm{Es} C_x (F^- - F^+) \end{cases} \quad (65)$$

The same system stands in cylindrical coordinates, replacing F by rF and z by r.

These equations clearly indicate that the total available local energy is given by $4\pi J = F^+ - F^-$ and the net radiative flux in the positive direction is then (the norm of vector $\mathbf{F}$) $|\mathbf{F}| = F = F^+ - F^-$ [46].

Table 3. Analytical approximized one-dimensional solutions for the radiative transfer equation in different geometries

Geometry	Rectangular reactor illuminated on one side	Rectangular reactor illuminated on both sides	Cylindrical reactor radially illuminated
Dimensionless variables	$Z = z/L$ (66) $\alpha = [Ea/(Ea + Es)]^{1/2}$ (67) $\delta = (Ea + Es)\alpha C_X L$ (68)	$Z = z/L$ (74) $\alpha = [Ea/(Ea + Es)]^{1/2}$ (75) $\delta = (Ea + Es)\alpha C_X L$ (76)	$Z = r/R$ (82) $\alpha = [Ea/(Ea + Es)]^{1/2}$ (83) $\delta = (Ea + Es)\alpha C_X R$ (84)
Solutions for available radiant energy profiles	$\frac{4\pi J_z}{F_0} = 2\frac{(1+\alpha)e^{-\delta(Z-1)} - (1-\alpha)e^{\delta(Z-1)}}{(1+\alpha)^2 e^{\delta} - (1-\alpha)^2 e^{-\delta}}$ (69)	$\frac{4\pi J_z}{F_0} = 2[e^{\delta Z}[(\alpha+1) + e^{-\delta}(\alpha-1)] + e^{-\delta Z}[(\alpha-1) + e^{\delta}(\alpha+1)]]/[(1+\alpha)^2 e^{\delta} - (1-\alpha)^2 e^{-\delta}]$ (77)	$\frac{4\pi J_r}{F_R} = \frac{1}{Z}\frac{2\cosh(\delta Z)}{\cosh(\delta) + \alpha\sinh(\delta)}$ (85)
Volumetric rate of radiant energy absorbed	$\frac{\langle \mathscr{A} \rangle}{(F_0/L)} = 1 - \left[\frac{4\alpha + (1-\alpha^2)(e^{\delta} - e^{-\delta})}{(1+\alpha)^2 e^{\delta} - (1-\alpha)^2 e^{-\delta}}\right]$ (70) $\lim_{C_X, \delta \to \infty}\left[\frac{\langle \mathscr{A} \rangle}{(F_0/L)}\right] = \frac{2\alpha}{1+\alpha}$ (71)	$\frac{\langle \mathscr{A} \rangle}{(2F_0/L)} = 2\alpha\left[\frac{e^{\delta}(\alpha+1) - e^{-\delta}(\alpha-1) - 2}{(1+\alpha)^2 e^{\delta} - (1-\alpha)^2 e^{\delta}}\right]$ (78) $\lim_{C_X, \delta \to \infty}\left[\frac{\langle \mathscr{A} \rangle}{(2F_0/L)}\right] = \frac{2\alpha}{1+\alpha}$ (79)	$\frac{\langle \mathscr{A} \rangle}{(2F_R/R)} = \frac{2\alpha\sinh(\delta)}{\cosh(\delta) + \alpha\sinh(\delta)}$ (86) $\lim_{C_X, \delta \to \infty}\left[\frac{\langle \mathscr{A} \rangle}{(2F_R/R)}\right] = \frac{2\alpha}{1+\alpha}$ (87)
Working illuminated volume	$\gamma = Z_2 = \frac{L_2}{L}$ (72) $2\frac{(1+\alpha)e^{-\delta(Z_2-1)} - (1-\alpha)e^{\delta(Z_2-1)}}{(1+\alpha)^2 e^{\delta} - (1-\alpha)^2 e^{-\delta}} \frac{-4\pi J_{lim}}{F_0} = 0$ (73)	$\gamma = \frac{L_2'}{L} + \frac{L - L_2}{L} = Z_2' + (1 - Z_2)$ (80) $2[e^{\delta Z_2}[(\alpha+1) + e^{-\delta}(\alpha-1)] + e^{-\delta Z_2}[(\alpha-1) + e^{\delta}(\alpha+1)]]/[(1+\alpha)^2 e^{\delta} - (1-\alpha)^2 e^{-\delta}] - \frac{4\pi J_{lim}}{F_0} = 0$ (81)	$\gamma = \frac{R_2'^2}{R^2} + \frac{R^2 - R_2^2}{R^2} = Z_2'^2 + (1 - Z_2^2)$ (88) $\frac{1}{Z_2}\frac{2\cosh(\delta Z_2)}{\cosh(\delta) + \alpha\sinh(\delta)} - \frac{4\pi J_{lim}}{F_R} = 0$ (89)

This system of ordinary differential equations was integrated in different systems of coordinates with boundary conditions expressing that a rectangular reactor was illuminated on one or two sides with a homogeneous incident radiant energy flux F_0 or that a cylindrical reactor was radially illuminated with homogeneous flux F_R [51].

The obtained solutions giving the available radiant energy profiles are summarized in Table 3 by Eqs. (66–69, 74–77, 82–85) and plotted in Fig. 4. The mean volumetric rate of radiant energy absorbed in the medium can be expressed from Eqs. (70, 78, 86) in Table 3. This ratio reaches an asymptotic value as the biomass concentration increases into the medium given by Eqs. (71, 79, 87) confirming that light energy transfer may be a limiting step when biomass concentrates in the medium [46].

The illuminated fraction of the reactor γ can be deduced from Eqs. (72, 80, 88) for different geometries by taking the roots of Eqs. (73, 87, 89) respectively [51].

Equations (47, 48) coupled to Eqs. (66–73) or (74–84) for rectangular geometries and to Eqs. (82–89) for cylindrical geometries may be used to calculate volumetric biomass growth rates in different operating conditions, when the one-dimensional approximation is made. It must be noticed that situations in Table 3 are not exhaustive and that the system of Eq. (65) can be integrated with a wide range of boundary conditions.

3.4.3 The P–N and Discrete Ordinate Methods for Solving the One-dimensional Radiative Transfer Equation

If the incident radiant light energy distribution and the scattering characteristics of the medium are well known, the one-dimensional radiative transfer equation can be solved with high accuracy without restrictive hypotheses as above. A thorough treatment of numerical methods is available in the literature [37, 53–56, 65, 71]. In this case, the one-dimensional equation of radiative transfer is solved in terms of specific intensity I and written in the form [36]

$$-\mu \frac{1}{(Ea + Es)C_x} \frac{dI_z}{dz} = I_z - \frac{1}{2} \bar{\omega}_0 \int_{-1}^{+1} p^0(\theta, \theta') I_z \, d\mu' \tag{90}$$

with $\mu = \cos\theta$, θ being the angle with the normal vector, $\bar{\omega}_0$ the albedo of single scattering equal to $Es/(Ea + Es)$ and the phase function

$$p^0(\theta, \theta') = \frac{1}{2\pi} \int_0^{2\pi} p(\theta, \phi, \theta', \phi') \, d\phi'. \tag{91}$$

Fig. 4a–c. Profiles in normalized radiant energy available $4\pi J/F_{0,R}$, vs the dimensionless thickness of the medium; $\alpha = 0.655$ (corresponding to *Spirulina*), δ as parameter: **a** rectangular reactor illuminated on one side; **b** rectangular reactor illuminated on both sides; **c** cylindrical reactor radially illuminated. (Permission from Elsevier)

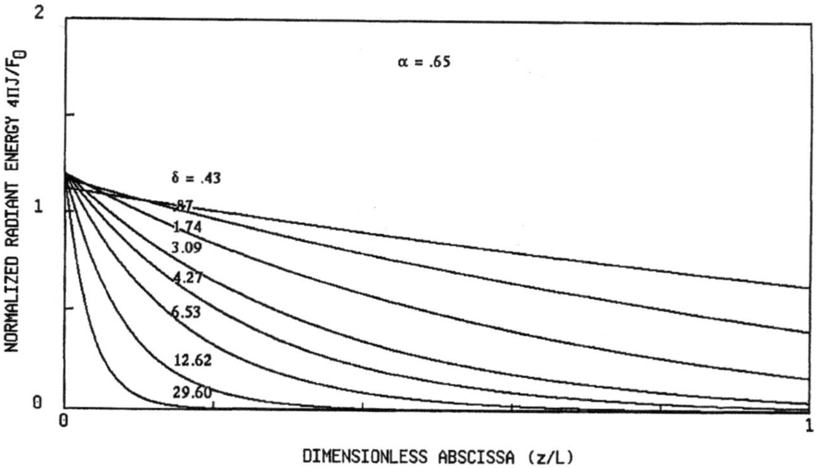
α = .65
δ = .43
.87
1.74
3.09
4.27
6.53
12.62
29.60
NORMALIZED RADIANT ENERGY 4ΠJ/F0
DIMENSIONLESS ABSCISSA (z/L)

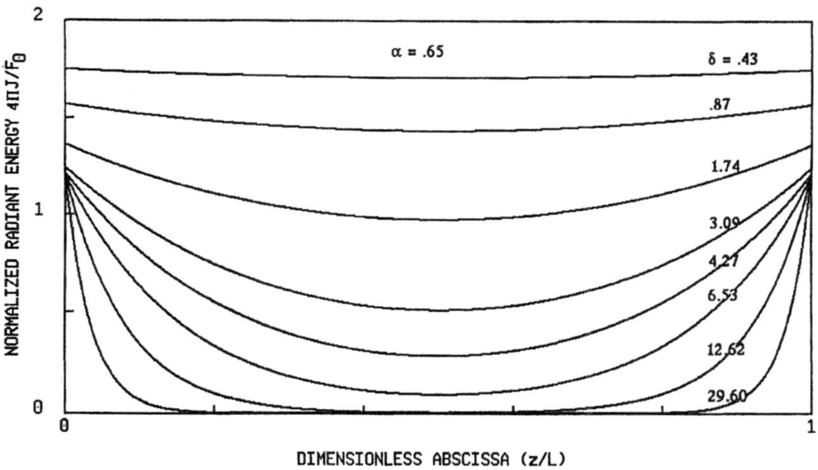
α = .65
δ = .43
.87
1.74
3.09
4.27
6.53
12.62
29.60
NORMALIZED RADIANT ENERGY 4ΠJ/F0
DIMENSIONLESS ABSCISSA (z/L)

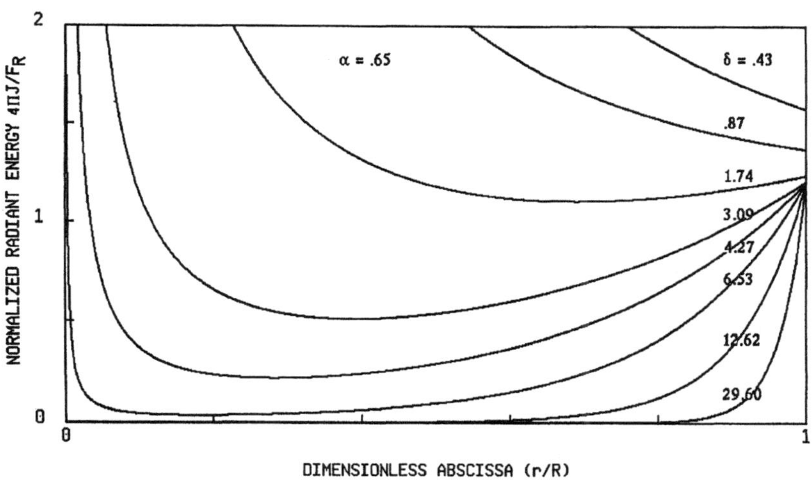
α = .65
δ = .43
.87
1.74
3.09
4.27
6.53
12.62
29.60
NORMALIZED RADIANT ENERGY 4ΠJ/FR
DIMENSIONLESS ABSCISSA (r/R)

In this phase function, ϕ is the azimuthal angle.

If the radiation field is considered as isotropic, the phase function is equal to 1 and Eq. (90) reduces to

$$-\mu \frac{1}{(Ea + Es)C_x} \frac{dI_z}{dz} = I_z - \frac{1}{2} \bar{\omega}_0 \int_{-1}^{+1} I_z \, d\mu' \tag{92}$$

3.4.3.1 P–N Methods

In such an approximation, Eq. (92) is reduced with a finite set of moment equations. The intensity I is expanded in an orthogonal series of spherical harmonics truncated after a finite number of terms N [37].

3.4.3.2 The Discrete Ordinate Method

This is an extension of the two-flux method of Schuster discussed above. The general equation at Eq. (92) is represented by a discrete set of equations for the average intensity over a finite number of ordinate directions. Integrals over solid angles are replaced by sums over the ordinate directions. For the direction μ_i Eq. (92) becomes [36, 37]

$$-\mu_i \frac{1}{(Ea + Es)C_x} \frac{dI_z(\mu_i)}{dz} = I_z(\mu_i) - \frac{1}{2} \bar{\omega}_0 \sum_{j=1}^{m} a_j I_z(\mu_j) \tag{93}$$

The a_j's coefficients are weighting coefficients, functions of the divisions of the interval [− 1; + 1]. Gaussian quadrature is usually used for their calculation. The discrete ordinate method is currently considered as a reference method for testing accuracy of results obtained with other methods [37, 65].

In these numerical methods, the local available radiant energy $4\pi J_z$ is obtained by the following double integral:

$$4\pi J_z = \int_0^{2\pi} \int_{-1}^{+1} I_z \, d\mu \, d\phi \tag{94}$$

The working illuminated length L_2 cannot be obtained from an analytical expression, but must be numerically calculated from the value of the compensation point ($1 \, W.m^{-2}$).

3.4.4 Radiative Transfer Theory for Three-Dimensional Applications: General Case

The general equation of radiative transfer in terms of specific intensity may be used on a finite medium such as a photobioreactor. In the general case, a one-dimensional approximation is inappropriate and the equation must be solved in its three-dimensional form. This affords a high accuracy for the calculation of the local available energy and at this level, the dependence of the wavelength on this calculation may be taken into account. The general form of the radiative transfer equation for direction **u** and wavelength λ is then written as [36, 38, 39, 50, 68]

$$\mathbf{u}\cdot\nabla I_\lambda = -[Ea(\lambda) + Es(\lambda)]C_x I_\lambda + \frac{Es(\lambda)}{4\pi} C_x \int_{\Omega'} p(\mathbf{u}, \mathbf{u}')I_\lambda \, d\omega' \tag{95}$$

Ea(λ) and Es(λ) are respectively the absorption and scattering mass coefficients for wavelength λ; C_x is the biomass concentration in the reactor. The first term on the right hand side of Eq. (95) denotes loss of radiant energy by absorption and scattering along direction **u**, and the second term denotes gain of radiant energy scattered from all directions of space with a phase function p(**u**, **u**′).

Equation (95) may be solved by numerical gridding techniques. Generally, the Monte Carlo method is used [37, 39, 50], but we have developed a new method with finite elements. It uses the explicit algorithm of the characteristic curve [72], the convergence of which is particularly convenient for solving Eq. (95).

From this numerical gridding of the reactor in specific intensity I_λ, the available radiant energy at each point of the reactor 4πJ can be obtained by the integrations given by Eqs. (40) and (41). The mean volumetric rate of radiant energy absorbed is then obtained by

$$\langle \mathscr{A} \rangle = \frac{1}{V}\int_V \int_0^\infty \int_\Omega Ea(\lambda)C_x I_\lambda \, d\omega \, d\lambda \, dV \tag{96}$$

This general approach for modeling radiative transfer must be restricted to simulation and design of photobioreactors as sophisticated numerical tools are required. However, it is the only way to obtain the local available energy in the reactor and the mean volumetric rate of energy absorbed with high accuracy and close agreement with experimental results. As discussed above in the introduction, this first quantity is of prime importance for studying the coupling between light transfer and growth kinetics, and the second allows energy balances to be performed.

3.5 Growth Kinetics of Spirulina platensis in Photobioreactors Under Light-Limiting Conditions

3.5.1 Physical Characteristics of Culture Media with Spirulina platensis

The absorption and scattering mass coefficients for each wavelength over the visible spectrum characterizing the *S. platensis* culture have been determined spectrophotometrically with the hypothesis of isotropic phase function [38] and

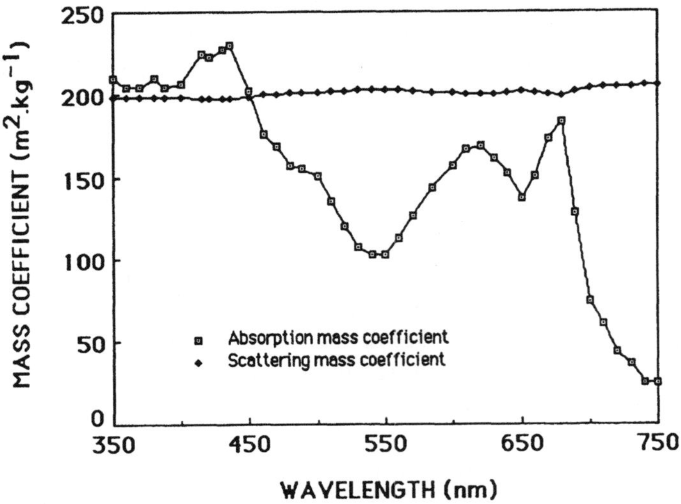

Fig. 5. Absorption and scattering mass coefficients for each wavelength in the range of visible spectrum for *S. platensis* culture medium. (Permission from AIChE)

are plotted in Fig. 5. The scattering mass coefficient is independent of wavelength in the range 350–750 nm because of the large size of filamentous microorganism compared to visible wavelengths. The main photosynthetic pigments, chlorophyll *a* and phycocyanin, are responsible for the values of the absorption coefficient, which appear similar to the absorption spectrum for the micro-organism concerned. From these results, the values obtained for the mean absorption and scattering coefficients are as follows [38]:

$$Ea = 150\,m^2.kg^{-1}$$

$$Es = 200\,m^2.kg^{-1}$$

These values show that the scattering mass coefficient can never be neglected compared with the absorption mass coefficient. This justifies choosing the radiative transfer theory to describe the photonic phase rather than Lambert's theory. However, the mean Lambert-Beer extinction coefficient in Eq. (53) has been determined and is equal to [7]:

$$m = 240\,m^2.kg^{-1}$$

3.5.2 Comparison Between Lambert, One- and Three-Dimensional Models for Obtaining the Mean Volumetric Rate of Radiant Energy Absorbed

The knowledge of absorption and scattering mass coefficients and extinction coefficient of Lambert-Beer allows the analytical calculation of the normalized ratio $\langle \mathscr{A} \rangle / \langle \mathscr{A}_0 \rangle$, i.e. the volumetric rate of radiant energy absorbed divided by the potential maximum volumetric rate of radiant incident energy. This has been done on rectangular photoreactors to compare results obtained from the exact three-dimensional solution and from various one-dimensional approximations.

Experimental determinations of the ratio $\langle \mathscr{A} \rangle / \langle \mathscr{A}_0 \rangle$ for two rectangular reactors of different optical thicknesses (respectively 0.05 and 0.08 m) and for a homogeneous incident flux of 40 $W.m^{-2}$ (fluorescent lamps) vs the biomass concentration have been performed using the opalescent plate method [39].

These experimental results are plotted in Fig. 6, the solid and dotted lines representing the calculated values obtained by the Lambert-Beer approximation (Eq. 56) for the two optical thicknesses. Clearly, the Lambert-Beer law, disregarding scattering phenomena, implies that the ratio $\langle \mathscr{A} \rangle / \langle \mathscr{A}_0 \rangle$ tends to 1 as the biomass concentration increases in the reactor and is inefficient for modeling radiative transfer in absorbing and scattering media.

In Fig. 7 the solid and dotted lines represent the results of the analytical solution provided by the one-dimensional approximation of Schuster (Eq. 70) at two different optical thicknesses. The asymptotic value of the ratio $\langle \mathscr{A} \rangle / \langle \mathscr{A}_0 \rangle$ calculated by Eq. (71) for high biomass concentration is added. For *S. platensis*, the α parameter is equal to 0.655 and Eq. (71) gives a maximum value for this ratio of 0.79. Equation (59) thus provides a fair approximation for the ratio $\langle \mathscr{A} \rangle / \langle \mathscr{A}_0 \rangle$, when the biomass concentration is high. However, an analytical solution of Eq. (70) is not sufficiently accurate to predict the volumetric rate of radiant energy absorbed for biomass concentration below 1 $kg.m^{-3}$. For high biomass concentration, only a fraction of the reactor is illuminated because

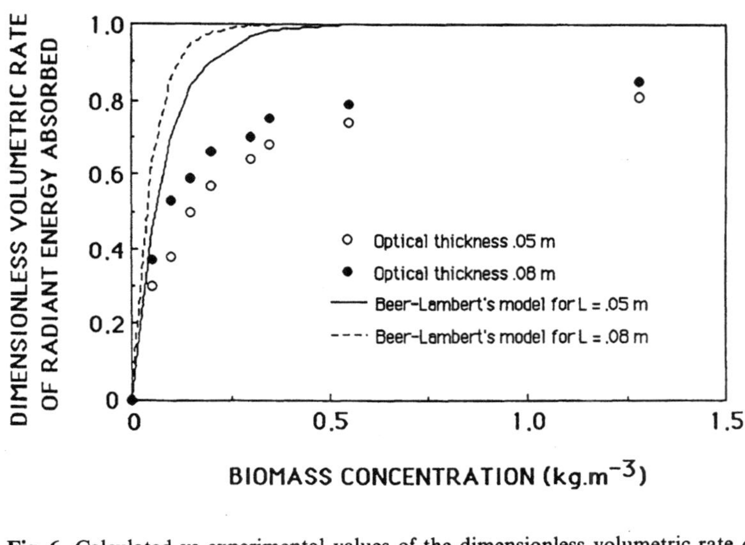

Fig. 6. Calculated vs experimental values of the dimensionless volumetric rate of radiant energy absorbed vs *Spirulina* biomass concentration obtained in two rectangular photobioreactors of different optical thicknesses. *Solid and dotted lines* stand for the analytical solution given by the Lambert-Beer approximation (Eq. 56). (Permission from AIChE)

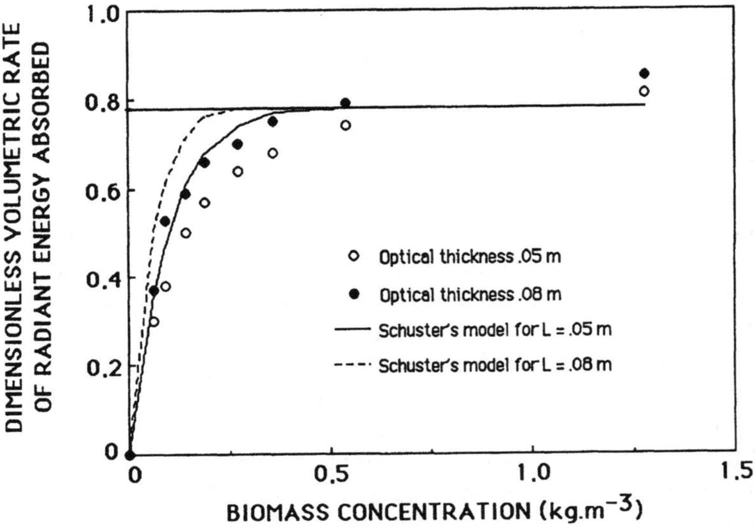

Fig. 7. Calculated vs experimental values of the dimensionless volumetric rate of radiant energy absorbed vs *Spirulina* biomass concentration obtained in two rectangular photobioreactors of different optical thicknesses. *Solid and dotted lines* stand for the analytical approximized one-dimensional solution for the volumetric rate of radiant energy absorbed (Eq. 70). Asymptotic value for the ratio $\langle \mathscr{A} \rangle / \langle \mathscr{A}_0 \rangle$ given by Eq. (71) is also presented. (Permission from AIChE)

of the absorption of radiant light energy by high pigment density, which justifies the concept of the working illuminated volume defined above. As this volume decreases, the loss of scattered radiation from the reactor in the two dimensions of space neglected also decreases and approaches zero. In this case, for high biomass concentration, the one-dimensional approximation is fully justified as seen in Fig. 7. Conversely, for low biomass concentration, a large proportion of the total incident energy is scattered from the total volume of the reactor and the one-dimensional approximation is inappropriate.

In Fig. 8 the same experimental results are shown, but here the solid and dotted lines have been obtained by numerical calculation from a gridding algorithm of finite elements (Eq. 95) [38]. Experimental values and numerical calculations are in fairly close agreement, irrespective of the biomass concentration or the optical thickness in the reactor.

Figure 9 shows the results of a numerical calculation of the ratio $\langle \mathscr{A} \rangle / \langle \mathscr{A}_0 \rangle$ vs the biomass concentration for a cylindrical reactor. The incident homogeneous flux was 40 W.m^{-2} and was supplied by halogen lamps to half the total annular area. This ratio reaches an asymptotic value at about 0.6 for the same biomass concentration as in the rectangular reactors. Also, there is a difference of up to 20% between the calculation using global coefficients and that using coefficients for each wavelength. This means that it is more accurate to use the absorption and scattering coefficients for each wavelength when the emission

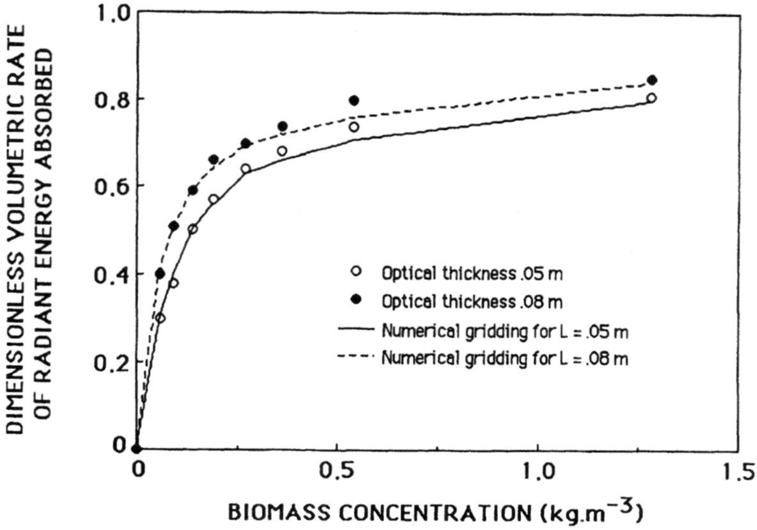

Fig. 8. Calculated vs experimental values of the dimensionless volumetric rate of radiant energy absorbed vs *Spirulina* biomass concentration obtained in two rectangular photobioreactors of different optical thicknesses. *Solid and dotted lines* stand for the numerical three-dimensional gridding for the determination of the volumetric rate of radiant energy absorbed with a finite element method (resolution of the general form of the equation of radiative transfer, Eq. 95). (Permission from AIChE)

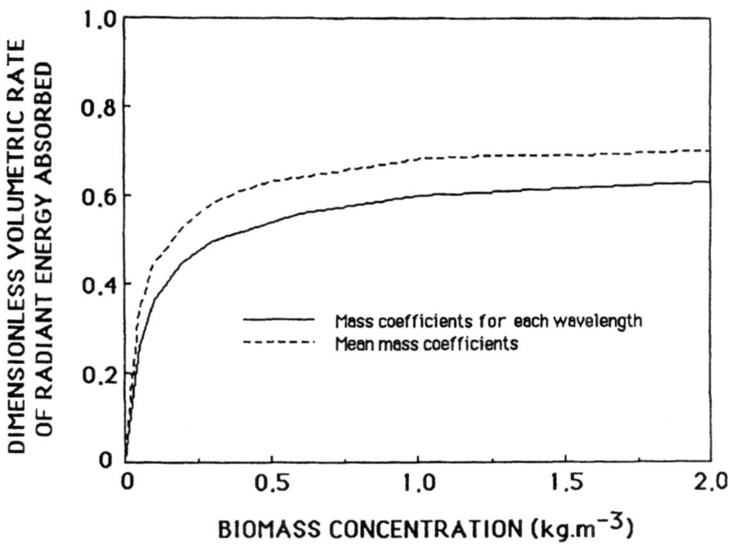

Fig. 9. Numerical three-dimensional calculation of the dimensionless volumetric rate of radiant energy absorbed vs *Spirulina* biomass concentration for a cylindrical photobioreactor. Comparison between the utilization of absorption and scattering mass coefficients for each wavelength (*solid line*), and mean global mass coefficients (*dotted lines*). (Permission from AIChE)

Table 4. Experimental conditions and results for batch cultures of *Spirulina platensis* in rectangular photobioreactors illuminated on one side. The reactor volumes were 1 l and 4 l for optical thicknesses of 0.05 and 0.08 m respectively (Permission from Wiley)

Incident radiant energy flux F_0 (W/m^2)	Optical thickness of the medium L.100 (m)	Mass volumetric rate in biomass $\langle r_X \rangle . 10^3$ (kg.m^{-3}.h^{-1})
4	5	1.5
7	5	1.6
8	8	1.7
11	8	1.8
11	8	1.8
12	8	2.1
18	8	2.5
20	8	3.2

spectrum of the light supplying system displays changes in energy distribution with wavelength. This is particularly important for halogen lamps used for illuminating reactors which supply most of their energy in the red range.

3.5.3 *Batch Cultures in Rectangular Photobioreactors*

Eight batch experiments were performed in rectangular photobioreactors illuminated on one side, 0.05 and 0.08 m thick (volume of 1 l and 4 l respectively) with *Spirulina platensis* for incident light fluxes F_0 ranging between 4 and 20 W.m^{-2} in the visible spectrum [73]. The pH was equal to 9.5 and temperature was 36 °C. For each experiment, a linear growth phase was attained, the rate of which is given in Table 4, indicating a strong influence of F_0. The one-dimensional two-flux model based on the assumptions of Schuster [57] (i.e. Eqs. 47, 48, 66–73 in Table 3) has been integrated for these eight batch cultures and compared with the experimental results obtained given the general mass balance equation from Eq. (2) [73]:

$$\frac{dC_x}{dt} = \langle r_x \rangle \tag{97}$$

The two biological parameters μ_M and K_J were obtained by an identification procedure giving

$$\mu_M = 0.073\,h^{-1}$$

$$K_J = 20\,W.m^{-2}$$

These values may be considered reliable within a 5% confidence interval in relation to the light transfer and physiological hypotheses presented. It should be noted that the simulations obtained in rectangular photobioreactors closely agree with experimental results [73].

Table 5. Comparison between the *Spirulina* mass volumetric growth rate in biomass calculated by the proposed model and experimentally determined on continuous cultures in a cylindrical photobioreactor for different biomass concentrations and incident radiant energy fluxes (Permission from Elsevier)

Biomass Concentration C_X (kg.m^{-3})	Incident radiant energy flux F_R (W.m^{-2})	Volumetric growth rate in biomass (from model) $\langle r_x \rangle$ (kg.m^{-3}.h^{-1})	Volumetric growth rate in biomass (experimental) $\langle r_x \rangle$ (kg.m^{-3}.h^{-1})	Standard deviation (%)
0.77	193	2.1 10^{-2}	1.9 10^{-2}	+ 10
0.70	195	1.8 10^{-2}	1.9 10^{-2}	− 5
0.47	23	7.7 10^{-3}	8.4 10^{-3}	− 9
0.47	68	1.3 10^{-2}	1.3 10^{-2}	0
0.46	45	1.1 10^{-2}	1.1 10^{-2}	0
0.46	21	7.2 10^{-3}	6.5 10^{-3}	+ 10
0.24	25	6.0 10^{-3}	6.6 10^{-3}	− 10
0.15	23	4.2 10^{-3}	5.0 10^{-3}	− 16
0.09	31	2.9 10^{-3}	2.6 10^{-3}	+ 10

3.5.4 Continuous Cultures in Cylindrical Photobioreactors

Nine experiments were performed in a radially illuminated cylindrical photobioreactor with a radius of 0.05 m (volume of 7 l) and incident light fluxes F_R ranging between 20 and 200 W.m^{-2} in the visible spectrum [51]. The pH was equal to 9.5 and the temperature was 36 °C.

The previous model for the growth of *Spirulina platensis* in a radially illuminated cylindrical reactor (i.e. Eqs. 47, 48, 82–89 in Table 3) has been integrated for these nine continuous cultures and compared with the experimental results obtained given the general mass balance equation in steady state:

$$\langle r_x \rangle = DC_x \tag{98}$$

where D is the culture dilution rate.

The model calculation of the mean volumetric biomass growth rate has been done using the μ_M and K_J values identified on rectangular photoreactors by Cornet et al. [46, 73] and recalled above. The results are given in Table 5. The model calculations and experimental data agree closely for biomass concentrations ranging between 0.09 and 0.77 kgm^{-3}; the mean deviation for the nine experiments is less than 10%.

Special attention must be paid to the predictivity of the proposed models since the cylindrical photoreactor simulations were obtained with the same parameters identified for rectangular reactors. Thus, parameters determined with incident fluxes ranging between 4 and 20 W.m^{-2} remain valid between 20 and 200 W.m^{-2}, representing a domain of variation of a factor of 50.

3.6 Dark Cycle: Respiration

Equation (38) enables one to calculate the decrease rate in *Spirulina* biomass by respiration in the dark at 36 °C, which is equal to $9 \cdot 10^{-4}$ kg.kg biomass^{-1}.h^{-1}. This rate has been confirmed by experiments in bioreactors. It has been shown that the micro-organism can be maintained for 24 h in darkness without the growth rate in the light being subsequently impaired [7]. Equation (38) can then be used if *Spirulina* is cultivated under dark cycle conditions.

4 Growth Kinetics Under Mineral Limitations

4.1 Physiological Effects of Nitrogen, Sulphur and Phosphorus Mineral Limitations on Spirulina Growth

4.1.1 Nitrate Limitation

Initial studies on the cyanobacterium *Anacystis nidulans* [74] showed that during nitrate depletion, growth stops, apoproteins of phycocyanins are degraded without significant changes in chlorophyll [75, 76] and carotenoid, and intracellular glycogen concomitantly accumulates [77]. Re-addition of nitrates to deficient cultures reverses these trends. These observations were confirmed in *Synechocystis* [78] and *Agmenellum quadruplicatum* [79]. Allen et al. [80] recently suggested that some inactivation of photosystem II could result in a decrease in photosynthesis efficiency.

The cyanobacterium *Spirulina* seems to follow this general response to nitrogen deprivation [81] though carbohydrate accumulation has not been extensively studied. Little information is available on intracellular glycogen [6] and exocellular polysaccharides [9], which are both sulfated.

Figure 10 shows a batch culture performed in a rectangular photoreactor with nitrate limitation. Growth and accumulation of major cell components in the presence of nitrates are initially exponential and rapidly become linear when light limitation is introduced by increasing shadowing. As soon as nitrates are exhausted, total amounts of proteins and chlorophyll *a* in the culture stabilize to steady-state levels. As expected, phycocyanins begin to be degraded and are used as a nitrogen reserve for the continued synthesis of other proteins, the amount of total proteins remaining constant. Intense synthesis of intracellular sulfated glycogen [6, 77–80] results in a marked accumulation of residual biomass. This intracellular glycogen accumulation is accompanied by an abundant excretion of sulfated exopolysaccharide.

Nitrate replenishment induces a rapid restoration of high levels of total proteins, which originates in a very strong and rapid increase of phycocyanins,

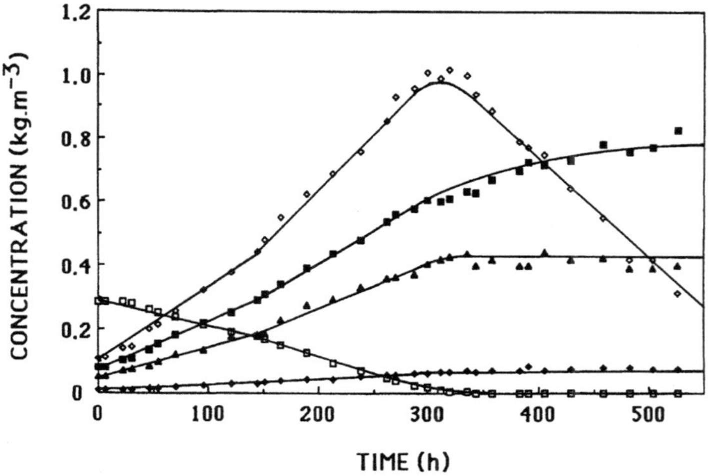

Fig. 10. Batch culture of *S. platensis* in a rectangular photobioreactor under NO_3^- depletion. Comparison between experimental data and the simulation obtained by the mathematical model presented in Sect. 4: (■) biomass concentration ($kg.m^{-3}$), (□) nitrate concentration ($kg.m^{-3}$), (▲) protein concentration ($kg.m^{-3}$), (◇) phycocyanin concentration ($kg.m^{-3} \times 10$), (◆) chlorophyll *a* concentration ($kg.m^{-3} \times 10$); the incident radiant energy flux F_0 is 8 $W.m^{-2}$ before 144 h and 12 $W.m^{-2}$ after. (Permission from Wiley)

followed, to a lesser extent, by other cell proteins. This does not lead to any appreciable biomass accumulation since the level of total sugars correlatively falls consequent to metabolization of accumulated glycogen for rapid protein synthesis.

4.1.2 Sulfate Limitation

The sulphur limitation was also reported by Cohen-Bazire and Bryant [82], and Wanner et al. [83] to impair phycobiliprotein content and composition in cyanobacteria. In a batch culture in a rectangular reactor with sulfate limitation, the effects of sulphur and nitrogen limitations are very similar (Fig. 11). Phycocyanins are degraded while carbohydrates largely accumulate, leading to abundant residual biomass formation. The main difference lies in the fact that the level of total proteins decreases, as the result of the degradation of phycocyanins, but also of other proteins. Most cell proteins are thus evidently used as sulphur reserves for the synthesis and accumulation of sulfated glycogen and exopolysaccharides. After sulfate re-addition, phycocyanins are synthesized more slowly than other proteins, suggesting that essential proteins are degraded during sulphur limitation and are reconstituted first.

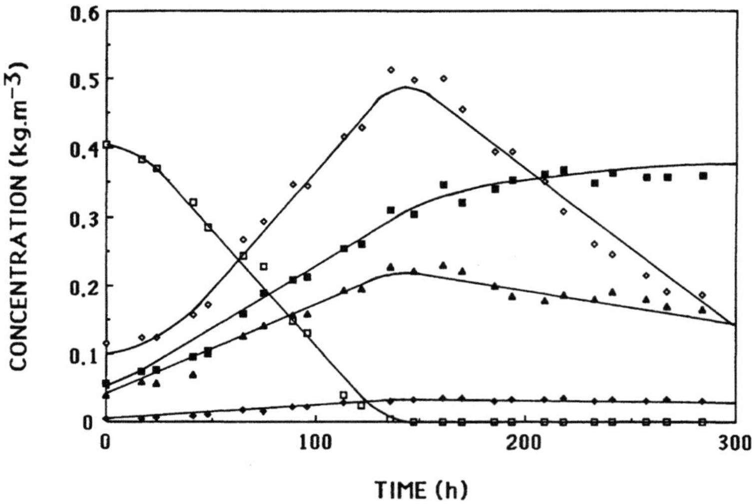

Fig. 11. Batch culture of *S. platensis* in a rectangular photobioreactor under SO_4^{2-} depletion. Comparison between experimental data and the simulation obtained by the mathematical model presented in Sect. 4: (■) biomass concentration ($kg.m^{-3}$), (□) sulfate concentration ($kg.m^{-3} \times 50$), (▲) protein concentration ($kg.m^{-3}$), (◇) phycocyanin concentration ($kg.m^{-3} \times 10$), (◆) chlorophyll *a* concentration ($kg.m^{-3} \times 10$); the incident radiant energy flux F_0 is 18 $W.m^{-2}$. (Permission from Wiley)

4.1.3 Phosphate Limitation

Much less is known about phosphate limitation in cyanobacteria. From continuous cultures under phosphate limitation, Healey and Hendzel [84] reported that an intracellular accumulation of polyosides appears with a concomitant decrease in phosphate, nitrogen, proteins, nucleic acids and chlorophyll contents.

Figure 12 shows the results obtained from a batch culture of *S. platensis* in a rectangular photobioreactor with phosphate limitation. Clearly, when the phosphate is exhausted, growth stops for biomass and main cell components, without their relative proportions changing. However, if the experiment is continued after this stationary phase, all the biomass components are subsequently re-synthesized at the same rate, probably from polyphosphate intracellular reserves, until these are totally exhausted. At this stage, one can observe a minor increase in the polyoside synthesis compared with protein synthesis. These experimental results on *Spirulina* conflict with the previous literature on other cyanobacteria, since the pigments, proteins or polyosides contents seem unaffected by phosphate starvation [84].

4.2 General Laws for Limiting Mineral Substrates

The main problem in formulating general kinetic laws in limiting conditions is to find simple macroscopic laws relating growth rate to extracellular concentration

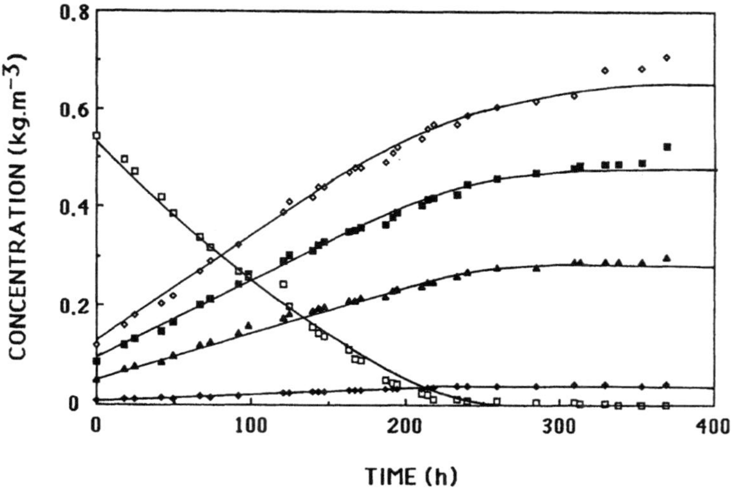

Fig. 12. Batch culture of *S. platensis* in a rectangular photobioreactor under HPO_4^{2-} depletion. Comparison between experimental data and the simulation obtained by the mathematical model presented in Sect. 4: (■) biomass concentration ($kg.m^{-3}$), (□) phosphate concentration ($kg.m^{-3} \times 50$), (▲) protein concentration ($kg.m^{-3}$), (◇) phycocyanin concentration ($kg.m^{-3} \times 10$), (◆) chlorophyll *a* concentration ($kg.m^{-3} \times 10$); the incident radiant energy flux F_0 is 11 $W.m^{-2}$

in limiting substrates. This requires a complete understanding of the mechanisms involved in the limiting process studied at different levels (metabolism, physiology, etc.). When photosynthetic micro-organisms are concerned, the problem is complicated because two or more limiting factors interact, since these micro-organisms are always cultivated in light-limiting conditions. In this case, it is particularly important to distinguish two different behaviours:

– the limiting substrates impair synthesis rates whatever the rate imposed by the availability of radiant light energy, i.e. the effects of the two (or more) limitations are additive. The limiting mineral substrate then acts on the growth rates throughout the working illuminated volume of the culture;

– the limitation is characterized only by a single limiting step, represented by either radiant light availability or mineral substrate concentration, which is assumed to be homogeneous in the culture medium, i.e. the effects of the two (or more) limitations are independent. The working illuminated volume must then be divided into two parts: a part in which the availability of radiant light energy is high with a constant growth rate limited by mineral concentration, and a part in which the growth rate results in local rates with low availability of radiant light energy.

This second approach has been used by Curless [85] for modeling light and mineral limitation in *Spirulina* cultures. However, it has been shown elsewhere [7, 80] that the *Spirulina* photosynthesis rate was impaired by mineral limitations whatever the value of the available radiant energy (the saturation curves of oxygen evolution and the CO_2 fixation rates are affected as well as the mineral limitation increases and the phycocyanin content decreases in the cells). This confirms the first hypothesis, which can be used for modeling both light and mineral limitations.

If limitation by light only is assumed, Eqs. (47) and (48) can be used to obtain the volumetric growth rate of biomass from knowledge of the working illuminated volume. These equations can be rewritten in the following simplified form (the mass volumetric biomass growth rate is denoted $\langle R_x \rangle$ in reference to the light limitation only):

$$\langle R_x \rangle = \langle \mu \rangle \gamma C_x \tag{99}$$

$$\langle \mu \rangle = \frac{1}{V_2} \int_{V_2} \mu_M \frac{4\pi J}{K_J + 4\pi J} dV \tag{100}$$

If mineral limitations appear, and assuming a Monod law for each limiting substrate i, the following general kinetic law can be applied for light and mineral limitations, according to the first hypothesis discussed above (Aiba law):

$$\langle r_x \rangle = \langle R_x \rangle \prod_{i=1}^{n} \frac{C_i}{K_i + C_i} \tag{101}$$

More complex kinetic equations may be required for modeling complex physiological behaviour as is the case for nitrate and sulfate limitations on *Spirulina* growth. However, Eq. (101) is widely used in more complex models, especially if compartment models are introduced.

4.3 Compartment Model for Spirulina Growth under Mineral Limitation by N, S and P

4.3.1 Compartment Models

Compartment models [1] are particularly useful in modeling mineral limitations since some biomass components, such as pigments, crucially affect the behaviour of the culture. If the concentration of biomass in the culture is C_x, the state of the culture is described by an overall chemical state vector **C**, which includes the concentrations of the compounds present in the biotic and abiotic phases. This state vector can be subdivided into a biotic state vector **X** and an abiotic vector **Y**. The different components of the biotic and abiotic phases that are suitable for use have been defined by Cornet et al. [73]. The biomass X stands for the total biomass (including exopolysaccharide) as long as no mineral limitation occurs, and the total biomass XT includes the additional intracellular glycogen appearing during mineral limitation. For convenience and because it is easier to work with experimentally measured variables, the biotic state vector will be

$$\mathbf{X} = (C_{PC}, C_{Pr}, C_{CH}, C_X, C_G)$$

and the abiotic state vector will be

$$\mathbf{Y} = (C_N, C_S, C_P)$$

The state vector is therefore

$$\mathbf{C} = (C_{PC}, C_{Pr}, C_{CH}, C_X, C_G, C_N, C_S, C_P)$$

where PC, Pr, CH, X, G, N, S, P denote respectively the phycocyanins, proteins, chlorophylls, biomass, glycogen, nitrates, sulfates and phosphates. In addition, we can define the mass biotic fraction of a component i as $z_i = C_i/C_X$ and the total biomass as $C_{XT} = C_X + C_G$.

Before the appearance of the limitation by sulfur or nitrogen, the previous experimental results indicate that the biotic mass fractions of proteins (z_{Pr}), phycocyanins (z_{PC}), and chlorophylls (z_{CH}) remain constant whatever the energy flux, in exponential or linear growth phases:

$$z_{Pr} = 0.63\,\mathrm{kg\,Pr.kg\,X^{-1}}$$

$$z_{PC} = 0.15\,\mathrm{kg\,PC.kg\,X^{-1}}$$

$$z_{CH} = 0.009\,\mathrm{kg\,CH.kg\,X^{-1}}$$

Some assumptions can be made here [73]:

– the C, H, O, N, S and P elemental composition of the biomass X remains constant; this assumption was experimentally confirmed for exponential and linear growth phases – under mineral limiting conditions, i.e. for stationary phases, this assumption is taken to remain valid;

– the biotic mass fraction of proteins remains constant under nitrate and phosphate starvation but decreases under sulphur starvation;

– the biotic mass fraction of chlorophylls remains constant under nitrate, sulfate, or phosphate starvation;

– the biotic mass fraction of phycocyanins decreases under nitrate and sulfate starvation, but remains constant under phosphate starvation.

In this case, the stoichiometric equation associated with growth is given by Eq. (30) or Eq. (31) if ionic species are considered. It should be noted that this equation implies a constant elemental formula for biomass with 10% of exopolysaccharide, i.e. for incident radiant energy fluxes down to $20\,\mathrm{W.m^{-2}}$. The conversion yields for nitrate, sulfate and phosphate are given by Eqs. (33–35). For higher incident fluxes, a new single stoichiometric equation must be constructed from different percentages of exopolysaccharide obtained by Cornet et al. [86] in continuous cultures and confirmed by Lu [87] in batch cultures. These results are summarized in Table 6. Such metabolic deviations with varying incident fluxes have been also observed on other photosynthetic micro-organisms [88].

Table 6. Modifications in the ratio proteins/exopolysaccharide (EPS) for different conditions of incident radiant light fluxes on the photobioreactor cultivating *S. platensis*

Incident light flux ($W.m^{-2}$)	Batch cultures [87]		Continuous cultures [86]	
	Proteins (%w)	EPS (%w)	Proteins (%w)	EPS (%w)
10 ± 1	–	–	65	10
50 ± 5	59	16	–	–
120 ± 10	57	19	56	20
160 ± 15	–	–	48	27
200 ± 20	43	33	42	34
300 ± 30	41	35	–	–

4.3.2 Definition of Overall Absorption and Scattering Mass Coefficients

The definition of absorption and scattering coefficients introduced in Sect. 3 have to be modified to take into account the modifications of pigment concentrations and cell size that occur in batch cultures during mineral limitations. Scattering is related to the size of cells suspended in the medium and is thus proportional to the mass concentration of the total biomass concentration C_{XT}, whereas absorption depends on pigment concentrations; absorption is thus proportional to the phycocyanin and chlorophyll content of cells, i.e. to the sum $C_{PC} + C_{CH}$, called the antenna content. For a rectangular reactor illuminated on one side and assuming the one-dimensional approximation of Schuster, the parameters α and δ of the light transfer model (Table 3) can thus be expressed as

$$\alpha = [Ea(C_{PC} + C_{CH})/[Ea(C_{PC} + C_{CH}) + EsC_{XT}]]^{1/2} \tag{102}$$

$$\delta = [Ea(C_{PC} + C_{CH})[Ea(C_{PC} + C_{CH}) + EsC_{XT}]]^{1/2}L \tag{103}$$

The light transfer equations derived from Schuster's assumptions clearly depend on both the biomass concentration and the pigment content of the cells. It is therefore justified a posteriori to describe absorption and scattering phenomena, which play different roles in light diffusion inside the reacting volume, by two coefficients. These overall absorption and scattering mass coefficients may be rewritten with these definitions:

$$Ea = 950\,m^2.kg^{-1} \text{ of antenna}$$

$$Es = 182\,m^2.kg^{-1} \text{ of total biomass}$$

4.3.3 Kinetic Model [73]

Two additional assumptions are needed to establish the different kinetic equations for each component of the model.

1 – under nitrate or sulfate limitations, the mean volumetric rate of phycocyanin or protein uptake and the mean volumetric rate of biomass synthesis are both proportional to the mean linear volumetric rate of biomass under light limitation $\langle R_X \rangle$ alone (Eq. 99). This linear volumetric rate possesses two important characteristics:

– it is unaffected by mineral limitations, which induce the synthesis of intracellular glycogen and exopolysaccharides under nitrate or sulfate starvation;

– it remains constant during the time course of the culture, since the mean quantum yield does not change, even during mineral limitations.

2 – the mean growth rate $\langle \mu \rangle$ follows a Monod law for nitrate, sulfate and phosphate concentrations as expected in Eq. (101). In the absence of mineral limitations, the volumetric growth rate in biomass is assumed to be proportional to the phycocyanin content C_{PC} rather than the biomass concentration. This can be justified by the fact that under light-limiting conditions, the growth metabolism is governed by the energy input rather than by enzyme activities. Equations (99) and (100) must be rewritten as follows:

$$\langle R_X \rangle = \langle \mu' \rangle \gamma C_{PC} \tag{104}$$

$$\langle \mu' \rangle = \frac{1}{V_2} \int_{V_2} \mu'_M \frac{4\pi J}{K_J + 4\pi J} dV \tag{105}$$

with $\mu'_M = 0.49\,h^{-1}$

The mean mass volumetric rate in biomass under light, nitrate, sulfate or phosphate limitation is then given by

$$\langle r_X \rangle = \langle R_X \rangle \frac{C_N}{K_N + C_N} \frac{C_S}{K_S + C_S} \frac{C_P}{K_P + C_P} \tag{106}$$

where K_N, K_S, K_P are the Monod constants for nitrate, sulfate and phosphate respectively.

Under nitrate, sulfate or phosphate deprivation, the chlorophyll content of active biomass remains constant, so that the mean mass volumetric rate is given by

$$\langle r_{CH} \rangle = z_{CH} \langle r_X \rangle \tag{107}$$

The mean mass volumetric rate for phycocyanins is proportional to that for biomass in conditions of non-limitation by minerals and to that of mean linear biomass synthesis $\langle R_X \rangle$ during nitrate or sulfate starvation:

$$\langle r_{PC} \rangle = z_{PC} \langle R_X \rangle \left[\frac{C_N}{K_N + C_N} \frac{C_S}{K_S + C_S} \frac{C_P}{K_P + C_P} - \left(\frac{K_N}{K_N + C_N} + \frac{K_S}{K_S + C_S} \right) \right] \tag{108}$$

where the term on the right-hand side represents the decrease in phycocyanin concentration, which only occurs when concentrations of nitrates or sulfates are low.

The mean mass volumetric rate for proteins only decreases during sulfate deprivation, so that

$$\langle r_{Pr} \rangle = z_{Pr} \langle R_X \rangle \left[\frac{C_N}{K_N + C_N} \frac{C_S}{K_S + C_S} \frac{C_P}{K_P + C_P} - \zeta \frac{K_S}{K_S + C_S} \right] \tag{109}$$

where ζ is a proportionality coefficient with respect to the mean linear volumetric rate $\langle R_X \rangle$ and is an adjustable parameter.

The mean mass volumetric rate for uptake of substrates is obtained from the mass conversion yields. For nitrates, sulfates and phosphates respectively, we have

$$\langle r_N \rangle = - Y_{N/X} \langle r_X \rangle \tag{110}$$

$$\langle r_S \rangle = - Y_{S/X} \langle r_X \rangle \tag{111}$$

$$\langle r_P \rangle = - Y_{P/X} \langle r_X \rangle \tag{112}$$

Under nitrate and sulfate deprivation, the continued increase in total biomass concentration results from the accumulation of intracellular glycogen. However, if cells do not divide, growth stops when cells reach a maximum size. This is described by the term $C_{PC}/(K_{PC} + C_{PC})$. The mean mass volumetric rate of total biomass synthesis is then

$$\langle r_{XT} \rangle = \langle R_X \rangle \left[\frac{C_N}{K_N + C_N} \frac{C_S}{K_S + C_S} \frac{C_P}{K_P + C_P} + \frac{C_{PC}}{K_{PC} + C_{PC}} \left(\frac{K_N}{K_N + C_N} + \frac{K_S}{K_S + C_S} \right) \right] \tag{113}$$

This model for growth of *S. platensis* under light, nitrogen, sulfur or phosphate limitation uses only seven parameters – μ'_M, K_J, K_N, K_S, K_P, K_{PC} and ζ – which remain to be identified. It is emphasized that light absorption and scattering parameters Ea and Es were experimentally measured separately.

This model for growth of *S. platensis* has been integrated for batch cultures performed in rectangular photoreactors illuminated on one side and compared with the experimental results obtained on ten different experiments for the general mass balance equation (from Eq. 2):

$$\frac{dC_i}{dt} = \langle r_i \rangle \tag{114}$$

The five biological parameters of the model for mineral limitations, K_N, K_S, K_P, K_{PC} and ζ were obtained by minimizing a quadratic criterion using a Gauss-Newton algorithm. Biomass, nitrate, sulfate, phosphate, proteins, phycocyanins and chlorophyll *a* concentrations have been taken into account.

A one-dimensional approximation of Schuster has been assumed for the determination of the available radiant energy profiles in the medium to minimize the calculation time, allowing the identification of parameters μ'_M and K_J, which have been discussed above. The results of the identification procedure are as follows (the parameters for light limitation are recalled here) [73]:

$$\mu'_M = 0.49\,h^{-1}$$

$$K_J = 20\ W.m^{-2}$$

$$K_N = 5.3 \cdot 10^{-3}\ kg\,NO_3^-.m^{-3}$$

$$K_S = 2.5 \cdot 10^{-4}\ kg\,SO_4^{2-}.m^{-3}$$

$$K_P = 2.7 \cdot 10^{-4}\ kg\,HPO_4^{2-}.m^{-3}$$

$$K_{PC} = 0.06\,kg.m^{-3}$$

$$\zeta = 0.55$$

Solid lines in Figs. 10, 11 and 12 show the numerical simulations obtained for nitrate, sulfate and phosphate limitations respectively. For each culture, the percentage of N, S or P recovery (defined as the total mass of N, S or P recovered in the synthesized biomass divided by the mass of N, S or P contained in the respective mineral substrates, nitrate, sulfate, phosphate) was respectively 96%, 91%, and 97%. This confirms that the main species involved initially in the process were correctly identified.

For nitrate and sulfate limitations, the model is in close agreement with the experimental results, thereby confirming the ability of the model to provide satisfactory predictions of biological kinetics in the photobioreactor until phycocyanins have totally disappeared. These kinetic parameters, identified here with a simple one dimensional model, may be extended to more complex geometries, such as cylindrical photobioreactors, and to more complex physical models related to light limitation. This has been done for cultures performed in cylindrical reactor with high incident radiant energy flux and with nitrate limitation [7]. In this case, the light transfer model used the numerical gridding in specific intensity presented in Sect. 3 (Eq. 95) and the kinetic model used the value of K_N previously given. The results obtained by simulation and the experimental data agree closely (results not shown) confirming the validity of the previous kinetic parameters over a wide range of operating conditions.

For phosphate limitations the simulation results are less accurate in the limiting phosphate concentration range. This shows that the Monod law is not satisfactory for modeling phosphate limitation. This is probably due to the large intracellular reserve of polyphosphates that are mobilized in a complicated way, and to the key role played by phosphates in the energy metabolism of cells. Some authors have proposed kinetic models for phosphate limitation in which the extra- and intracellular phosphate concentrations are considered [39, 89]. The prime hypothesis advanced in this section concerning the additive effects of limitations should perhaps be reconsidered in this case.

5 Limitation by the Carbon Source

5.1 *Different Cases in CO_2 Transfer and HCO_3^- Consumption Rates*

At high pH values ($8 < pH < 11$), the total dissolved carbon ($CO_2 + HCO_3^- + CO_3^{2-}$) is essentially composed of bicarbonate HCO_3^- and carbonate CO_3^{2-} ions. For example, at pH 9.5, the concentration of carbonate and bicarbonate ions is about 2000 times higher than the CO_2 concentration. This is an important specificity compared with the classical approach used for gas transfer rate analysis, because this creates a large buffer reserve of total dissolved carbon in the culture media. This implies that the assumption of non-accumulation of dissolved gases in the medium is no longer valid and that the mass balance equations should not be written at steady state and should account for an accumulation term in the liquid medium. This entails solving non-stationary differential equations.

Cyanobacteria are known to concentrate intracellular bicarbonate if the extracellular concentration is low [11–14, 90, 91]. This concentration proceeds against the HCO_3^- concentration gradient but maintains the activity of the Rubisco at a high level. However, this active transport consumes part of the produced ATP and the energy yield of photosynthesis thus decreases, resulting in lower growth rates.

The study of the limitation by the carbon source and its modeling are therefore complex problems since information has to be obtained at three levels:

– the study of the CO_2 gas–liquid transfer and any limitation by the CO_2 transfer rate;

– the study of chemical equilibria in complex media with high ionic strength to calculate the CO_2–HCO_3^-–CO_3^{2-}-concentrations in the culture medium as function of pH;

– the study of HCO_3^- limiting concentrations on the physiology and metabolism of *S. platensis* to postulate a kinetic law.

Moreover, the general modeling of these coupled phenomena requires mass balances on O_2 and CO_2 in the gas phase to be performed to supplement the above information.

Two principal cases may be examined:

1 – limiting concentration of bicarbonate: in this case, the Rubisco controls the bioconversion rates (growth rate, glycogen and exopolysaccharide synthesis rates) which depend on the bicarbonate concentration. This bicarbonate concentration is related to the gas-liquid mass balance and to the dissociation equilibria. Either at steady state or in transient state, the bicarbonate concentration is obtained by solving the mass balance equations (Eqs. 9–15) enabling one to link bicarbonate concentration to process parameters, i.e. gas–liquid volumetric transfer coefficient k_La, gas flow rate, entrance molar fraction of CO_2, total pressure;

2 – non-limiting concentration of bicarbonate: The bioconversion rates depend on other parameters (light energy transfer, mineral limitations). However, the CO_2 balance remains to be solved because the total dissolved carbon concentration can either slowly increase or decrease, depending on the process parameters and on the physiological activity of the micro-organisms.

5.2 *Physical and Chemical Equilibria for the CO_2–HCO_3^-–CO_3^{2-} System*

5.2.1 *Gas–Liquid Equilibrium*

For the most commonly encountered conditions (low CO_2 partial pressure) in aeration of photosynthetic cultures, the CO_2 fugacity coefficient is equal to 1. The gas-liquid equilibrium for CO_2 in ideal mixtures is then given by the following relation (Henry's law):

$$C'^{*}_{CO_2} H' = y_{CO_2} P \tag{115}$$

where H' is Henry's constant, which depends on temperature and on the composition of the liquid medium, y_{CO_2} is the CO_2 molar fraction in the gas and P the total pressure.

5.2.2 *Chemical Equilibria*

When reacting with water, the CO_2 is converted to carbonic acid H_2CO_3 which instantly gives bicarbonate ion. For this reason, the first reaction of CO_2 with water is often written

$$CO_2 + H_2O \underset{k'_{-1}}{\overset{k'_1}{\leftrightarrows}} HCO_3^- + H^+ \tag{116}$$

$$\text{with } K_1 = \frac{k'_1}{k'_{-1}} = \frac{C'_{HCO_3^-} C'_{H^-}}{C'_{CO_2^-}} \tag{117}$$

The second chemical equilibrium involves bicarbonate and carbonate ions:

$$HCO_3^- \underset{k_{-2}}{\overset{k_2}{\leftrightarrows}} CO_3^{2-} + H^+ \tag{118}$$

$$\text{with } K_2 = \frac{k_2}{k_{-2}} = \frac{C'_{CO_3^{2-}} C'_{H^+}}{C'_{HCO_3^-}} \tag{119}$$

This reaction is often considered as instantaneous; the equilibrium between carbonate and bicarbonate is then always satisfied.

In order to complete the reaction scheme, the water dissociation equilibrium must be considered (ionic product Ke):

$$H^+ + OH^- \leftrightarrows H_2O \tag{120}$$

Furthermore, when the liquid medium contains calcium ions with no complexant, the solubility product for $CaCO_3$ must be considered:

$$Ca^{2+} + CO_3^{2-} \leftrightarrows CaCO_3 \tag{121}$$

$$\text{with } K_S = C'_{Ca^{2+}} C'_{CO_3^{2-}} \tag{122}$$

For synthetic culture media, the equilibrium constants K_1, K_2, Ke vs the temperature are available in the literature [92]. For example, at 25 °C, $K_1 = 4.38 \cdot 10^{-7}\ mol \cdot l^{-1}$, $K_2 = 4.65 \cdot 10^{-11}\ mol \cdot l^{-1}$, and $Ke = 10^{-14}\ mol \cdot l^{-1}$.

5.3 *Conservation Law for the CO_2–HCO_3^-–CO_3^{2-} System*

In the general case, in an aerated WTR at fixed pH with no presumption concerning batch or continuous processing, the following differential system, deduced from the above reaction scheme, may be written for the liquid phase:

$$\begin{cases} \dfrac{dC'_{CO_2}}{dt} = Ek_La(C'^*_{CO_2} - C'_{CO_2}) + D(C'^E_{CO_2} - C'_{CO_2}) - k'_1 C'_{CO_2} + k'_{-1} C'_{HCO_3^-} C'_{H^+} \\ \dfrac{dC'_{HCO_3^-}}{dt} = D(C'^E_{HCO_3^-} - C'_{HCO_3^-}) + k'_1 C'_{CO_2} + k_{-2} C'_{CO_3^{2-}} C'_{H^+} - k'_{-1} \\ \qquad C'_{HCO_3^-} C'_{H^+} - k_2 C'_{HCO_3^-} - \dfrac{Y_{C/XT}}{M_{HCO_3^-}} \langle r_{XT} \rangle . \\ \dfrac{dC'_{CO_3^{2-}}}{dt} = D(C'^E_{CO_3^{2-}} - C'_{CO_3^{2-}}) + k_2 C'_{HCO_3^-} - k_{-2} C'_{CO_3^{2-}} C'_{H^+} \end{cases} \tag{123}$$

$Y_{C/XT}$ is the mass conversion yield from bicarbonate to total biomass defined in the first section of this chapter, $M_{HCO_3^-}$ is the molar mass for bicarbonate and D is the dilution rate. E stands for the enhancement factor by chemical reaction for the CO_2 transfer. Generally, E may be taken to be close to one, except for low volumetric mass transfer coefficients ($k_La < 10\,h^{-1}$) or pH higher than 10 [7,93]. In these conditions, the enhancement factor E may be determined from the generalized Hatta number [94], rewriting Eqs. (116) and (120) in the form

$$CO_2 + OH^- \underset{k'_{-1}}{\overset{k'_1}{\leftrightarrows}} HCO_3^- \tag{124}$$

Then, depending on the pH and bicarbonate concentration, this reaction rate may be considered respectively as irreversible pseudo-first order or pseudo-second order or reversible.

This system of equations is a simplified reaction model compared with the complete system written by Curless [85] and Erickson et al. [95], but these approximations are justified by many authors. The equation providing the accumulation rate for the bicarbonate gives the HCO_3^- concentration time course in the medium with growth of *S. platensis.* If only light limitation is present, the mass volumetric rate $\langle r_{XT} \rangle$ is given by Eqs. (47) and (48); however, if a mineral limitation by HCO_3^- occurs, the model requires an additional kinetic equation to obtain $\langle r_{XT} \rangle$.

If the carbon dioxide/bicarbonate (Eq. 116) and bicarbonate/carbonate (Eq. 118) reactions are considered to be instantaneous and equilibrated, the system of equations at Eq. (123) can be simplified to a system of only one differential equation on the CO_2 transfer and two algebraic equations (Eqs. 117 and 119) giving the bicarbonate and carbonate concentrations at thermodynamic equilibrium [7].

To complete this kinetic model, the gas balance equation is required:

$$\frac{P}{RT}\frac{\partial y_{CO_2}}{\partial t} = -\,Ek_L a\,\frac{1-\varepsilon}{\varepsilon}\,(C'^{*}_{CO_2} - C'_{CO_2}) - \frac{\partial (G y_{CO_2})}{\partial V_G} \tag{125}$$

where y_{CO_2} is the molar fraction of CO_2 and G is the total molar flow rate. If the difference in molar fraction between the input and the output of the reactor is low, the gas phase can be considered as perfectly mixed so that $\partial(Gy_{CO_2})/\partial V_G \approx y^E_{CO_2}G^E - y_{CO_2}G$. Moreover, as previously discussed in Sect. 2, the accumulation term in Eq. (125) may often be neglected.

The system of equations at Eqs. (123) and (125) must be simplified for different operating conditions:

– for a continuous culture, the system of equations at Eq. (123) has to be solved entirely, except if a physical limitation by CO_2 transfer rate occurs or if a steady state is reached. In these cases, the accumulation terms in d/dt vanish;

– for a batch culture, the dilution rate D is equal to zero. In addition, for limitation by the CO_2 transfer rate, the terms in d/dt also vanish.

In other general cases, experiments showed that the time constant of such processes was very high [93] (three months at pH 10), and depended mainly on pH and the ratio G/V.

5.4 Effects on Spirulina Growth

As discussed above, the *Spirulina* growth rate is limited in two ways as regards the carbon source:

1 – the culture may be limited by the physical CO_2 transfer rate between the gas and liquid phases. In this case, the mass volumetric growth rate is fixed by the technical characteristics of the apparatus and is equal to

$$\langle r_{XT} \rangle = r_{XT} = \frac{M_{HCO_3^-}}{Y_{C/XT}}\,k_L a\,C'^{*}_{CO_2} \tag{126}$$

Table 7. Effect of bicarbonate limitations on the *Spirulina* mass volumetric growth rate in total biomass obtained in batch cultures vs the CO_2 amount in the gas phase. The maximum growth rate is obtained for an incident radiant energy flux of 200 W.m^{-2} in a cylindrical photobioreactor of 5 l

CO_2 amount in the incoming gas flow rate (ppm)	Mass volumetric rate in biomass $\langle r_{XT} \rangle . 10^3$ (kg.m^{-3}.h^{-1})	Ratio $\frac{\langle r_{XT} \rangle}{\langle r_{XT} \rangle_{max}}$
Maximum mass volumetric growth rate under only light limitation ($F_0 = 200$ W.m^{-2}) $\langle r_{XT} \rangle_{max}$	11.0	1
2800	9.0	0.82
2100	7.5	0.68
1400	5.6	0.51
700	3.3	0.30
350	1.9	0.17

2 – the culture may be limited by the bicarbonate concentration in the liquid medium. In this case, the problem is very similar to that of mineral limitations and this requires a kinetic equation relating the mass volumetric rate of biomass under light limitation $\langle r_{XT} \rangle$ and the bicarbonate concentration in the medium. Little information is available concerning work on this subject; Table 7 presents preliminary results obtained by Cornet [7] on bicarbonate limited cultures with the cyanobacterium *S. platensis.* Unfortunately, because of the complexity of the intracellular bicarbonate accumulation mechanism and of the non-stationary balance equations giving the extracellular bicarbonate concentration, it is impossible, at the present time, to propose a kinetic law for this phenomenon. Table 7 presents only the decrease in mass volumetric rate of biomass with the limiting CO_2 concentration in the gas phase.

6 Biochemically Structured Model of Growth

In the foregoing sections, the kinetics of the metabolic conversions occurring during the growth have been treated macroscopically without entering into the details of the metabolic pathways and energy transduction processes. The light energy transfer limitations have been explored via a single stoichiometric equation for growth, the rate of which is calculated using light energy distribution models of varying mathematical complexity. This was a so-called unstructured model. However, such an interpretation appears to be too oversimplified if the objective of modeling is also to account for exopolysaccharide synthesis which is associated with growth and depends on light energy availability.

In order to represent the variable composition of total biomass it is therefore necessary to treat active biomass and exopolysaccharide synthesis as independent anabolic events and no longer by a single stoichiometric equation. One way could be to correlate empirically each of these rates as functions of energy availability. The approach which is presented here uses, first, a detailed description of metabolic pathways for characterizing the two anabolic events, and second, an analysis of the thermodynamic efficiency of chemical energy producing machinery. This leads to a biochemically structured model.

6.1 Coupling photophosphorylations and Photosynthesis: ATP/2e⁻ Determination and Biomass Composition Prediction

6.1.1 Basic Formalisation for a Structured Model of S. platensis Growth

As previously stated in Sect. 2, to match the consumption of the reduced factors (considered here as NADPH, H^+) and the ATP consumption for an anabolic process, it is necessary to consider two additive stoichiometric equations: one for $NADP^+$ reduction (water photolysis, Eq. 28) and one for ATP regeneration (Eq. 29). This investigation of light energy influence upon rates and biomass composition will therefore need four stoichiometric equations: two for anabolic processes (Eqs. 25 and 26) and two for light energy conversion into chemical energy (ATP) and into reducing power (NADPH, H^+). Considering a pseudo-steady state assumption for ATP and cofactors, the four reaction rates must satisfy two constraints leaving two rates that must be independently correlated. By comparison, the previous approach, which uses a single stoichiometric equation for total biomass formation leads to one rate equation which has to be determined on the basis of the light transfer model, when light energy transfer is the rate limiting process.

Using the stoichiometric coefficients reported in Table 2 for active biomass and exopolysaccharide synthesis, the two rate constraints between the average rates ($\langle J_{XA}\rangle$, $\langle J_{EPS}\rangle$, $\langle J_{COF}\rangle$, $\langle J_{ATP}\rangle$) take the following form:

$NADP^+$ balance:

$$2.874\langle J_{XA}\rangle + 1.920\langle J_{EPS}\rangle = 2.779\langle J_{COF}\rangle \tag{127}$$

ATP balance:

$$3.568\langle J_{XA}\rangle + 3.330\langle J_{EPS}\rangle = 3.544\langle J_{ATP}\rangle \tag{128}$$

It must be noticed that the establishment of the stoichiometric equations responsible for the two anabolic events (J_{XA} and J_{EPS}) results from a thorough analysis of the operative metabolic pathways of *S. platensis* including Calvin cycle, amino-acids, nucleic acids, carbohydrate and lipid pathway synthesis. Two major oversimplifying assumptions have been systematically made: i) assimilation of all reducing power to NADPH, H^+ cofactors; ii) expression of all chemical energy consumption or production as ATP taking all triphosphate nucleotides to be equivalent and assuming that one monophosphate nucleotide production is equivalent to the production of two ADPs. A more refined technique would consider each metabolic reaction step separately leading to a large number of stoichiometries to be considered and a large number of metabolic intermediates that would have to be assumed at pseudo-steady state. This approach implies calculating all metabolic fluxes and lies outside the scope of this analysis.

Therefore, the biochemically structured model presented here involves four rates $\langle J_{XA} \rangle$, $\langle J_{EPS} \rangle$, $\langle J_{COF} \rangle$ and $\langle J_{ATP} \rangle$. The analysis of biochemical pathways, still simplified, leads to two structuring relations (Eqs. 127 and 128). Two relations remain to be established considering the light energy availability in order to express a kinetic model for active biomass and exopolysaccharide synthesis. Knowing that cofactor synthesis rate $\langle J_{COF} \rangle$ and ATP production rate $\langle J_{ATP} \rangle$ are intimately linked to light energy transfer for photosynthetic micro-organisms, this calls for a deeper analysis of the kinetics of photosynthesis and photophosphorylations. Such an approach affords a better understanding of the light energy limitation than an empirical evaluation of the exopolysaccaharide yield $\langle J_{EPS} \rangle / \langle J_{XA} \rangle$ and provides clues to a correct formulation of kinetic models for photosynthetic micro-organisms.

6.1.2 Bioenergetic Analysis of Coupling Photophosphorylations and Photosynthesis

A detailed description of metabolism provides valuable relations between the exopolysaccharide and active biomass production rates and the rates of the intracellular turn over of chemical energy and reducing power carriers. The basic idea here is to add a thorough description of the energy transducing process in order to assess the ratio $\langle J_{ATP} \rangle / \langle J_{COF} \rangle$ which is termed the $P/2e^-$ ratio.

6.1.2.1 Mechanistic and Biochemical Analysis

The central feature of photosynthesis is the conversion of light energy into redox energy, meaning that photon capture causes a component to change its redox potential from being moderately electropositive to being highly electronegative. The electrons released from this component serve to generate an electrochemical gradient, flowing through either a cyclic pathway back to rereduce the original component, or a non-cyclic pathway to reduce additional electron acceptors, i.e. $NADP^+$ in the case of thylakoid membranes [10]. In the non-cyclic pathway, electrons are extracted from water, pass through a proton-translocating-electron transfer chain and then a second reaction centre and are ultimately donated to $NADP^+$. The production of ATP involves a proton circuit that is closely analogous to the electron transport chain for respiration.

Without entering into the details of the biochemical mechanisms, these processes can be illustrated by the classical Z-scheme (Fig. 1) enabling one to summarize the overall reaction by one stoichiometric equation:

$$4\,h\nu + H_2O + NADP^+ + ADP + Pi \rightarrow NADPH, H^+ + ATP + H_2O + \tfrac{1}{2}O_2 \qquad (129)$$

This representation is, however, too simplistic to be used in a biochemically structured model for growth, as the $P/2e^-$ ratio remains equal to 1 which is obviously not the case when anabolic processes are considered. From Eqs. (25) and (26), the $P/2e^-$ ratio equals $3.568/2.874 = 1.24$ for active biomass synthesis, and $3.3/1.92 = 1.73$ for exopolysaccharide synthesis. In other words, NADPH, H^+ produced by non-cyclic electron flow is directed to the Calvin cycle, which fixes CO_2, and further to other biosynthesis pathways which always require more than 1 ATP per NADPH, H^+ produced. Thus the previous stoichiometry of ATP synthesis associated with electron flow leaves a shortfall ranging between 24% and 73% whether active biomass or exopolysaccharide synthesis are considered, which is evidence of a cyclic electron flow exclusively devoted to ATP synthesis.

Although the idea of cyclic electron flow has been accepted for some time, the pathway is, surprisingly, still not well understood [10]. A mechanistic scheme cannot be used as a basic tool for a mathematical model; only a phenomenological treatment can give insight into the coupling of the two processes of ATP and NADPH, H^+ synthesis.

6.1.2.2 Phenomenological Treatment of the Light Energy Transducing Process

The analysis proposed here consists of splitting the overall process into two coupled reactions, one responsible for ATP synthesis:

$$3h\nu + ADP + Pi \rightarrow ATP + H_2O \qquad (130)$$

and one responsible for water photolysis:

$$h\nu + H_2O + NADP^+ \rightarrow \tfrac{1}{2}O_2 + NADPH, H^+ \qquad (131)$$

The sum of these two reactions restores the global reaction (Eq. 129). The choice of $3h\nu$ for ATP synthesis and one $h\nu$ for water hydrolysis should be regarded as more or less arbitrary and does not correspond to any well-established mechanism. However, this stoichiometry can be justified considering that two light quanta at PS I and one light quantum at PS II are necessary to create an H^+ gradient over the thylakoid membrane while one light quantum only at PS II is necessary to initiate water splitting. Formally, this separation between the functioning capabilities of the two photosystems corresponds to the previous stoichiometric analysis distinguishing between ATP synthesis rate J_{ATP} (Eq. 29) and cofactor reduction rate J_{COF} (Eq. 28).

Let A_{ATP} and A_{COF} be the two affinities of two reactions:

$$A_{ATP} = 3h\nu + g_{ADP} + g_{Pi} - g_{ATP} - g_{H_2O} \tag{132}$$

$$A_{COF} = h\nu + g_{H_2O} + g_{NADP^+} - g_{NADPH} - g_{H^+} - \tfrac{1}{2} g_{O_2} \tag{133}$$

where the g_is stand for the partial molar Gibbs energies, i.e. the chemical potentials of the components in the physiological conditions and $h\nu$ the radiation energy of one mole of photons ($h\nu = 176.4\,kJ.mol^{-1}$ for a 680 nm wavelength radiation). Considering the values of Gibbs energy changes in standard conditions (pH = 7, 25 °C, ionic strength $0.2\,mol.l^{-1}$), $g_{ATP} + g_{H_2O} - g_{ADP} - g_{Pi} = 32.5\,kJ.mol^{-1}$ [96] and $g_{NADPH} - g_{NADP^+} = 21\,kJ.mol^{-1}$ [97], the affinities of the photoreactions can be calculated ($g_{H_2O} = -237.2\,kJ.mol^{-1}$, $g_{H^+} = -40.0\,kJ.\ mol^{-1}$ at pH = 7, $g_{O_2} = -1.9\,kJ.mol^{-1}$ at $p_{O_2} = 0.21$ atm) assuming a total conversion of radiant energy into chemical energy:

$$A_{ATP} = 3h\nu - 32.5 = 497\,kJ.mol^{-1} \tag{134}$$

$$A_{COF} = h\nu - 216.3 = -40\,kJ.mol^{-1} \tag{135}$$

These results indicate that the first reaction is exergonic (positive affinity) and the second reaction is endergonic (negative affinity) due to the fact that only one light reaction at photosystem II is not sufficient to encompass the redox span $H_2O/\frac{1}{2}O_2$ to $NADP^+/NADPH$.

These considerations suggest treating the coupling of these two rates as a free energy transduction process between an exergonic reaction and an endergonic reaction offering the possibility of calculating the coupling efficiency, which in turns corresponds to the rate ratio, in terms of the linear energy converter theory.

6.1.2.3 Determination of the $P/2e^-$ Ratio

As exopolysaccharide and active biomass synthesis rates vary when changing the light energy availability, the $P/2e^-$ ratio is also modified. The treatment of photophosphorylation and water photolysis in terms of a fixed stoichiometry is thus invalid. Also, the stoichiometric equations for ATP synthesis and for cofactors reduction are therefore evidently incompletely coupled leading to the overproduction of ATP which is the energy driving process. The theory of the linear energy converter in terms of linear thermodynamics of irreversible processes presents the perfect vehicle for handling this problem.

The rate of entropy production, i.e. the dissipation function that corresponds to the coupling of the two processes, takes the form

$$\sigma = J_{ATP}A_{ATP} + J_{COF}A_{COF} \tag{136}$$

By virtue of the second principle of thermodynamics σ must remain positive.

The theory of the linear energy converters states that the rates J_{ATP} and J_{COF} can be expressed as multilinear functions of the affinities according to a linear phenomenological relationship with the affinities A_{ATP} and

A_{COF} [1,28,98,99];

$$J_{ATP} = L_{AA}A_{ATP} + L_{AC}A_{COF} \tag{137}$$

$$J_{COF} = L_{AC}A_{ATP} + L_{CC}A_{COF} \tag{138}$$

It is stressed that the cross-coefficients L_{AC} must be equal according to the Onsager reciprocity relationships which are assumed to apply. This formalism applies to the near equilibrium domain which normally requires $A \ll RT$ for linearity. The free energy transduction between photophosphorylation and cofactor reduction obviously operates at far-from-equilibrium conditions. However, the linear relationships are still considered valid for some steady-state region of linear behaviour. This extension of the linear domain to far-from-equilibrium conditions is still debated [100] though Onsager reciprocity relationships are demonstrated to apply even at far-from-equilibrium conditions for an arbitrary steady state [101]. The most valid rationale is that, for a stable and adaptative functioning of cell machinery, the time-average values of the rates and affinities must follow the linear energy converter formalisation [28].

The classical treatment for such equations [1,98,100] involves reducing the dimensions of the system of equations (Eqs. 137 and 138), leading to the definition of the coupling coefficient and of the phenomenological stoichiometric coefficient, which are dimensionless coefficients. The coupling coefficient is given by

$$q = \frac{L_{AC}}{(L_{AA}L_{CC})^{1/2}} \tag{139}$$

and the phenomenological stoichiometric coefficient is expressed by

$$\chi = \left(\frac{L_{CC}}{L_{AA}}\right)^{1/2} \tag{140}$$

Introducing the generalized forces ratio $x (x < 0)$:

$$x = \chi \frac{A_{COF}}{A_{ATP}} \tag{141}$$

the following expressions are obtained:

$$J_{ATP} = L_{AA}A_{ATP}(1 + qx) \tag{142}$$

$$J_{COF} = \chi L_{AA}A_{ATP}(q + x) \tag{143}$$

$$\sigma = L_{AA}A_{ATP}^2(1 + 2qx + x^2) \tag{144}$$

The ratio of the two rates J_{ATP}/J_{COF}, which corresponds to the $P/2e^-$ ratio is given by

$$\frac{P}{2e^-} = \frac{1 + qx}{\chi(q + x)} \tag{145}$$

Therefore, the theory of linear energy converters enables one to express the ratio of the two rates as a function of the three variables (q, χ and x). The theoretical calculation of q, χ and x is then performed in two steps.

First, the fixed coefficients q and χ are assumed to be determined by the internal regulation of metabolism, which leads to a minimal rate of entropy production (which still remains positive) ensuring a maximum power output [1, 28, 102]. This gives the value of the degree of coupling $q = 0.910$ and the expression of the phenomenological stoichiometric coefficient:

$$\chi = q n_{AC} \tag{146}$$

where n_{AC} stands for the mechanistic stoichiometry, i.e. the stoichiometry which is deduced from the analysis of the reaction scheme. Typically in the case of coupling between photophosphorylation and water photolysis, $n_{AC} = 1$ mole $NADP^+$ reduced/mole ATP produced, leading to

$$\chi = q \tag{147}$$

The second step concerns the determination of force ratio x the value of which is calculated as a function of the external conditions applied to the system. More precisely, it is shown that for an isolated system, the stability principle [98, 99] directs the system towards its thermodynamic equilibrium states, i.e. to a zero affinities state. For an open system, the dissipation function is optimized considering the external constraints applied. This, in turn, leads to expressing the optimization of the dissipation function as a variational problem using a Lagrangian function $\mathscr{L} = \sigma - \lambda f$. The external constraint is expressed as $f(x, q, \chi) = 0$; the problem simply reduces to solving the system

$$\left(\frac{\partial \mathscr{L}}{\partial A_{ATP}}\right)_x = 0 \tag{148}$$

$$\left(\frac{\partial \mathscr{L}}{\partial x}\right)_{A_{ATP}} = 0 \tag{149}$$

$$f(x, q, \chi) = 0 \tag{150}$$

which leads to express the multiplicatory Lagrangian coefficient λ and the variables x and A_{ATP}.

In the case of photosynthesis, two different cases have to be considered, namely a non-limiting conditions case and the case of limitation of rates by the light energy transfer rate.

1) In the case of non-limitation by light energy transfer rate (saturation of photosynthesis), the product $J_{ATP}A_{ATP}$ is fixed by the non-limiting functioning conditions of the overall metabolism. In other words, the product $J_{ATP}A_{ATP}$ must remain completely independent of the characteristics of the energy transducing system. The f-function must be expressed as follows:

$$J_{ATP}A_{ATP} = L_{AA}A_{ATP}^2(1 + qx) = K \tag{151}$$

The resolution of the system of Eqs. (148), (149) and (150) gives the expression of x:

$$x = \frac{-1 + \sqrt{1 - q^2}}{q} \tag{152}$$

so that $x = -0.644$ for $q = 0.910$.

This value of x corresponds to $P/2e^- = 1.71$ (Eq. 145). Using Eqs. (134), (135) and (141) with $x = q = 0.91$ leads to recalculation of the affinities and the value of the effective radiant energy which is converted into chemical energy:

$$(h\nu)_{eff} = 76.6\,kJ.mol^{-1}$$

$$A_{ATP} = 197.3\,kJ.mol^{-1}$$

$$A_{COF} = -139.7\,kJ.mol^{-1}$$

The difference between the incident radiant energy $h\nu = 176.4\,kJ.mol^{-1}$ and the effective value $(h\nu)_{eff}$ represents the part of radiant energy dissipated into heat at the antenna level. It can be interpreted as an adaptative control of the photosystems when the light energy is provided in non-limiting conditions.

2) When the light energy transfer is the rate limiting process, the external constraint now concerns the rate J_{ATP}. The f-function is then expressed as follows:

$$J_{ATP} = L_{AA}A_{ATP}(1 + qx) = K \tag{153}$$

The solution of Eqs. (148), (149) and (150) gives $x = 0$ which in turns leads to $P/2e^- = 1.21$, $A_{COF} = 0$ and $(h\nu)_{eff} = 216.4\,kJ.mol^{-1}$. Obviously, the effective radiant energy cannot be greater than the incident radiant energy value which explains that for very limiting incident energy fluxes below the compensation point, chemical energy is supplied by respiration.

6.1.3 Comparison of Experimental and Theoretical Values of $P/2e^-$

The experimental determination of the biomass composition enables one to calculate the overall $P/2e^-$ ratio by the weighted sum of different $P/2e^-$ for the synthesis of the fractions of cell material. Importantly, such calculations are only feasible if a biochemically structured approach to metabolism is used. The values of $P/2e^-$ obtained for the synthesis of the different parts of the biomass are given in Table 2.

To test the applicability of the previous theoretical developments, the influence of three different cases of limitation on the biomass composition has been investigated.

6.1.3.1 Light Energy Transfer Limitation

The cultures performed at a low light energy incident flux are considered ($F_0 < 20\,W.m^{-2}$, non-limiting part of Figs. 10–12). As stated above, total biomass is composed of 10% exopolysaccharide and 90% active biomass. These values have already been used to establish the stoichiometric equation for the synthesis of total biomass under such incident flux conditions (Eqs. 27 and 30). For such conditions, the $P/2e^-$ value is equal to 1.27 (Table 2) which is in fairly good agreement with the theoretical value of 1.21 obtained in strict light energy limitation conditions.

If the incident light energy flux is increased up to 300 $W.m^{-2}$, experimental results show that the proportion of exopolysaccharide produced is sharply increased, corresponding to an increase in the $P/2e^-$ ratio which can be interpreted as a progressive removal of the light energy limitation. In the extreme case, the synthesis of the exopolysaccharide only corresponds to $P/2e^- = 1.73$ (Table 2), compared to the theoretical value for non limitation conditions of 1.71.

6.1.3.2 Nitrogen and Mineral Limitations

The experimental results presented in Sect. 4 indicate that biomass formed is composed of 92 wt% of carbohydrates (glycogen and sulfated exopolysaccharides) and 8 wt% of lipids considering that the constitutive proteins are synthesized from the amino-acids produced by the breakdown of the phycocyanins. The resulting stoichiometric equation for nitrate limited growth is written as follows:

$$CO_2 + 2.236H_2O + 0.015H_2SO_4 + 3.435ATP + 2.012(NADPH, H^+)$$

$$\rightarrow CH_{1.655}O_{0.860}S_{0.015} + 3.435(ADP + Pi) + 2.012NADP^+ \quad (154)$$

The corresponding $P/2e^-$ value is thus $3.435/2.012 = 1.71$, which is identical to the theoretical value obtained for non-limited light transfer conditions for the coupling between photophosphorylation and $NADP^+$ reduction. This means that a severe limitation by nitrogen nutrient in batch cultures may overcome the influence of the light energy transfer limitation and impose new regulation rules for the metabolism, i.e. towards an orientation of more ATP and less reducing power productions. This switch of the metabolism towards non-light energy transfer conditions can be macroscopically confirmed by the phycocyanin consumption indicating less need for antenna and light energy conversion devices.

6.1.3.3 Carbonate Limitation

The carbonate limitation induces, as expected, an inhibition of exopolysaccharide synthesis and an improvement of phycocyanins synthesis. The resulting $P/2e^-$ calculated from the stoichiometric equation (Eq. 25) is equal to 1.23 corresponding to strict light energy transfer conditions. In this case, the limited availability of CO_2 at a fixed rate imposes a fixed ATP production rate J_{ATP} whatever the light energy flux. This in turn produces the same effect as a light energy transfer rate, i.e. the lowest $P/2e^-$ obtainable equal to 1.21.

It may be concluded that the theoretical predictions of the $P/2e^-$ ratio are in close agreement with the experimental values obtained for different limitation conditions. At the present time, it should be stressed that the two asymptotic values of $P/2e^-$, 1.21 and 1.71, correspond strictly to the values obtained for carbonate and nitrate limitation conditions respectively. Furthermore, the changes in biomass compositions with light energy availability, i.e. increase in exopolysaccharide fraction with increase in incident light flux, are qualitatively represented by the theoretical model. This confirms the usefulness and the potential of such a biochemically structured approach associated with a thermodynamic analysis of energy transduction processes, for representing not only the overall production and consumption rates but also the internal composition of biomass and the variation of conversion yields with external conditions. In this respect, the thermodynamics of linear energy converters and the relevant optimization of the entropy production rate which must be considered as a function of the external constraints imposed on the system is a promising tool for handling the expressions of the conversions yields.

6.2 Thermodynamic Efficiency for Batch and Continuous Cultures

The above analysis implicitly leads to the determination of the macroscopic thermodynamic efficiency and of the enthalpy efficiency, the definitions of which are still being debated [103–110]. This point can now be clarified.

The conversion of radiant light energy into chemical energy at microscopic or macroscopic levels is described by three types of balance equations:

– mass balance (Sect. 2);

– photonic phase balance knowing the expression of the local volumetric rate of radiant energy absorbed (Sect. 3);

– energy balances accounting for the interchanges of energy between the material phase and the photonic phase.

The entropy balance is obtained from the Gibbs equation, splitting the rate of entropy accumulation into a source term and a convective term [38, 98, 99]. This last balance affords the expression for the dissipation function, and the thermodynamic treatment of energy conversion using the thermodynamics of irreversible processes which is concerned with the study or relations between rates and affinities. This enables one to establish the thermodynamic efficiency for photosynthesis in batch or continuous cultures, which is then defined by [38]

$$\eta_{th} = \frac{\sum_j \sum_p \nu_{pj} \langle J_j \rangle g_p}{\langle \mathscr{A} \rangle - \sum_j \sum_s \nu_{sj} \langle J_j \rangle g_s} \tag{155}$$

where subscripts s and p refer respectively to substrates and products ($\nu_s < 0, \nu_p > 0$).

Table 8. Calculation of thermodynamic efficiency for photosynthesis in batch and continuous cultures of *S. platensis* (values of $\langle \mathscr{A} \rangle$ in cylindrical photobioreactors were obtained by numerical determinations using finite element method [38]) (Permission from AIChE)

Mean volumetric rate of radiant energy absorbed $\langle \mathscr{A} \rangle$ ($W.m^{-3}$)	Reaction rate $\langle J_X \rangle$ ($kmol.m^{-3}.s^{-1}$)	Thermodynamic efficiency η_{th} (–)
80[a]	$1.95\,10^{-8}$	0.15
110[a]	$2.46\,10^{-8}$	0.14
140[a]	$3.02\,10^{-8}$	0.13
180[a]	$3.78\,10^{-8}$	0.13
330[a]	$5.52\,10^{-8}$	0.10
390[a]	$5.95\,10^{-8}$	0.09
480[a]	$6.33\,10^{-8}$	0.08
700[a]	$8.50\,10^{-8}$	0.08
750[a]	$8.50\,10^{-8}$	0.07
790[b]	$9.67\,10^{-8}$	0.07
980[a]	$1.10\,10^{-7}$	0.07
1150[a]	$1.32\,10^{-7}$	0.07
1280[b]	$1.42\,10^{-7}$	0.06
1680[b]	$1.47\,10^{-7}$	0.05

[a] batch cultures
[b] continuous cultures

If the arbitrary datum level for energy (reference state) has been chosen sufficiently close to the actual conditions, the chemical potentials of the different species can be considered to be equal to the reference chemical potentials g_i^o, i.e. the Gibbs energy of formation of component i.

Therefore, the calculation of the thermodynamic efficiency η_{th} requires the knowledge of a set of stoichiometric equations associated with their kinetic models and the non-accumulation constraints of metabolic intermediates, and the knowledge of a light energy transfer model to assess the terms $\langle \mathscr{A} \rangle$ and $\langle J_j \rangle$.

Such calculations have then been performed for *S. platensis* cultures in photobioreactors. The experimental results for several batch and continuous cultures in different photobioreactors and with different incident illuminations are collected in Table 8. In this table the volumetric rates of radiant energy absorbed have been calculated using the numerical finite elements technique of the characteristic curve presented in Sect. 3. The standard free enthalpies used for the calculation of the dissipation function of entropy are defined at a pH of 7 for HCO_3^- and liquid water, and unit molality for all other reactants, a temperature of 308 K and atmospheric pressure. These data for biomass and other species are provided by Roels [1]. The biomass standard chemical potential has been taken as $6.1\ 10^8\ J.kmol^{-1}$ [1].

It appears that the thermodynamic efficiency decreases as the volumetric rate of radiant energy absorbed increases in the reactor. This decrease is correlated with modifications in the preliminary stages of photosynthesis and particularly in electron transfer efficiency, which is affected by a high flux of photons. The highest value of 0.15 in energy yield was found for low volumetric rates of radiant energy, i.e. for low light energy inputs. In these cases, the values obtained for the mean radiant energy available in the reactor $\int_0^\infty 4\pi J_\lambda \, d\lambda$ are less than 10 $W.m^{-2}$, which corresponds to the maximum quantum yield for photosynthesis.

It should be emphasized that the general behaviour of the thermodynamic efficiency can be represented solely as a function of the volumetric rate of radiant energy absorbed whether batch or continuous cultures are concerned. This quantity appears to be a key variable for correlating the conversion yield of light energy into chemical affinity.

7 Conclusion

Given the obvious difficulties in representing photosynthetic micro-organisms growth kinetics, it is useful to consider the overall strategy that can be applied.

Clearly, the modeling of growth can be systematically forced using a stoichiometric approach which insures the elemental conservation balances are satisfied. At the simplest level, i.e. a single stoichiometric equation, the determination of a fixed elemental composition for biomass enables one to calculate the stoichiometric coefficients corresponding to biomass growth and thereby to predict, without requiring any experimental data, all the conversion yields that are assumed to remain constant during the growth process.

This stoichiometric model can be used to improve the physical light transfer models responsible for light energy distribution through an absorbing and scattering dense medium. The models presented are of varying mathematical complexity. The simplest Lambert-Beer model fails to represent correctly the culture behaviour with a fixed physical coefficient (light attenuation coefficient) and constant physiological coefficients. Schuster's model, which has been explored for different bioreactor geometries, has an acceptable mathematical complexity without excessive loss of accuracy in representing the physics of the problem. The concept of working illuminated volume affords a tractable system of equations that enables computer simulations and the identification of the physiological coefficients from experimental data. The most elaborate model for the description of the light energy transfer involves numerical gridding techniques for solving the radiative transfer equations. The main point is certainly that it is possible to represent the behaviour of *Spirulina platensis* cultures over a wide range of experimental conditions (rectangular and cylindrical photobioreactors, batch and continuous cultures, incident light flux ranging between 5 and 300 $W.m^{-2}$) with constant physiological coefficients provided the light transfer problem is correctly handled.

The influence of mineral limitations (nitrogen, sulphur, phosphorus) has been studied via a compartment model allowing a variable biomass composition to be represented as a function of the depletion of the different nutrients. The mathematical model involves independent kinetic rates for proteins, phycocyanins and other cell mass components (carbohydrates, nucleic acids, lipids) and absorption and scattering coefficients depending on the phycocyanin content of the cells. The model is able to represent the consumption of pigments under nitrogen or sulphur deprivation and relative accumulation of glycogen. The effect of phosphate deprivation is less well understood, probably due to large intracellular reserve pools.

Very little reliable experimental data is available for studying the influence of CO_2–HCO_3^-–CO_3^{2-} limitation. The qualitative presentation reported here shows that this complex problem involves physiological metabolic regulation of the Calvin cycle, gas–liquid transfer characterization inside the medium, chemical equilibria between the different forms of dissolved carbon, strongly dependent on the pH value and, to a lesser extent, total ionic strength.

Biochemically structured models are promising tools for characterizing the orientations of the metabolism by the determination of the $P/2e^-$ ratio. The theoretical calculations, which involve the theory of linear energy converters, enable the effects of the limitations to be satisfactorily interpreted at a metabolic level. As predicted by the theory, the macroscopic entropy and enthalpy balances established for the growth of *Spirulina platensis* show that the global thermodynamic efficiency for light energy conversion decreases when light energy availability in the medium is increased, which corresponds to a highest value of 15% for low energy inputs. This clearly demonstrates the promising features of such analysis for the determination of the yields and of the reaction rates during growth and/or metabolite production processes of photosynthetic micro-organisms.

Acknowledgements. The authors are grateful to the European Space Agency (ESA) for their financial support through the MELiSSA project, and the Centre National d'Etudes Spatiales (CNES France).

Some experimental determinations were performed at the CNRS, Laboratoire de Biochimie Fonctionelle des Membranes Végétales (Gif sur Yvette-France). We are especially indebted to G. Dubertret and A. Tremolières.

8 References

1. Roels JA (1983) Energetics and kinetics in biotechnology. Elsevier Biomedical Press
2. Van Eykelenburg C (1979) Antonie van Leeuwenhoek 45: 369
3. Ciferri O (1983) Microbiol Rev 47: 551
4. Clement G (1975) Nutr Rev Alim 29: 477
5. Fox RD (1986) Algoculture: la Spirulina, un espoir pour le monde de la faim. Edisud
6. Sekharam KM, Venkataraman LV, Salimath PV (1989) Food Chemistry 31: 85
7. Cornet J-F (1992) Thesis n° 1989, University of Paris XI, Orsay, France
8. Vincenzini M, Sili C, De Philippis R, Ena A, Materassi R (1990) J Bacteriol 172: 2791
9. Filali-Mouhim R, Cornet J-F, Fontaine T, Fournet B, Dubertret G (1993) Biotech Letters 15: 567
10. Nicholls DG, Ferguson SJ (1992) Bioenergetics 2. Academic Press
11. Miller AG, Colman B (1980) J. Bacteriol 143: 1253
12. Coleman JR, Colman B (1981) Plant Physiol 67: 917

13. Badger MR, Andrews TJ (1982) Plant Physiol 70: 517
14. Kaplan A, Zenvirth D, Reinhold L, Berry JA (1982) Plant Physiol 69: 978
15. Volokita M, Zenvirth D, Kaplan A, Reinhold L (1984) Plant Physiol 76: 599
16. Badger MR, Bassett M, Comins HN (1985) Plant Physiol 77: 465
17. Gonzales de la Vara L, Gomez-Lojero C (1986) Photosynthesis research 8: 65
18. Myers J, Kratz KA (1955) J Gen Physiol 39: 11
19. Cheng KH, Colman B (1974) Planta 115: 207
20. Lloyd NDH, Canvin DT, Culver DA (1977) Plant Physiol 59: 936
21. Birmingham BC, Coleman JR, Colman B (1982) Plant Physiol 69: 259
22. Nilsen S, Johnsen O (1982) Physiol Plant 56: 273
23. Smith AJ (1983) Ann Microbiol 134B: 93
24. Carr NG, Whitton BA (1973) The biology of blue green algae. Botanical Monographs, Blackwell Scientific Publications, vol 9
25. Fogg GE, Stewart WDP, Fay P, Walsby AE (1973) The blue green algae. Academic Press
26. Carr NG, Whitton BA (1982) The biology of blue green algae. Botanical Monographs, Blackwell Scientific Publications, vol 19
27. Fay P, Van Baalen C (1987) The cyanobacteria. Elsevier
28. Dussap CG (1988) PhD Thesis, University B. Pascal, Série E, n° 409, Clermont-Ferrand, France
29. Oura E (1983) In: Rehm HJ and Reed G (Eds) Biotechnology 3: 1
30. Stouthamer AH (1973) Antonie Van Leeuwenhoek 39: 545
31. Tremolières A (1990) CNRS-IBP Orsay. Personal communication
32. Zarrouk C (1966) Thesis, University of Paris, France
33. Lee HY (1986) Master thesis, Manhattan University, Kansas, USA
34. Vonshak A, Guy R, Poplawsky R, Ohad I (1988) Plant Cell Physiol 2: 721
35. Peschek GA (1987) In: Fay P and Van Baalen C (Eds) The Cyanobacteria. Elsevier, p 118
36. Chandrasekhar S (1960) Radiative transfer. Dover publications Inc.
37. Siegel R, Howell JR (1992) Thermal radiation heat transfer, Third edition. Hemisphere publishing corporation, USA
38. Cornet J-F, Dussap CG, Gros J-B (1994) A.I.Ch.E. Journal 40: 1055
39. Aiba S (1982) Ad Biochem Eng 23: 85
40. Smith RA (1980) Ecological Modeling 10: 243
41. Cosby BJ, Hornberger GM (1984) Ecological modeling 23: 1
42. Ogawa T, Terui G (1970) J Ferm Technol 48: 361
43. Ogawa T, Kozasa H, Terui G (1971) J Ferm Technol 50: 143
44. Ten Hoopen HGJ, Roels JA, Van Gemert JM, Nobel PJ, Fuchs A (1981) In: Advances in Biotechnology, Int Ferm Symp Pergamon Press, 315
45. Frohlich BT, Webster JA, Attai MM, Shuler ML (1983) Biotech Bioeng Symp 13: 331
46. Cornet J-F, Dussap CG, Dubertret G (1992) Biotech Bioeng 40: 817
47. Fredrickson AG, Brown AH, Miller RL, Tsuchiya HM (1961) ARS Journal 1429
48. Bird RB, Stewart WE, Lightfoot EN (1960) Transport phenomena. J Wiley and Sons Inc
49. Leifer A (1988) The kinetics of environmental aquatic photochemistry, theory and practice. ACS Professional Reference Book, USA
50. Spadoni G, Bandini E, Santarelli F (1978) Chem Eng Science 33: 517
51. Cornet J-F, Dussap CG, Gros J-B, Binois C, Lasseur C (1995) Chem Eng Science 50: 1489
52. Stramigioli C, Spiga G, Santarelli F (1982) Ind Eng Chem Fundam 21: 119
53. Incropera FP, Thomas J-F (1978) Solar Energy 20: 157
54. Daniel KJ, Laurendeau NM, Incropera FP (1979) ASME J Heat Transfer 101: 63
55. Chu CM, Churchill SW (1955) J Phys Chem 59: 855
56. Boffi VC, Santarelli F, Stramigioli C (1977) J Quant Spectrosc Radiat Transfer 18: 189
57. Schuster A (1905) Astrophys J 21: 1
58. Shibata K (1958) J Biochem 45: 599
59. Kurata H, Furusaki S (1993) Biotechnol Prog 9: 86
60. Cornet J-F (1996) Technical note 23.4 ESA, PRF 141315 MOU ECT/FG/CB/ 95.205
61. Van de Hulst (1981) Light scattering by small particles. Dover Publications Inc.
62. Duntley SQ (1963) Journal Opt Soc Am 53: 214
63. Wyatt PJ (1968) Applied Optics 7: 1879
64. Göbel F (1978) Biochim Biophys Acta 588: 593
65. Houf WG, Incropera FP (1980) J Quant Spectrosc Radiat Transfer 23: 101

66. Alfano OM, Negro AC, Cabrera MI, Cassano AE (1995) Ind Eng Chem Res 34: 488
67. Cabrera MI, Alfano OM, Cassano AE (1995) Ind Eng Chem Res 34: 500
68. Cassano AE, Martin CA, Brandi RJ, Alfano OM (1995) Ind Eng Chem Res 34: 2155
69. Pasquali M, Santarelli F (1996) AIChE Journal 42: 532
70. Cornet J-F, Marty A, Gros J-B (1997) Biotechnol Prog 13: 408
71. Viskanta R, Toor JS (1972) Wat Res Research 8: 595
72. Pironneau O (1989) Finite element methods for fluids. J Wiley and Sons Inc.
73. Cornet J-F, Dussap CG, Cluzel P, Dubertret G (1992) Biotech Bioeng 40: 826
74. Allen MM, Smith AJ (1969) Arch Microbiol 69: 114
75. Lau RH, Mac Kenzie MM, Doolittle FW (1977) J Bacteriol 132: 771
76. Yamanaka G, Glazer AN (1980) Arch Microbiol 124: 39
77. Lehman M, Wober G (1976) Arch Microbiol 111: 93
78. Allen MM, Hutchison F (1980) Arch Microbiol 128: 1
79. Stevens SE, Balkwill DL, Paone DAM (1981) Arch Microbiol 130: 204
80. Allen MM, Law A, Evans EH (1990) Arch Microbiol 153: 428
81. Boussiba S, Richmond AE (1980) Arch Microbiol 125: 143
82. Cohen-Bazire G, Bryant DA (1982) In: Carr NG and Whitton BA (Eds) The biology of cyanobacteria. B.A. Botanical Monographs, Blackwell Scientific Publication 19: 143
83. Wanner G, Henkelman G, Schmidt A, Kost HP (1986) Z Naturforsch 41: 741
84. Healey FP, Hendzel LL (1975) J Phycol 11: 303
85. Curless CE (1986) Master Thesis, Manhattan University, Kansas, USA
86. Cornet J-F, Dussap CG, Gros J-B (1993) Technical note 19.2 ESA, PRF 130820
87. Lu X (1996) Master Thesis, University of Bruxels, FAME, Belgium
88. Tsygankov AA, Laurinavichene TV (1996) Biotech Bioeng 51: 605
89. Okada M, Sudo R, Aiba S (1982) Biotech Bioeng 24: 143
90. Lehman JT (1978) J Phycol 14: 33
91. Miller AG, Turpin DH, Canvin DT (1984) Plant Physiol 75: 1064
92. Edwards TJ, Maurer G, Newman J, Prausnitz JM (1978) AIChE Journal 24: 966
93. Cornet J-F, Dussap CG, Gros J-B (1995) Technical note 19.4 ESA, PRF 130820
94. Sherwood TK, Pigford RL, Wilke CR (1975) Mass Transfer. Mc Graw Hill
95. Erickson LE, Curless CE, Lee Y (1987) Annals NY Academy of Sciences 308
96. Alberty RA, Goldberg RN (1992) Biochemistry 31: 10610
97. Miller SL, Smith-Magowan D (1990) J Phys Chem Ref Data 19: 1049
98. Prigogine I (1967) Thermodynamics of irreversible processes 3rd edition. Interscience publishers
99. Glansdorff P, Prigogine I (1971) Thermodynamic theory of structure, stability and fluctuations. J. Wiley Interscience
100. Walz D (1990) Biochim Biophys Acta 1019: 171
101. Hill TL (1977) Free energy transduction in biology. Academic Press
102. Stucki JW (1980) Eur J Biochem 109: 269
103. Aiba S, Ogawa T (1977) J Gen Microbiol 102: 179
104. Ogawa T, Aiba S (1978) J Appl Chem Biotechnol 28: 515
105. Ogawa T, Fujii T, Aiba S (1978) Biotech Bioeng 20: 1493
106. Lee YK, Pirt SJ (1982) Biotech Bioeng 24: 507
107. Aiba S, Ogawa T (1983) Biotech Bioeng 25: 2775
108. Lee YK, Erickson LE, Yang SS (1984) Biotech Bioeng 26: 926
109. Iehana M (1987) J Ferment Technol 65: 267
110. Lee YK, Erickson LE, Yang SS (1987) Biotech Bioeng 29: 832

Received May 1997

Apoptosis and Cell Culture Technology

Mohamed Al-Rubeai
Centre for Biochemical Engineering, School of Chemical Engineering, University of Birmingham, Birmingham B15 2TT, UK

The importance of apoptosis in cell culture has now been widely recognised. Recent work has shown that this process is widely manifested during the in vitro cultivation of commercially important mammalian cell lines.

In this review I summarise what is now known of the characteristics, significance and regulatory mechanisms of apoptosis. As the process of cell proliferation and cell death are now considered intimately related, particular attention is paid to highlight the progress and opportunities in the field of cell culture engineering. The strategies that have been undertaken to prevent the induction of apoptosis in cell culture and those which have been suggested as possibilities to improve culture productivity through the apoptosis route are discussed with given examples.

Advances in Biochemical Engineering/
Biotechnology, Vol. 59
Managing Editor: T. Scheper

1 Introduction

Today, the biotechnology industries are searching for and executing strategies to sustain themselves as they enter the new millennium. The survival time for biotech companies is getting shorter, which means fewer of them can afford the time and money involved in bringing new products to the market. There are more than 400 human health care products in development and over 40 therapeutic biologics have been approved and a significant number of these products are produced in animal cells. It is therefore not surprising to see animal cell culture continuing to play an increasingly important role in the production of biopharmaceuticals.

Significant strides have been made over the past several years in the development of process strategies for animal cells that provide optimal growth conditions and maximise production (Table 1). The understanding of the regulation of cell growth, cell death, and synthesis and secretion of protein products has been increasingly utilised for effective design and operation of both simple and intensive production processes. Nevertheless, the scale up and optimisation of animal cell cultures to meet production demands has been problematic due to the low productivity and instability of the cell lines used. There are substantial variations in most aspects of physiological response and the specific kind of differentiation which yields the most product. Animal cells are also sensitive to the hydrodynamic forces generated by mechanical agitation or by bubble areation [1–4]. Even oxygen from bubble or bubble-free aeration can have a detrimental effect on the cells [5]. Another important factor which disadvantage animal cell culture processes is the complex medium requirement that normally contains serum and growth factors. Considering that nutrient limitation is often the rate-limiting step for increasing cell number, increasing viability and prolonging culture duration, many strategies have been adopted to overcome this problem, such as the use of controlled feeding technique using single or multi-nutrient feeds, and the application of rational medium design protocols.

Many of the problems associated with growth, metabolism and production in recombinant cell lines can now be dealt with through a new approach which

Table 1. Process strategies for the production of biopharmaceutical products from animal cells

Process strategy	Example
Maximisaton of viable cell number	Nutrients feeding
Improvement of medium formulation	Serum and protein free media
Implementation of high-performance reactor configurations	Spin filter perfusion system
Maximisation of product production rate per cell	Efficient expression system, cell cycle control

Table 2. Control and improvement of product expression

1. Transfection and expression
2. Metabolic engineering
   a. Design or redirection of metabolic pathways
      1. Reduction of toxic metabolic production
      2. Increase in the rate of substrate uptake and product formation
      3. Creation of novel pathways
   b. Controlling proliferation/improving adaptability
   c. Controlling cell death/increasing viability
3. Process engineering

Table 3. Increased cell viability and survivability: The advantages

1. Increased number of cell factories
2. Enhanced adaptation to nutrient limited conditions and protein-free medium
3. Reduced DNA and proteases contamination
4. Efficient control of cell division
5. Prolonged and increased viral replication
6. Improved efficacy and fidelity of products

involves the direct modification of the function of metabolic networks. The improvement of cellular activities by metabolic engineering of enzymatic, transport and regulatory functions of cells have been widely accepted as a way forward for integrated process development [6]. The use of rDNA technologies to achieve controlled proliferation, enhanced survival and improved cellular responses to the external environment by manipulating the regulatory pathways of cell death and cell division should enable more effective and more specific strategies for product improvement (Table 2).

In animal cell culture, viability remains the single most important parameter by which culture success can be judged. Proteases are released from necrotic cells, and their presence in the culture supernatant may result in the degradation or alteration of the desired product. Furthermore, the presence of debris during the declining phase of the culture can adsorb product thus highly complicating the purification process [7] The advantages of increasing viability and survivability of cells in culture is outlined in Table 3. An important aspect of cell viability is that the post-translational modification including glycosylation of protein products can be altered during cell death with negative implications for the pharmaceutical industry. It has been reported that in batch cultures of CHO cells glycosylation of interferon-γ deteriorates reproducibily with time with non-glycosylated IFN-γ reaching up to 25% of the total product during the mid death phase [8].

2 Growth, Survival and Death

Large scale production of biopharmaceuticals normally utilise "transformed" cell lines that are characterized by a finite lifespan, less dependency on surface adherance and a lower requirement for serum. These continuous cell lines have the advantages of greater growth rates, higher cell densities, higher yield and general ease of maintenance in simple media [9]. Generally, in all cell lines, the growth rate and growth extent are determined by the cell cycle duration, number of cell divisions and cell deaths. The molecular machinery of the cell cycle and the checkpoints that help to ensure proper progression of cells from G1 to M have been extensively studied in the last few years [10–13]. The progression from G1 to M is controlled at least in part by a family of enzymes called the cyclin-dependent kinases (CDKs) which, in turn, are activated by a group of positive regulators called cyclines [14–15]. Several of these cell cycle regulators have been shown to play an important role in the development of cancer, a disease characterised by loss of cellular growth control. Like cancer cells, "transformed" or continue cell lines do not stop dividing for the purpose of repairing lesions but cell division continue as long as there is sufficient nutrients and that other suitable chemical and physical conditions for survival are maintained. Depletion of nutrients results in a reduction of DNA and protein synthesis, thereby including cell death. The cells respond to chemical and physical stress by altering their morphology and metabolism as well as invoking repair mechanisms (Fig. 1). In normal tissue a delicate balance exists between and within cell populations which is controlled by the regulation of proliferation, differentiation and death. During development of animal organs both cell proliferation and cell death processes interact to regulate and maintain the proper number and type of cells. In cell culture, such a balanced state is absent and cells therefore continue to divide forming a physiologically and morphologically uniform population until death is triggered. Attempts to separate cell growth from cell death for the purpose of maintaining nondividing cells for high differentiated product production have been unsuccessful.

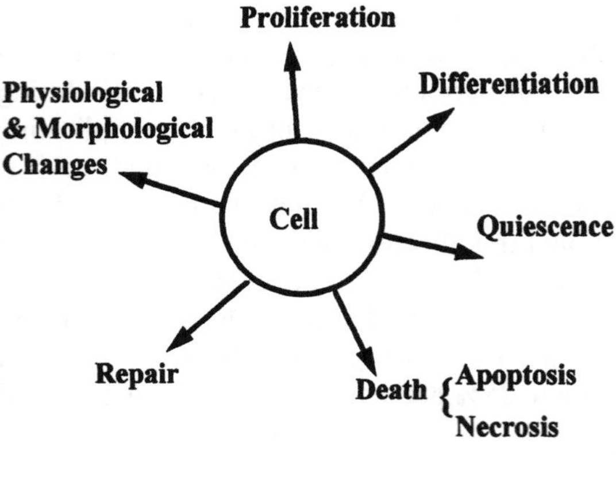

Fig. 1. Cellular response to environmental stimuli

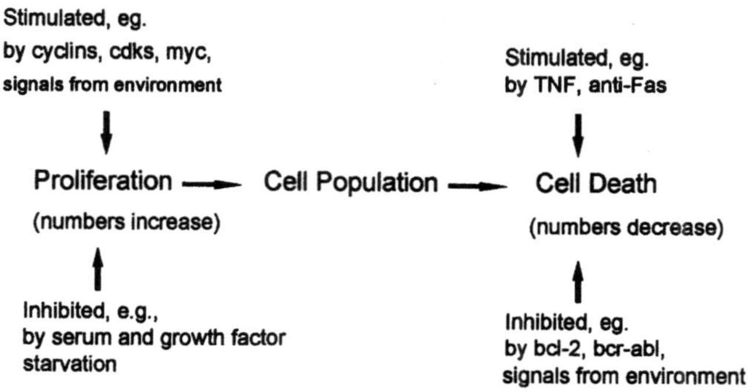

Fig. 2. Interrelationship between cell proliferation and cell death in determining population size (based on: Williams GT (1991) Cell 65: 1097).

In contrast to normal cells or finite cell lines, the continuous cell line cannot be kept quiescent and viable over an extended period. In the former type of cells, characterised by the occurrence of a contact-inhibited quiescent state, recent evidence suggests that the signalling for cell death may have something in common with early cell cycle events, specifically the G1 phase. A molecular key appears to determine whether cells go on to the S phase and survive, or to leave the cell cycle and die. Continuous cell lines, however, lack such mechanisms which arrest the cell in G1 and consequently invokes cell death in the absense of survival factors. Instead it has been observed that cell death occurs during all phases of the cell cycle and is not restricted to any particularly "quiescence or non-growing" phase [16]. Although in some cell types as thymocytes induced death normally occurs after cell arrest at the G1/S border [17], it has been found that arrest at G1 is not a prerequisite and that the cells can die at many different points in the cycle [18]. Uncoupling cell death and cell proliferation remains an intriguing task for animal cell technologists which, if it is achieved, would result in a significant cost reduction and would allow the operators to establish an efficient integration of process operations of high predictability and reliability. In Fig. 2, we summarise the essential elements of the interrelationship between cell proliferation and cell death with illustration of the various factors involved in the regulation of both processes, some of them will be dealt with in more detail in this review.

3 Programmed Cell Death

Cells may die in two ways, necrosis and apoptosis (Fig. 3). Necrosis is an accidental cell death process in which the cell typically exhibits distinctive morphological and physiological characteristics [19] including changes in mitochondrial morphology and function and in the ability of the plasma

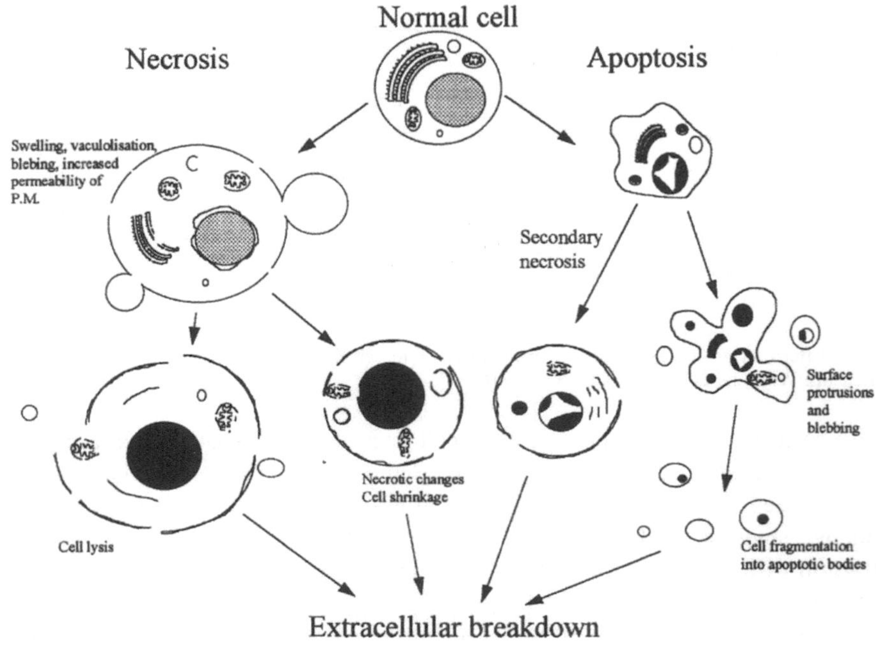

Fig. 3. Morphological features of the different stages in apoptosis and necrosis

membrane to regulate osmotic pressure. Initially a progressive hydration of the cytoplasm occurs followed by membrane and organelle disruption allowing the leakage of the lysosomes into the cytoplasm and finally nuclear disintegration and complete disruption of the cell take place. The injury to membranes as described by Buja et al. [19], involves three stages: 1) discrete alterations in ionic transport systems of membranes, 2) a non-specific increase in membrane permeability and 3) physical membrane disruption. Necrosis often involves cell swelling rather than cell shrinkage; however, the mode depends on the speed of injury. Coagulation necrosis (sometimes called oncosis), which results from denaturation and coagulation of cellular proteins, is pathological cell death that can be triggered by factors from outside the cell such as hypoxia, hyperthermia, physical damage and toxic injury. It is characterized by cellular swelling, blebbing, increased membrane permeability due to the failure of the ionic pumps, karyolsis and DNA fragmentation in a nonspecific fashion [20]. Other types including colliquative necrosis have also been described based on the specific feature of the disease process and the tissue involved. A consequent inflammatory reaction in adjacent tissue usually results due to the leakage of cellular contents into the extracellular space before the cell rapidly disintegrates.

Apoptosis is, by contrast, a regulated and controlled process, determined in a generally pre-programmed manner. There is increasing evidence that this self-demise process occurs in most animal cell types and is controlled by a putative "yin-yang" mechanism involving the activation of killer genes and the deactivation of survival genes. Clearly, this raises the possibility that modifying the apoptotic response of cells in the bioreactor may suppress or delay the induction of death in response to chemical and physical stress. This may lead to the generation of more robust cells lines, which are able to survive more extreme bioreactor operation strategies.

3.1 *Morphological Features of Apoptosis*

Apoptosis is characterised by features which contrast markedly with those of necrosis and, because of its significance in the process of tissue regulation, it is now the subject of intense research activity. Important features seen during apoptosis in cultured cells (Fig. 4) are the condensation, margination and fragmentation of chromatin, loss of volume and radical changes in the cellular morphology which includes the modification of the cytoskeleton. Following these events the plasma membrane convolutes and losses its microvilli and

a

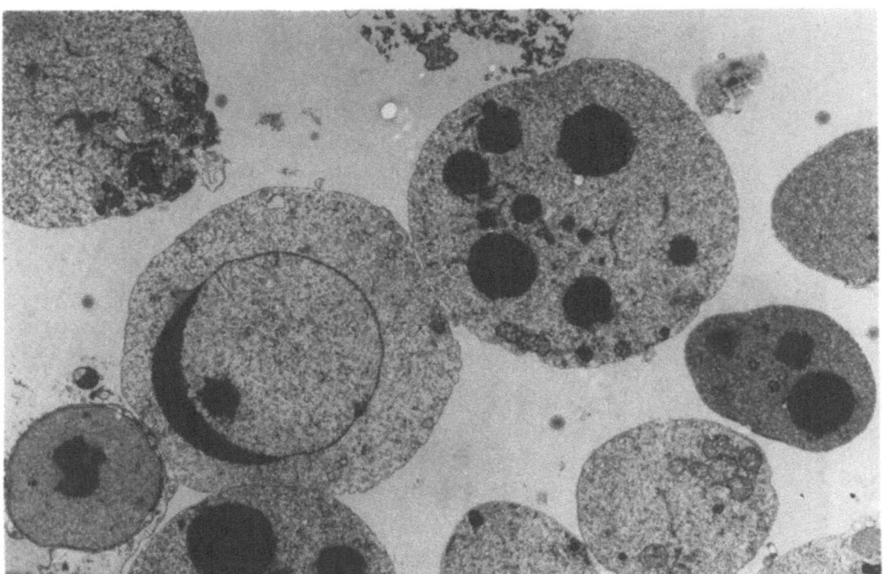

Fig. 4. a Transmission electron micrograph of apoptotic Burkitt lymphoma cells (from: Milner et al., 1996 [22]). **b** The morphological features of apoptosis and necrosis in in vitro culture in mammalian cells visualised by staining with acridine orange and propidium iodide

b Non-fixed cells stained with a mixture of Acridine Orange (A O) and Propidium Iodide (PI)

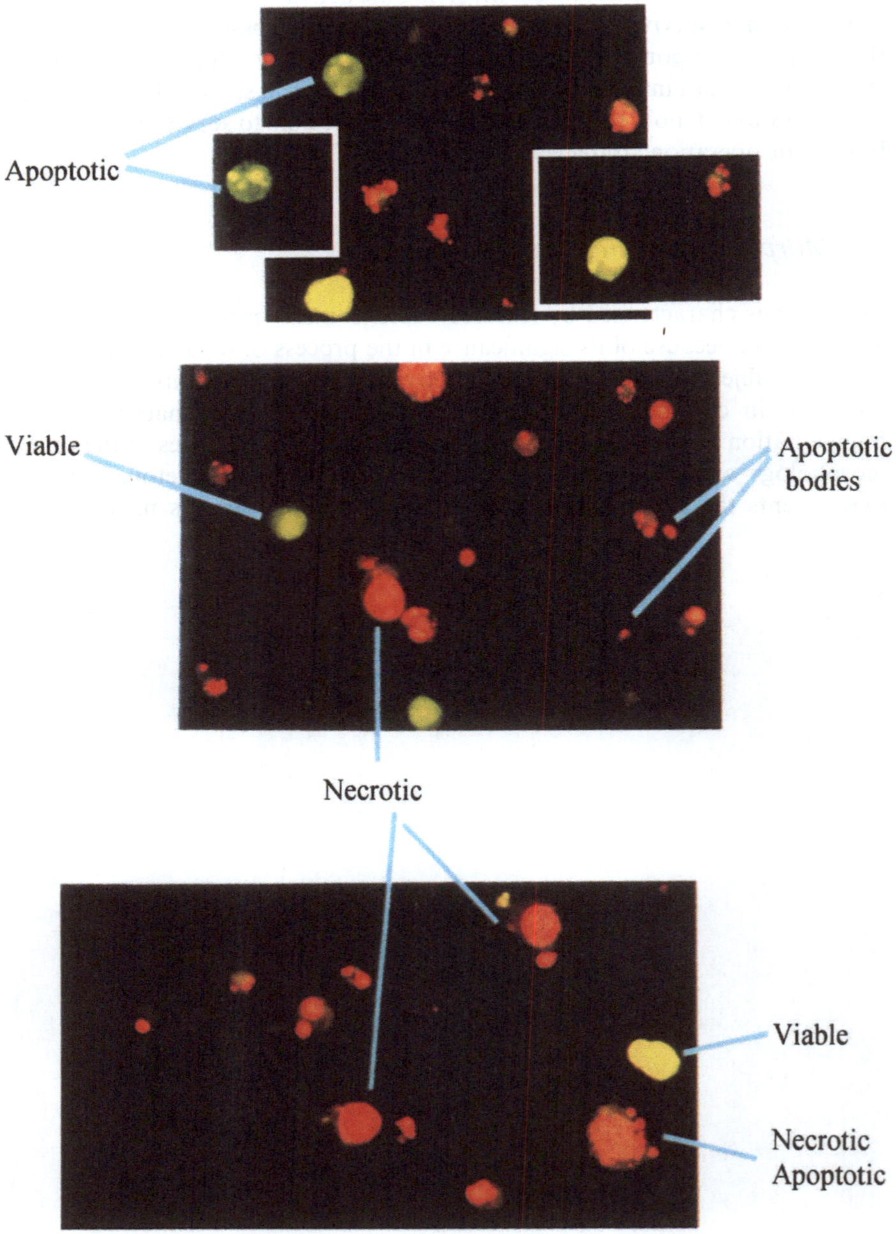

Fig. 4. continued

eventually the cell fragments to form apoptotic bodies that contain cytoplasmic organelles and, usually, nuclear fragments. In the body, the apoptotic bodies are immediately recognised and engulfed by phagocytic or nearby cells. The recognition of apoptotic cells by phagocytes has been suggested as due to the movement of membrane phospholipids such as phosphatidylserine from the inner side to the outer side of the plasma membrane.

One of the major distinguishing feature of apoptosis, which serves as the most characteristic biochemical feature of the process, is the extensive damage to nuclear chromatin so that extracted DNA runs as a characteristic ladder of oligonucleosomal fragments. The DNA fragmentation appears to take place in an orderly fashion. First, large fragments of 300–750 kb are formed and these are degraded into smaller fragments (50–70 kb) followed by the formation of nucleosome size pieces of around 180 bp (Fig. 5). Although the internucleosomal fragmentation is always associated with apoptosis, this type of cell death is not strictly defined by internucleosomal fragmentation and may involve other types of DNA fragmentation.

Apoptosis, especially in non-haematopoietic cell types, can have atypical features which make its identification difficult if not sometimes, impossible (Z. Darzynkiewicz, personal communication). Some of the possibilities include lack of DNA ladder on gel or even lack of chromatin condensation and other morphological features. Under conditions of batch culture cultivation, we failed to find apoptosis in chinese hamster ovary (strain K1) and *Spodoptera frugiperda* (Sf9) cell lines using gel electropheresis of DNA and acridine orange staining techniques [21]. Recently, we have identified at the death phase of a recombinant CHO batch culture, apoptotic cells with slightly different morphology from those of hybridoma and plasmacytoma cell lines. It is possible that due to the simultaneous occurrence of apoptosis and necrosis, a situation which often occurs during the death phase of batch cultures as a result of the combination of nutritional and physical triggers, the typical morphological characteristics of apoptosis are concealed. Darzynkiewicz suggests that atypical apoptosis may take place when one or more pathways of apoptosis are blocked while other pathways are not, resulting in, for example, cessation of DNA cleavage at the 50–300 kb. This may be the case in some cell lines grown under specific conditions.

3.2 Methods for the Detection and Analysis of Apoptosis

Many methods are now available for the examination of apoptosis in various types of cells and tissues. Recently, articles which describe these methods comprehensively have been published by Milner et al. [22] and Darzynkiewicz et al. [23]. A book describing procedures for studying cell growth and apoptosis has also been published [24].

3.2.1 Microscopically Based Methods

The low incidence of the occurrence of apoptotic cells in tissues and their rapid disappearance due to phagocytosis makes them difficult to detect. In contrast, in vitro cell culture provides a process for significantly increasing the number of spontaneous or induced transient apoptotic cells. Microscopy can be employed to detect and examine the morphology of apoptotic cells either indirectly after processing and staining of a section or directly by in situ examination of unstained cells. Evan and co-workers used time-lapse video microscopy to monitor and study apoptotic events induced by *c-myc* in serum-deprived cells [25]. However, stains such as toluidine blue, acridine orange and silver stain have been used to demonstrate the condensation of chromatin [21, 26–28]. We use dual staining with acridine orange and propidium iodide of unfixed cells to differentiate between necrotic, apoptotic and viable cells (Fig. 4b). The technique is employed to obtain quantitative information on the kinetics of cell death in various commercially important cell lines and to assess the influence of medium components and the physical environment on cell growth and death in bioreactors. It is worth noting that this technique can recognise those cells which have undergone apoptosis and subsequently lost their membrane integrity to become 'secondary necrotic' as well as both types of apoptotic bodies- with or without chromatin. The clearest way to demonstrate the morphological characteristics of apoptosis remains electron microscopy (Fig. 4a). Although this technique is not suitable for quantitation of apoptosis, it can illustrate detailed morphological features of the process, such as the state of cytoplasmic organelles, cellular membrane and nuclear structure. Al-Rubeai et al. [29] demonstrated the occurrence of apoptosis in hybridoma cell culture and its ultrastructural features by transmission and scanning electron microscopy. Their work indicated that the cell membrane undergoes structural changes during the late stages of batch culture with the disappearance of microvilli and the appearance of blebs and deep indentations. Further ultrastructural features of apoptosis has been shown to include very late cytoplasmic damage despite the good preservation of the plasma membrane and cytoplasm [30].

3.2.2 DNA Fragments

DNA fragmentation into oligonucleosome-size fragments occurs simultaneously with the manifestation of morphological changes such as chromatin condensation. The oligonucleosomal fragments can be visualized on electrophoretic agarose DNA gel as a distinctive ladder of multiples of 180–200 bp (Fig. 5). In contrast to apoptosis, DNA fragments produced during necrosis are of irregular sizes, resulting in a smear on agarose gel. Although the ladder pattern of DNA fragments has been considered as the biochemical hallmark of apoptosis, in many cases the classical morphological characteristics of apoptosis can be manifested in the absence of DNA fragmentation [31, 32]. Moreover, it is

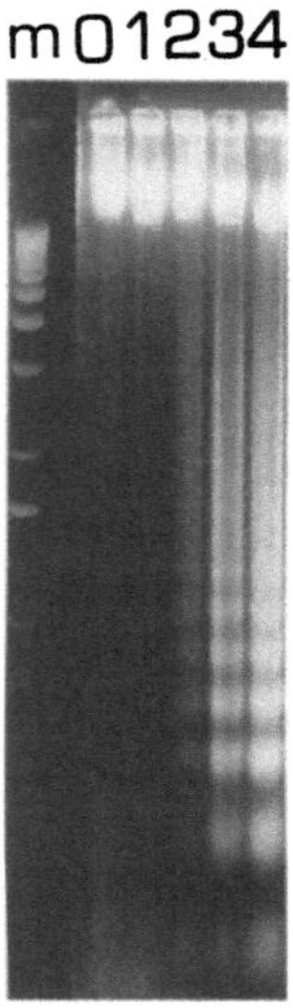

Fig. 5. DNA fragmentation in apoptosis during batch culture of hybridoma cells. *m*: molecular markers, 0–4: days in cultivation

now apparent that DNA fragmentation is much more complex than initially thought and most, if not all cells also produce much larger DNA fragments [33].

3.2.3 Flow Cytometric Methods

Flow cytometry is one of the most suitable quantitative techniques for the study of apoptosis is heterogenous cells [34]. This technique has been the centre of considerable attention, as it enables the quantitation of virtually any cell-associated property or cell organelle for which there is a fluorescent probe such as the DNA structure and metabolic parameters associated with cell death. Most of the cytometric techniques focus on two features of apoptosis which distinguished this process from necrosis: specific DNA fragmentation and intact membrane integrity.

The changes in the scattering properties of cells undergoing apoptosis has been exploited in the analysis of this processes in Burkitt lymphoma cells when deprived of serum [35]. Inducing these cells to undergo apoptosis resulted in a clear shift of the high forward light scatter (FLS)/low side light scatter (SLS) population to the low FLS/high SLS population. However, we found that the decrease in FLS (cell size) and the increase in SLS (cytoplasmic granularity) is always associated with the decline phase in batch cultures of hybridoma, CHO and myeloma cell lines where the dead cell population is composed of apoptotic, necrotic and secondary necrotic cells. In addition, light scatter measurements take no account of the apoptotic bodies which have a low FLS/low SLS. It is clear that light scatters cannot always provide adequate discrimination of apoptosis and necrosis in heterogeneous cell cultures.

Apoptotic cells are characterised by a reduction in DNA stainability with fluorescent dyes. The level of DNA staining of apoptotic cells is normally below that of G0/G1, which results in the appearance of a distinct cell cycle region on the DNA histrogram (sub-G1 peak). In hybridoma batch culture the sub-G1 peak is clearly evident on day 2 when the cell viability as measured by the trypan blue exclusion is 95% (Fig. 6). The size of this peak increases with the age of the culture and its distinction as a peak becomes less visible as the viability decreases. The increasing sub-G1 population merges into the cell debris which suggests that, during the death phase of suspension batch culture, the cells begin to disintegrate into apoptotic bodies and small fragments. Indeed, the sub-G1

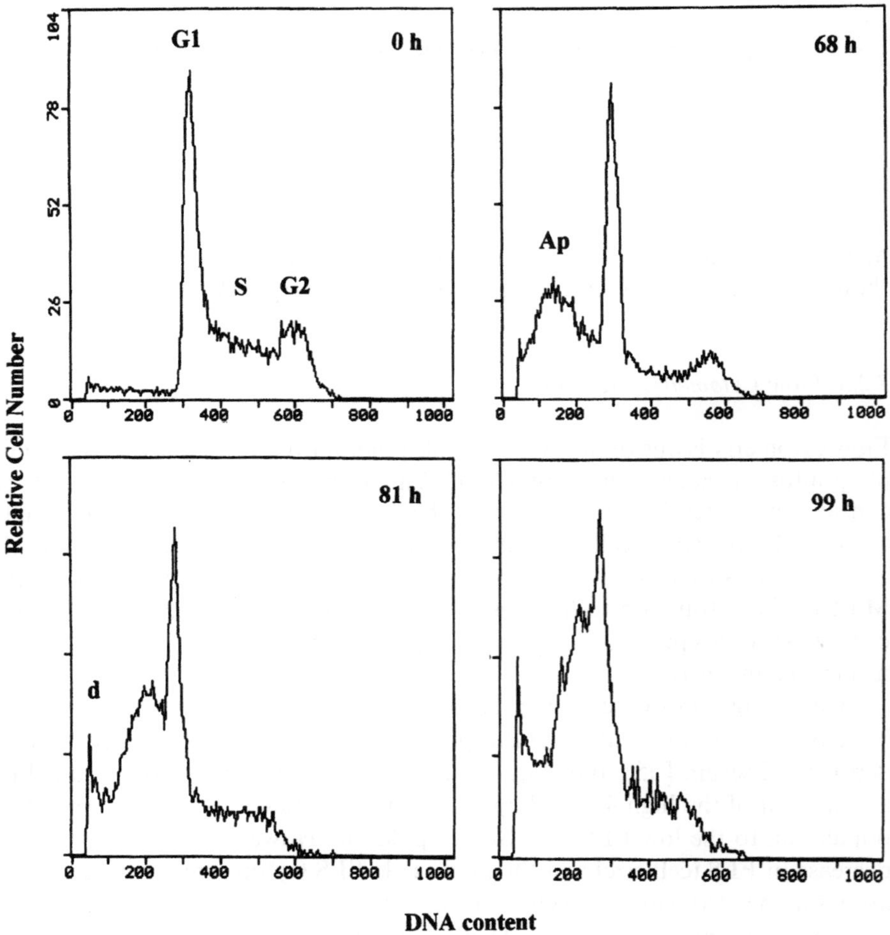

Fig. 6. Demonstration of apoptosis in hybridoma batch culture by flow cytometric measurement of DNA distribution. *Ap*: apoptotic cell population; *d*: nuclear debris

population may represent fragments of apoptotic and necrotic nuclei, individual apoptotic bodies containing various amounts of condensed chromatin and individual chromosome or chromosomal aggregates from used mitotic cells.

Labelling of DNA strand breaks generated during DNA fragmentation can be accomplished by labelling of the 3'OH termini at sites of DNA cleavage with biotin conjegated nucleotides in a reaction catalysed by exogenous TdT [36, 37] or DNA polymerase (nick translation) [38]. Gavrieli et al. [36] have given the name 'TUNEL' which stands for 'Terminal transferase Utilizing Nick End Labelling' for the above method of detecting apoptotic DNA strand breaks. A kit called ApoptTag for the detection of apoptosis based on this technique is now available from ONCOR. The method was later simplified by a single step labelling DNA strand breaks with deoxynucleotides directly conjugated with a fluorescent molecule, using exogenuous TdT [39]. Using flow cytometry, this method proved satisfactory for the detection of apoptosis in HL-60 and lymphoma cultures.

There are several other flow cytometric techniques in use for assessing cell viability. These techniques are based on the assessment of the structural and functional integrity of the plasma membrane using, for example, the exclusion of propidium iodide (PI) and cellular retention of fluorescin diacetate. The exclusion of PI is also used in combination with uptake of the cationic flurochrome Rh123, normally used to study mitochondrial activity. Live cells with active mitochondria exclude PI and retain Rh123. While these methods are adequate to identify necrotic cells they cannot discriminate between viable and apoptotic cells nor between necrotic and secondary necrotic cells. Generally they can be functionally considered as similar to the non-flurochrome based cytotoxicity assays such as trypan blue exclusion, release of ^{51}Cr or LDH from damaged cells, uptake of neutral red into lysosomes and conversion of MTT to formazan via the mitochondrial dehydrogenase activity.

Finally, it is important to recognise that although flow cytometry is quantitative, reproducible and can be applied to a variety of cell types, the observed morphological characteristics of cells undergoing apoptosis vary not only between cell types, but also between different triggers of apoptosis and can be absent or misleading in some cases.

4 Mechanisms and Death Signals

The activation of the apoptotic pathways invovles a diverse variety of physiological and non-physiological triggers determined by the presence or absense of extracellular signals. Examples of the signals which trigger apoptosis are γ-irradiation [40], glucocorticoides [41], interaction with a surface receptor, e.g. Fas or APO-1 [42] and removal of essential growth factors and/or survival factors [e.g. 43], (see Table 4).

Table 4. Apoptosis induced by the removal of growth factors, mitogens and hormones (based on Collins KL and Lopez R 1993 TIBS 18, 307; Vermes I. and Haanen C. 1994 In: Advances in Clinical Chemistry, Vol 31, p 177, Academic Press)

Cell type	Factor
Myeloid progenitor	Granulocyte colony stimulating factor
T lymphoctye	Interleukin 2, 4
B lymphocyte	Antigen
Megakaryoblasts	Interleukin 11
Eosinophil	Interleukin 5
Neurons	Nerve growth factor
Endothelium cells	Fibroblast growth factor
Erythroid progenitor cells	Erythropoietin
Prostate	Testostrone
Adrenal cortex	Adrenocorticotropic hormone
Liver cells	Phenobarbital, cyproterone
Fibroblast constitutively expressing *c-myc*	serum

A cell invariably responds to such "death signals" in different ways depending on the nature of the signal and on cell behaviour [44]. Glucocorticoids have been shown to activate Ca^{2+} independent protein kinase C (PKC), while the protective signals is provided by an increase in the cytosolic Ca^{2+} level and proper activation of PKC [45]. Iwata suggested that the mechanism of glucocorticoid-induced apoptosis is thymocytes may have some relation to the mechanism of thymic selection that determine the T cell reprertoire. Moreover, the apoptotic reaction of thymocytes and in the immune system can be the consequence of the "unbalancing" of many different transduction signals which may act upon particular switch and effector [46]. A large number of studies supports the view that Ca^{2+} plays an important role as a signaling molecule in apoptosis [47, 48]. However, Ca^{2+} is not involved in all forms of apoptosis nor in the regulation or inhibition of the process [48, 49]. Calcium can activate a number of latent enzymes that are involved in apoptosis such as the endonuclease, protease, phospholipase, and transglutaminase. The activation of endonuclease, probably via a mechanism that involves the function of a nuclear Ca^2 pump, leads to the cleavage of DNA into oligonucliotide fragments. Activation of the transglutaminase enzyme in apoptotic cells results in the assembly of a highly cross-linked intracellular protein network [50]. It may be that the apoptotic cell strengthens its membranes against the risk of lysis by the expression of the transglutamine which explains why apoptotic cells are robust and sealed with the intact osmotic gradients, while necrotic cells are fragile and easy to break up.

The interaction with a certain class of surface receptor can trigger an efficient apoptosis pathway. The Fas antigen (also called CD95 or APO-1), a member of the low affinity nerve growth factor (NGF) superfamily of proteins which also includes tumor necrosis factor (TNF), is thought to be an important molecule in the regulation of normal immune responses. Interaction with anti-Fas antibody or with cells expressing Fas ligand (FasL) induces apoptosis in many lymphoid and non-lymphoid cell types [51–53]. However, due to the inability of anti-Fas antibody to induce apoptosis in all myeloma cells it has been suggested that variability in Fas expression may determine sensitivity to apoptosis [54].

The apoptosis pathway may be regulated in a number of different ways, depending on the stress signal, the expression of several genes and the microenvironment at the site of activation. Apoptosis has been shown to require the active synthesis of RNA and protein. In a number of circumstances the suppression of mRNA or protein by inhibitors such as actinomycin D and cycloheximide has led to the activation of apoptosis [55–59]. By contrast, these inhibitors in some other situations either have no blocking effect or can actually induce apoptosis [60, 61]. To explain such an apparent contradiction, it has been suggested that if protein synthesis is inhibited, for apoptosis to continue the cell must have all the necessary machinery for invoking the process of cell death [41]. It is also suggested that cells may contain regulatory proteins that can either inhibit or promote cell death and that catabolism of such proteins determines whether stopping RNA or protein synthesis suppresses or induces apoptosis [62]. It is increasingly clear that the signalling molecules in apoptosis are also involved in other phenotypic expression such as proliferation and differentiation and therefore it is difficult to separate the action of these molecules into distinct functional mechanisms.

4.1 Apoptosis Induced by Removal of Growth Factors and Nutrient Deprivation

Cells survive and proliferate in culture when they are provided with adequate amounts of nutrients and a suitable physical environment. The extracellular environment must meet the essential requirement of cells to survive and whenever these conditions cease to exist, the cells begin to die. Growth factors and energy suppliers are considered the most significant factors that influence the biological activities of cells. The depletion of these nutrients may not only block proliferation but also can induce apoptosis. It is now widely accepted that both proliferation and apoptosis are controlled by an integrated mechanism and that the signals which trigger or prevent one process also influence other aspects of cellular activity. For example, the presence of mitogens to induce cell proliferation and survival factors to inhibit apoptosis are both necessary for maintainenance of cell viability [63]. Various types of growth factors, such as insulin-like growth factor, interleukin 2, platelet-derived growth factors and macrophage colony stimulating factor, have been reported to inhibit apoptosis in many cell types (see Table 4). It is suggested that all animal cells are dependent on the presence of at least one growth factor [64]. However, the main effect of these growth factors is not necessarily to enhance both survival and proliferation as some of them are considered poor mitogens but potent survival factors and vice versa. In cell culture technology, where continuous cell lines are used, survival factors which provide anti-apoptic properties are probably the most important molecules required to maintain both high viability and growth.

Serum is essential for the growth and survival of mammalian cells. It provides growth factors, protein, lipids and other nutrients to cells in culture. The removal of serum has been shown to induce apoptosis in many cell lines such as hybridomas and plasmocytomas [21]. However, when serum or plasma is present in the medium at a high concentration, apoptotic cell death is induced, although such death can be prevented by the addition of thiol compounds such as L-cyst(e)ine and L-tryptphan [65].

The mechanism by which the removal of growth factors induces apoptosis has been shown to involve the expression of the protooncogene *c-myc* gene product. Fibroblasts that express *c-myc* grow well in serum supplemented medium but undergo apoptosis if the serum is removed [66]. It was also observed that, in the absense of growth factors and down regulation of *c-myc*, cell arrest in G_0 occurs. This suggests that *c-myc* may activate a regulatory programme that leads either to proliferation or to apoptosis.

Apoptosis also occurs when cells are deprived of energy suppliers such as glutamine and glucose. An important characteristic of apoptosis in continuous cell lines of haemopoeitic origin is that it is induced by removal of glutamine or glucose [21, 67]. The role of these molecules in the apoptotic pathways is not yet clear but since both glutamine and glucose levels regulate both the cell growth and catobolic pathway operation, their deprivation may prevent DNA synthesis and block cells into G1, thereby triggering apoptosis. This suggestion, however, does not explain why cells in S, G2 and M phases also undergo apoptosis. It is more likely that apoptosis is initiated when maintenance energy required to produce ATP to drive the biosynthesis of cellular material is reduced to a critical level, resulting in the activation of a regulatory protein that promotes apoptosis.

4.2 Genetic Regulation of Apoptosis

Although there have been suggestions that apoptosis may not be controlled at the gene expression level, there is today a large body of data indicating that the process of cell death is under genetic control and that the basic machinery to induce apoptosis is present in all mammalian cells. However, the regulatory pathway of programmed cell death is influenced by the type and origin of the signal which in turn either inhibits or activates certain genes. Some of the earliest evidence of the presence of the so called 'death genes' came from the studies on the nematoed *Caenorhabdities elegans*. In this organism, 3 genes have been identified: *ced-3*, ced-*4* and *ced-9* genes, the first two of which are required for all somatic cell death to occur while the third gene is required to protect cells that should survive from undergoing cell death (for review see [68]). The *ced3* gene product is similar to a family of mammalian cysteine proteases which includes interleukin-1β-converting enzyme (ICE). Many observations suggest that these proteases play important roles in apoptosis. For example, the overexpression of the murine *Ice* gene induces apoptosis in several cell lines [69]. Moreover, the cowpox virus *crmA* gene product, a potent inhibitor of ICE, has been shown to protect cells against apoptosis induced by growth factor removal [70]. Several other protease inhibitors have been reported to block apoptosis in many cells types [37]. However, using TPCK (serine protease inhibitor) and BTEE (trypsin and chymotrypsin-like enzymes inhibitor) we have been unable to obtain any positive inhibition of either spontaneous or induced apoptosis in hybridoma cultures. Many members of the ICE family have been recently identified and it seems that more than one gene is activated to ensure effective cell elimination (for review see [72]). To what extend the activation of ICE family proteases is involved in apoptosis is still unknown. Evan et al. [25] asked two questions which need to be answered to understand the possible mechanism by which ICE acts: a) How do the genes that suppress or predispose cells to apoptosis and survival factors regulate the activities of the ICE proteases? b) What are the consequences of ICE protease activity and what substrates are cleaved and to what biological end?

Some of the genes involved in the regulation of cell proliferation have been also implicated in the control of apoptosis. For examples, the *p53* tumor suppressor gene which plays a central role in the regulation of cell proliferation, has been shown to be required for mediating apoptosis induced in DNA-damaged mice [72]. *p53* functions by arresting the cell in G1 phase after DNA damage, which allows time for repair to proceed. If damage remains unrepaired the cell will then undergo apoptosis. This action of *p53* has been ascribed to its ability to induce expression of *WAF1/CIP1*, which encodes an inhibitor of G1 cyclin-dependent kinases [73]. However, *p-53* is not involved in the control of apoptosis during the normal development of mice.

Another gene that stimulate cell proliferation and is involved in apoptosis is *c-myc* (see above). Research by Gerard Evan and co-workers indicated that the *c-myc* gene product gives the cell two option: proliferate or die but the outcome depends on external survival signals such as the presence of growth factors (for review see [25]). In addition, inhibition of *c-myc* expression by antisense oligonucleotide in T-cell hybridoma cells inhibits activation-induced apoptosis [74]. The requirement of *c-myc*, like that of *p-53*, is not universal for the induction of apoptosis. Moreover, the effect of both genes in inducing apoptosis is cell type and stimuli specific.

4.3 Inhibition of Apoptosis

The other class of genes invovled in the regulation of apoptosis represents those genes which confer resistance to apoptosis. The best known example of these 'survival genes' is the mammalian protooncogene *bcl-2*, a functional equivalent to the *ced-9* gene in *C. elegans*. The action of *bcl-2* was demonstrated by cloning breakpoint in the chromosome translocation, t(14; 18); characteristic of the majoirty of follicular lymphomas. Originally it was found that *bcl-2* promoted the survival, but not proliferation of IL-3 dependent myeloid cell lines after the removal of IL-3 [75]. Moreover, transgenic mice expressing *bcl-2* develop follicular hyperplasia with large excess of B lymphocytes with enhanced survival [76]. Further studies have shown that *bcl-2* inhibits apoptosis in a variety of cell types in response to a variety of signals including chemothrapeutic drugs, ionizing radiation, cytotoxic lymphokines, growth factor deprivation, nutrient depletion, hyperoxia, hypoxia and some viruses. However, its inhibitory effect on apoptosis is not universal, as it has been shown that it does not prevent apoptosis induced by a variety of triggers including cytotoxic T-lymphocytes [77, 78].

The precise molecular function of *bcl-2* remains unclear. It has been suggested that the association between *bcl-2* and the 23 kDa p23 R-*ras* gene product may indicate a possible role of this member of the *ras* gene family in the regulation of apoptosis [79]. Studies on human breast cancer have indicated that *p53* can down regulate the expression of *bcl-2* [80], but it can upregulate the expression of *bax* which encodes a dominant inhibitor of the *bcl-2* protein that forms heterodimers with it [81]. It is probable that the ratio of Bcl-2 to Bax proteins may predetermine the apoptotic response of some cells to death triggers – higher Bcl-2 leads to survival while higher Bax leads to death [82].

Bax is a member of the Bcl-2 extended family of proteins that also includes bcl-X_s, bcl-X_L, Bad, Bak and Mcl-1. Very little is known about the interaction among these proteins but their increasing number and their functional similarities suggest that there may be a duplication of function and a common molecular pathway that controls the suppression or predisposition of mammalian cells to apoptosis.

The localisation of the *bcl-2* gene at several intracellular sites including the nuclear membrane, endoplasmic reticulum, mitochondrial and plasma membrane and the subsequent aggregation under stressed conditions suggest that *bcl-2* might target itself to specific sites following the exposure of the cell to an apoptotic inducing agent [83]. It has been suggested that the principal product of *bcl-2*, which is a 26 kDa integral membrane protein containing a stretch of hydrophobic amino acids at the carboxyl terminus, is used for post-translational insertion into the cellular membrane [84].

Hockenbery et al. [85] and Kane et al. [86] proposed that apoptosis involves the generation of reactive oxygen species and that the survival enhancing activity of bcl-2 may be mediated through an oxidative pathway. This idea was suggested because of the belief that many of the agents that trigger apoptosis generate reactive oxygen species and that superoxide ion and H_2O_2 are produced mainly around mitochondria where bcl-2 is predominantly located. However, several recent studies provided evidence against this hypothesis showing that reactive oxygen species are not a central part of the apoptotic regulatory pathways [87, and Cotter TG, personal communication]. Furthermore, recent data suggests that bcl-2 exerts its anti-apoptotic activity by a mechanism that does not involve oxidative reactive pathways [87–89, and our unpublished results].

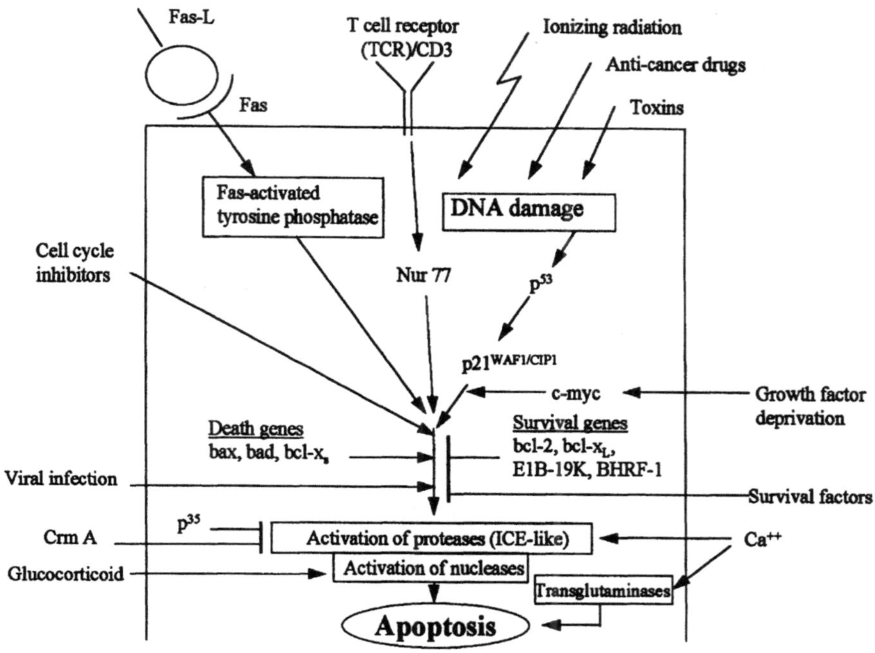

Fig. 7. Regulation of apoptosis in mammalian cells

The interaction between *bcl-2*, *c-myc* and growth factors has been proposed as a possible mechanism to explain the intricacy of the control of apoptosis, proliferation and survival. According to this model, the presence of cytokines that are found in serum results in survival and proliferation of cells. *c-myc* which is a transcriptional factor, can induce apoptosis when cells are deprived of cytokines or treated with cytotoxic drugs [25]. However, expression of *bcl-2* protein inhibits *c-myc-* induced apoptosis independent of any mitogenic activity, thus promoting survival. From this, we can conclude that any changes in the supply of growth factors could lead to changes in the regulatory pathways of growth genes with important implications for tumor progression/inhibition. In addition to the relevance of these finding to cancer therapy, it also important for conditions where cell survival is desirable, including production of biopharmaceuticals from animal cells. The present knowledge about the regulation of apoptosis is outlined in Fig. 7.

5 Exploiting Apoptosis in Cell Culture Technology

The study of apoptosis has now been recognised as an area which should have tremendous promise and potential value in cancer, neurodegeneration, heart diseases, viral diseases and autoimmunity, but the implications of this activity outside the medical field have also been recently realised. In particular, the rapid advances of the last few years have heightened interest in the mechanisms of cell death in the bioreactors used for the production of biopharmaceuticals. The question as to why there should be any interest in the mechanism of cell death in animal cell culture technology can be answered by illustrations from experience in the culture of hybridoma and myeloma, two cell lines which are used extensively for the large scale production of antibodies. Apoptosis in hybridoma cells was first reported by Al-Rubeai et al. [29] in a study of the cellular ultrastructre during batch culture and by the Franek group [90, 91] in protein-free media. A recent kinetics study demonstrated an association between DNA fragmentation and cell density during batch cultivation of 24 hybridoma cell lines with the presence of short fragments of 100 bp at the mid exponential phase and unresolved long fragments at the late death phase [92]. In the physical and chemical environment of a bioreactor, it has been recently shown that apoptosis plays a critical role in the determination of growth and survival of both hybridoma and myeloma cell lines [21]. It has also been demonstrated that apoptotic responses can be caused by serum, glutamine and glucose depletion, cytotoxic metabolite accumulation and by the stressful hydrodynamic conditions often found in bioreactors [93–95]. Significant levels of cell death occur in high cell density cultures, specifically when process intensification by means of cell retention or immobilisation is employed. For example, in fixed bed systems the cells are exposed to oxygen limitation within the porous carriers [96]. Depreviation of hybridoma cells of oxygen in bioreactor culture for 45 h resulted in 70% dead cells, of which 90% were apoptotic [97]. The induction of apoptosis under anoxic conditions was attributed to a critical reduction in the intracellular energetic balance and protein synthesis and probably also to the production of stress proteins similar to those synthesised following heat shock.

In seeking to implement higher performance bioproduction strategies it is necessary to identify ways to reduce apoptotic cell death of cultured cells. Prior research strongly suggests that such apoptotic cell death might be reduced by specific genetic modification of the cells. Changes in bioreactor configuration and/or operating protocols may further reduce apoptosis.

The growing appreciation of the central role of apoptosis in controlling cell number and viability in the cultures of cell lines that are used for the production of biopharmaceuticals has stimulated interest in metabolic engineering of cell survival. The most obvious application is the transfection of cell lines with anti-apoptotic genes, such as *bcl-2*, *bcr-abl*, *v-abl* and *E1B* to inhibit cell death, thereby prolonging culture duration. The successful transfection and expression of these genes in cell lines producing products of commercial importance should lead to the development of more robust cell lines which are able to survive in sub-optimal conditions, including nutrient limitation, and in stressful physical environments. The feasibility of such objectives was demonstrated in my laboratory by showing that, unlike unmodified cells, a lymphoblastoid cell line transfected with *bcl-2* can grow in agitated suspension culture without prior adaptation, suggesting that the *bcl-2* gene protects the cells from death by apoptosis induced by hydrodynamic forces [98]. This work also demonstrated that the suppression of apoptosis by *bcl-2* following growth arrest allowed the maintenance of cells in a growth arrested state for much longer than would otherwise be possible. Moreover, the ability of *bcl-2* to enhance growth in serum free media suggests that the anti-apoptotic effect may be important in the design of simpler and more effective serum/protein free media.

The expectation of increasing survival and consequent increase in culture productivity through transfection with anti-apoptotic genes has stimulated a rush of cell manipulation programmes which have prevention of apoptosis as their primary goal. Amongst other properties, overexpressed anti-apoptotic genes in industrially important cell lines such as hybridoma, myeloma and CHO has been generally shown to increase culture productivity and to enhance survival in nutrient-depleted media (e.g. [99] and our unpublished data) (Table 5). However, Murray et. al. [100] reported that *bcl-2* fails to suppress apoptosis during the decline phase of NSO myeloma culture and also in response to serum depreviation. Moreover, *bcl-2* may not be universally useful for the improvement of product productivity even though its expression can lead to enhancement of survivability. An important example here relates to chimeric antibody production, which has been shown recently to be reduced in growth-enhanced myeloma cells transfected with *E1B* adenovirus genes [101].

There are also indications that the overexpression of *bcl-2* may enhance cell survival in hypoxic and hyperoxic conditions. We have overexpressed *bcl-2* in hybridoma cells and found a significant increase in culture duration when subjected to hypoxic conditions by gassing the culture with 90% nitrogen and 10% carbon dioxide (our unpublished data). The prolongation of culture duration resulted from the suppression of apoptosis but eventually necrosis reached very high levels and thus prevented the incubation beyond 50 h. The inhibition of apoptosis in such anerobic conditions by *bcl-2* provides further evidence that the protective mechanism may not be through the regulation of the levels of reactive oxygen species. On the other hand, growth in hyperoxic conditions was only slightly increased due to the high levels of necrosis. It is

Table 5. Commercial cell lines transfected with anti-apoptotic genes

Cell line	Gene transferred	Reference
Hybridoma	*bcl-2*	Itoh et al. (Biotech. Bioeng. 1995)
Hybridoma	*bcl-2*	Simpson & Al-Rubeai (AIChE meeting, 1995)
NS0	*bcl-2*	Murray et al. (SGM meeting, 1995)
NS0	*E1B*	Mercille et al. (Cell Culture Engineering V, 1996)
CHO	*bcl-2*	Reddy et al. (Cell Culture Engineering V, 1996)
CHO	*bcl-2*	Goswami et al. (Cell Culture Engineering V, 1996)

possible that although the product of *bcl-2* gene can repress apoptosis in conditions of excess oxygen, the high level of necrosis resulting from the severe injury to cellular structures can lead to an apparent reduction of cell growth promotion.

The benefit for cell culture from the overexpression of anti-apoptotic genes can, therefore, only be realised in normal conditions or conditions of mild stress which do not lead to high level of necrosis or cell destruction. Such conditions are normally encountered in high cell density cultures where cells are continuously under nutrient limitation leading to reduction in growth rate and also cell death with consequent problems arising from cell lysis and the release of DNA, proteases, etc.

The large scale production of vaccines in animal cells requries a relatively sophisticated processes utilising a suitable cell-virus system and appropriate production technology. With the tremendous demand on producing viruses as rapidly, cheaply and efficiently, as possible there is a need for extending the cell life cycle to allow more viral replication cycles. It has been reported that apoptosis occurs on infection of the cells by many viruses including the chicken anemia virus [102], human immunodeficiency virus type 1 [103], influenza A and B viruses [104] and insect baculovirus [105]. The hypothesis was presented that interference by the virus in the intracellular and intercellular signalling could cause abnormal induction of an activation-associated cell-suicide process by apoptosis. On the other hand, a number of other viruses have been found to possess genes, such as p35, which can prevent apoptotic cell death in their host cells [105, 106].

Recognising the importance of apoptosis in virus infection may have dramatic consequences for the conception of new strategies of research and development for enhancement of viral yield (and of expressed product in the case of recombinant virus) in vaccine production and gene therapy processes. In order to prevent apoptosis, antiapoptotic genes should be overexpressed in either the viruses or the infected cells, thereby allowing the cells to survive longer and the viruses to multiply further.

In closing, I would like to point out that the first half of the 1990s has resulted in an explosion of knowledge and information about the characteristics and occurrence of apoptosis as well as the physiological and non-physiological triggers, molecular pathways and genes involved in the death and survival process. The second half of the 1990s will, undoubtedly, witness a giant leap towards the realisation of our goals to develop drugs that inhibit apoptosis induced by infection and drugs that stimulate apoptosis for the treatment of cancers. In addition to the obvious relevance of this knowledge to human therapies, they may also be of importance for the industrial production of biologics. In this respect, the development by genetic manipulation of robust cell lines superior in their growth and productivity as well as the optimisation of the physiological environment for the large scale cultivation of cell lines should lead to efficient and safe processing of protein products and reduction of contaminated products such as DNA and proteases.

Acknowledgements. This work is supported by the European Community, Framework IV Programme (Biotechnology- 950207) and the Biotechnology and Biological Sciences Research Council, UK.

6 References

1. Handa-Corrigan A, Emery AN, Spier RE (1989) Enzyme Microb. Technol. 11: 230
2. Kioukia N, Nienow AW, Emery AN, Al-Rubeai M (1992) Trans IChemE 70: 143
3. Kunas KT, Papoutsakis ET (1990) Biotechnol Bioeng 36: 476
4. Al-Rubeai M, Oh SKW, Musaheb R, Emery AN (1990) Biotechnol Lett 12: 323
5. Emery AN, Jan DC-H, Al-Rubeai M (1995) Appl Microbiol Biotechnol 43: 1028
6. Bailey JE (1991) Science 252: 1668
7. Kearns MJ (1990) Bio/Technol 8: 409
8. Curling EM, Hayter PM, Baines AJ, Bull AT, Gull K, Strange PG, Jenkins N (1990) Biochem J 272: 333
9. Freshney RI (1989) Animal cell culture, a practical approach, IRL Press, Oxford
10. Heichman KA, Roberts JM (1994) Cell 79: 557
11. King RW, Jackson PK, Kirschner MW (1994) Cell 79: 563
12. Nurse P (1994) Cell 79: 547
13. Sherr CJ (1994) Cell 79: 551
14. Murray A (1994) Current opinion in cell biology 6: 872
15. Sherr CJ, Roberts JM (1995) Gene & Development 9: 1149
16. Al-Rubeai M, Emery AN, Chalder S, Jan DC (1992) Cytotech 9: 85
17. Ashwell JD, Cunningham RE, Noguchi PD, Hernandez D (1987) J Exp Med 165: 173
18. Cotter TG, Lennon SV, Glynn JM, Green DR (1992) Cancer Res 52: 997
19. Buja LM, Eigenbrodt ML, Eigenbrodt EH (1993) Arch Pathol Lab Med 117: 1208
20. Manjo G, Joris I (1995) American J Pathol 146: 1
21. Singh RP, Al-Rubeai M, Gregory CD, Emery AN (1994) Biotechnol Bioeng 44: 720
22. Milner AE, Wang H, Gregory CD (1966) In: Al-Rubeai M, Emery AN (eds) Flow Cytometry Applications in Cell Culture, Marcel Dekker, New York
23. Darzynkiewicz Z, Li X, Gong J (1994) In: Darzynkiewicz Z, Robinson JP, Crissman HA (eds) Methods in Cell Biology, Vol 41 Flow Cytometry, 2nd edition, Academic Press
24. Studzinski GP (1995) Cell growth and apoptosis: A practical approach, IRL Press
25. Evan GI, Harrington E, McCarthy N, Gilbert C, Benedict MA, Nunez G (1995) In: Thomas NSB (ed) Apoptosis and cell cycle control in cancer, BIOS, Oxford
26. Bursch W, Paffe S, Putz B, Barthel G, Schlte-Hermann R (1990) Carcinogenesis 11: 847
27. DelVecchio M, Leoncini L, Buerki K, Kraft R, Megha T, Barbini R, Tosi P, Cottier H (1991) Int J Cancer 47: 38
28. Moser B (1995) Microscopy and Analysis 50: 21
29. Al-Rubeai M, Mills D, Emery AN (1990) Cytotechnol 4: 13
30. Falcieri E, Gobbi P, Zamai L, Vitale M (1994) Scanning Microscopy 8: 653
31. Oberhammer F, Fritsch G, Pavelka M, Froschl G, Tiefenbacher R, Purchio T, Schulte-Hermann R (1992) Toxicol Lett 64–65: 701.
32. Cohn GM, Sun X-M, Snowden RT, Dinsdale D, Skilleter DN (1992) Biochem J 286: 331
33. Walker R (1995) In: CHI Conference on Commercial Prospects of Apoptotic Research, Cambridge, MA, May 1–3
34. Al-Rubeai M, Emery AN (1993) Bio/Technol 11: 572
35. Dive C, Gregory CD, Phipps DJ, Evans DL, Milner AE, Wyllie AH (1992) Biochimica et Biophysica Acta 1133: 275
36. Gavrieli Y, Sherman Y, Ben-Sasson SA (1992) J Cell Biol 119: 493
37. Gorczyca W, Bruno S, Darzynkiewicz RJ, Gong J, Darzynkiewicz Z (1992) Int J Onclogy 1: 639
38. Gold R, Schmied M, Rothe G, Zischler H, Breitschopt H, Wekerle H, Lassman H (1993) J Histochem Cytochem 41: 1023

39. Li X, Traganos F, Melamed MR, Darzynkiewicz Z (1995) Cytometry 20: 172
40. Cohen JJ (1991) Adv Immunol 50: 55
41. Wyllie AH (1980) Nature 284, 555
42. Itoh N, Yonehara S, Ishii A, Yonehara M, Mizushima S, Sameshima M, Hase A, Seto Y, Nagata S (1991) Cell 66: 233
43. Collins MKL, Marvel J, Malde P, Lopez-Rivas A (1992) J Exp Med 176: 10430
44. Schwartzman RA, Cidlowski A (1993) Endocrine Rev 14: 133
45. Iwata M (1995) In: Kroemer G, Martinez AC (eds) Apoptosis in Immunology, Springer Berlin Heidelberg New York
46. McConkey DJ, Orrenius S, Jondal M (1990) Immunol Today 11: 120
47. Bellomo G, Perotti M, Taddei F, Finardi G, Nicotera P, Orrenius S (1992) Cancer Res. 52: 1342
48. McConkey DJ, Orrenius S (1995) In: Kroemer G, Martinez AC (eds) Apoptosis in Immunology, Springer, Berlin Heidelberg New York
49. Lennon SV, Kilfeather SJ, Hallet MB, Campbell AK, Cotter TG (1991) Clin Exp Immunol 87: 465
50. Fesus L, Davies PJA, Piacentini M (1991) Eur J Cell Biol 56: 170
51. Owen-Schaub LB, Yonehara S, Crump III WL, Grimm E (1992) Cell Immunol. 140: 197
52. Klas C, Debatin KM, Jonker RR, Krammer PH (1993) Int Immunol 5: 625
53. Nagata S, Golstein P (1995) Science 267: 1449
54. Yoshino T, Kondo E, Cao L, et al. (1994) Blood 83: 1856
55. Cohen JJ, Dukle RC (1984) J Immunol 132: 38
56. Wyllie AH, Morris RG, Smith AL, Dunlop DJ (1984) J Pathol 142: 67
57. Sellins KS, Cohen JJ (1987) J Immunol 139: 3199
58. Shi Y, Szalay MG, Paskar L, Boyer M, Singh B, Green DR (1990) J Immunol 144: 3326
59. Crompton T (1991) Growth Factors 4: 109
60. Takano YS, Harmon BV, Kerr JFR (1991) J Pathol 163: 329
61. Brach MA, de Vos S, Gruss HJ, Hermann F (1992) Blood 80: 2920
62. Carson D, Ribeiro JM (1993) The Lancet 341: 150
63. Cotter TG, Al-Rubeai M (1995) TIBTECH 13: 1251
64. Raff MC (1992) Nature 356: 397
65. Kurita T, Namiki H (1994) J Cellular Physiology 161: 63
66. Evan GI, Wyllie AH, Gilbert CS, Littlewood TD, Land H, Brooks M, Waters CM, Penn LZ, Hancock DC (1992) Cell 69: 119
67. Mercille S, Massie B (1994) Biotechnol Bioeng 44: 1140
68. Steller H (1995) Science 267: 1445
69. Miura M, Zhu H, Rotello R, Hartwieg EA, Yuan J (1993) Cell 75: 653
70. Wang L, Miura M, Bergeron L, Zhu H, Yuan J (1994) Cell 78: 739
71. Yuan J (1995) Current Opinion in Cell Biology 7: 211
72. Lane DP (1993) Nature 362: 786
73. El-Deiry WS (1995) In: Thomas NSB (ed) Apoptosis and cell cycle control in cancer, BIOS, Oxford
74. Shi Y, Glynn JM, Guilbert L, Cotter TG, Bissonette R, Green DR (1992) Science 257: 212
75. Vaux DL, Cory S, Adams JM (1988) Nature 335: 440
76. McDonnell TJ, Korsmeyer SJ (1991) Nature 349: 254
77. Hockenbery D, Nunez G, Millman C, Schreiber RD, Korsmeyer SJ (1990) Nature 348: 334
78. Gregory CD (1992) Rev Med Virol 2: 205
79. Ferandez-Sarabia M, Bischoff J (1993) Nature 366: 274
80. Halder S, Negrini M, Monne M, Sabbioni S, Croce CM (1994) Cancer Res 54: 2095
81. Reed JC (1994) J Cell Biol 124: 1
82. Oltvai ZN, Korsmeyer SJ (1994) Cell 79: 189
83. Halder S, Jena N, Cross R, Sedar AW, Wachsberger PR, Beatty C, Croce CM (1994) Cell Death and Differentiation 1: 109
84. Chen-Levy S, Cleary MS (1990) J Biol Chem 265: 4929
85. Hockenbery D, Oltvai Z, Yin X, Milliman C, Korsmeyer S (1993) Cell 75: 241
86. Kane D, Safafian T, Anton R, Hahn H, Gralla E, Valentine J, Ord T, Bredesen D (1993) Science 262: 1274
87. Muschel R, Bernhard E, Garza L, Mckenna E, Koch C (1995) Cancer Res 55: 995
88. Shimizu S, Eguchi Y, Kosaka H, Kamiike W, Matsuda H, Tsujimoto Y (1995) Nature 374: 811
89. Jacobson MD, Raff MC (1995) Nature 374: 814

90. Franek F, Dolnikova J (1991) Febs Letters 248: 285
91. Franek F, Vomanstek T, Dolnikova J (1992) Cytotechnol 8: 55
92. Solis-Recendez MG, Perani A, D'Habit B, Stacey GN, Maugras M (1995) J Biotechol 38: 117
93. Mercille S, Massie B (1994) Biotechnol Bioeng 44: 1140
94. Al-Rubeai M, Singh RP, Goldman MH, Emery AN (1995) Biotechnol Bioeng 45: 463
95. Franek F (1995) Biotechnol Bioeng 45: 86
96. Racher AJ, Looby D, Griffiths JB (1990) J Biotechnol 15: 129
97. Mercille S, Massie B (1994) Cytotechnol 15: 117
98. Singh RP, Emery AN, Al-Rubeai M (1996) Biotechnol Bioeng 52: 166
99. Itoh Y, Ueda H, Suzuki E (1995) Biotechnol Bioeng 48: 118
100. Murray K, Ang C-E, Gull K, Hickman JA, Dickson AJ (1995) In: Abstract, Society for General Microbiology meeting, University of Bath, April 3–6
101. Mercille S, Gervais C, Paquette D, Tardif AB, Martin L, Massie B (1995) In: Cytotechnology 17 (Abstrct, Cell Culture Engineering V meeting, San Diego)
102. Jeurissen SHM, Wagenaar F, Pol JMA, van der Eb AJ, Noteborn MHM (992) J Virol 66: 7383
103. Gougeon M, Montagnier L (1993) Science 260: 1269
104. Hinshaw VS, Olsen CW, Dybdahl-Sissoko N, Evans D (1994) J Viorol 68: 3667
105. Chem RJ, Fechheimer M, Miller LK (1991) Science 254: 1388
106. Crook NE, Clem RJ, Miller LK (1993) J Virol 67: 2168

Received June 1996

Author Index Volume 1-59

Author Index Vols. 1-50 see Vol. 50

Subject Index

MIX
Papier aus verantwortungsvollen Quellen
Paper from responsible sources
FSC® C105338

If you have any concerns about our products,
you can contact us on
ProductSafety@springernature.com

In case Publisher is established outside the EU,
the EU authorized representative is:
Springer Nature Customer Service Center GmbH
Europaplatz 3, 69115 Heidelberg, Germany

Printed by Libri Plureos GmbH
in Hamburg, Germany